AF660074

Post Harvest Management and Technology

NIPA® GENX ELECTRONIC RESOURCES & SOLUTIONS P. LTD.
New Delhi-110 034

About the Authors

Dr. Dev Raj is *Professor and Head (Post Harvest Technology),* in the Department of Post Harvest Technology, ASPEE College of Horticulture, Navsari Agricultural University, Navsari, Gujarat India. He is working as Head of Department since 2012. He also worked as *Scientist (Food Technology),* in the Department of Food Science and Technology, Dr. YS Parmar University of Horticulture and Forestry Nauni, Solan (HP) 173230, India. He is having more than 20 year's experience in research, teaching and extension experience in the field of Post Harvest Technology and Food Science & Technology. He is graduated in Horticulture from Dr. YS Parmar University, Solan; MSc in Post Harvest Technology of Horticultural Crops from University of Agricultural Sciences, GKVK, Bangalore and PhD in Post Harvest Technology of Horticultural Crops from Dr YS Parmar University, Solan. He has been awarded with ICAR Junior Research Fellowship during MSc and merit Scholarship during PhD. He also served at Directorate of Mushroom Research (Indian Council of Agricultural Research), Chambaghat as Senior Research Fellow for about 2 years. He has a very good experience in the field of processing, preservation and value addition of fruits and vegetables. He has to his credit 5 patents, more than 70 research papers published in journals of national and international repute along with more than 70 popular articles/technical articles, 32 book chapters, 16 books related to Food Processing. He was Co-PI and project leader for different projects under All India Coordinated Research Project on Post Harvest Technology (ICAR). He also handled externally funded project of NOVOD Board, GOI as Co-PI. He has more than 30 research recommendations for the farming communities. He handled more than 30 research experiments as PI. He has guided 9 MSc and 8 PhD Students as major guide. He is recipient of more than 25 awards including Fellow award of ISHRD, Kejriwal award (2008) and Dr J.S. Pruthi Memorial Award (2010) for their research contribution in the interest of food industries development by All India Food Processor's Association, New Delhi. He is member for many professional societies. He is referee of prestigious Journal of Food Science and Technology. He is also appointed as paper setter cum evaluator by many agricultural universities.

Dr. Rajendrakumar Maganlal Naik is *Principal and Dean,* ASPEE College of Horticulture and N. M. College of Agriculture, Navsari Agricultural University, Navsari, Gujarat - India. He also worked of Director Student Welfare, Navsari Agricultural University, Navsari, Gujarat. He also worked up to the position of Professor and Head in the Department of Agriculture Extension, N. M. College of Agriculture Navsari Agricultural University Navsari – 396450 (Gujarat) India. He is having more than 35 years' experience in research, teaching and extension experience. He is graduated in Agriculture from GAU Gujarat and completed PhD in Agriculture Extension and Agronomy from N. M. College of Agriculture Navsari Agricultural University Navsari. He has a very good experience with respect to extension of the developed technologies up to farmer level. He has to his credit more than 30 research papers published in journals of national and international repute along with more than 150 popular articles/technical articles, 28 books / chapters Booklets, 10 book folders, 4 practical manuals. He organised more than 110 farmer's trainings and workshops. He has delivered good numbers of recommendations for the farming communities. He has guided 11 MSc and 2 PhD Students as major guide along with member / minor guide in Advisory committee of numerous students. He is recipient of various awards including Academic Achievement awards by Shri Shuklcshwar Mahadev Mandal, Anawal (2007) and Best Extension Education thesis award by GAAS, Ahmedabad (2008) for his contribution. He is member for many professional societies. He is also appointed as paper setter cum evaluator by many agricultural universities. He was also appointed as a member of Peer review Team for accreditation of Agricultural College and University by ICAR.

Dr. Z. P. Patel is **Vice-Chancellor** of Navsari Agricultural University, Navsari (Gujarat). He born on 1st May, 1957, native of Navsari, Gujarat has did his B.Sc. (Zoology) from South Gujarat University, Surat; M.Sc (Agri.) in Agricultural Entomology from erstwhile Gujarat Agricultural University, Navsari and Ph.D in Agricultural Entomology from GAU, Anand (Gujarat) in 1994. He reached to the highest position of Vice-Chancellor, after serving various positions viz. JRF, SRF, Senior Research Assistant, Assistant Professor, Associate Professor, Professor. He was first Principal of College of Agriculture, Waghai and worked as Dean, Faculty of Agriculture at NAU Navsari. He also established accredited NABL Food Quality Testing Laboratory at Navsari during his tenure as Professor. For all these efforts, he is undoubtedly titled as ***'Man of Infrastructure Development'*** in the University. He possessed about 40 years of experience in Research, Education, Extension & Administration and 27 years' hands-on experience in Tribal belt. As a Vice Chancellor, now he has taken the Name and Fame of the Navsari Agricultural University to new heights by improving the standards of research, teaching, extension, infrastructure development and administrative improvements in every aspect. He had Major contributions in the Pest Management; and has important research achievements in the field of Entomology and worked specially to control Fruit flies in Fruit crops and Vegetables. He developed and commercialized one kind of first fruit fly trap (**Nauroji Stone House Fruit Fly Trap**) in the country which is very effective, economic, eco-friendly and sustainable for the management of fruit flies enhancing the export potential and is patented technology. His work on fruit fly gave him recognition as **"Man of Fruit Fly"** in academic and farming fraternity. He also developed **Crop Pest Map** in Sapota which guided farmers to implement Integrated Pest Management (IPM) strategies more efficiently in time to protect the crop and reap higher income. His contribution for control of **Sapota Seed/Fruit Borer;** an insect named *Trymalitis margarias* found first time in the world is of paramount importance and saved the crop in Gujarat as well as in country from its destruction. He is also known for his contribution for **Bio-Control of Sugarcane Pyrilla** which was one of the most devastating pest of sugarcane in South Gujarat and has been controlled by a very effective natural enemy, *Epiricania melanoleuca* which is proved as most successful example of Bio-control in Gujarat. He has guided total **17** students (12 M.Sc & 5 Ph.D) in Agricultural Entomology as Major advisor. He is also a good extension educationist and wholeheartedly participated in Krishi Mahotsav and successfully convinced farmers for the adoption of new technologies. He also delivered numerous TV talks and radio talks. He was successful in creating large scale awareness among farmers regarding use of fruit fly trap. Dr. Patel has in his credits about more than 5 patents, 47 research papers published in referred international and national journals, 03 Books published, 04 papers presented in International seminar, 12 paper presented in National seminar, 10 Research technologies developed for farmers, 38 popular articles and booklets, 02 lead lectures delivered in National conference, 04 projects completed as Principal Investigator and 06 projects initiated as Principal. He has acted as Chairman/Member/Convenor in many research, extension, academic forums of NAU, Navsari before joining as Vice-Chancellor NAU, Navsari. During the tenure of his service, he received Awards, recognitions and appreciations and is a proud recipient of the most coveted Research Excellence Awards from two consecutive Hon. Chief Ministers of Gujarat State (i) First by Shri Keshubhai Patel during 2000 (Sardar Patel Agricultural Research Award, GoG) for the Crop-Pest map in Sapota (ii) Shri Narendrabhai Modi in 2012 (Federation of Gujarat Industries Award) for the Fruit fly trap. Other awards include **Significant Achievement Award** by Rotary International, Gandevi in 2001 and Publication of success story "Integrated Pest Management for Higher Fruit Production" of RKVY in "Coffee Table Book" – a publication of GoI in 2012. He also has foreign exposure as attended 25 days international training at Israel and delivered lecture in UC Davis University of California, USA. He is always a man of different thoughts and his *Jarahatke approach* made him a versatile person in all his work. He is a good team leader, punctual, visionary, and a quick decision-maker. A true son of the soil, Dr. Z. P. Patel has always thought for the welfare of farmers and students; and his untiring efforts, due diligence, and perseverance have set an example for the scientific and teaching fraternity.

Post Harvest Management and Technology
An Illustrative Guide

Dev Raj
Professor & Head (PHT)
Department of Post Harvest Technology
ASPEE College of Horticulture
Navsari Agricultural University
Navsari – 396450, Gujarat, India

R. M. Naik
Principal and Dean
ASPEE College of Horticulture
Navsari Agricultural University
Navsari – 396450, Gujarat, India

Z. P. Patel
Vice Chancellor
Navsari Agricultural University
Navsari, Gujarat, India

NIPA® GENX ELECTRONIC RESOURCES & SOLUTIONS P. LTD.
New Delhi-110 034

NIPA® GENX ELECTRONIC RESOURCES & SOLUTIONS P. LTD.

101,103, Vikas Surya Plaza, CU Block
L.S.C. Market, Pitam Pura, New Delhi-110 034
Ph : +91-11-43860225, Mob.: +91 9717133558, 9540816132
E-mail: newindiapublishingagency@gmail.com
Website: www.nipaersources.com

© 2025, Publisher

Print ISBN: 978-93-58874-37-2

ebook ISBN: 978-93-58874-68-6

All rights reserved. No part of this publication may be reproduced, stored in a retrieval system or transmitted in any form or by any means, including electronic, mechanical, photocopying recording or otherwise without the prior written permission of the publisher or the copyright holder.

This book contains information obtained from authentic and highly reliable sources. Reasonable efforts have been made to publish reliable data and information, but the author/s, editor/s and publisher cannot assume responsibility for the validity, accuracy or completeness of all materials or information published herein or the consequences of their use. The work is published with the understanding that the publisher and author/s are not attempting to render any professional services. The author/s, editor/s and publisher have attempted to trace and acknowledge the copyright holders of all material reproduced in this publication and apologize to copyright holders if permission and/or acknowledgements to publish in this form have not been taken. If any copyrighted material has not been acknowledged, please write to us and let us know so that we may rectify the error, in subsequent reprints.

Trademark Notice: NIPA®, the NIPA® logos and their presentations (the way they are written/presented) in this book are the trademarks of the publisher and hence may not be used without written permission, if copied or used without authorization, the infringer will be prosecuted as per law.

NIPA® also publishes books in a variety of electronic formats. Some content that appears in print may not be available in electronic books, and vice versa.

Composed and Designed by NIPA®.

Preface

Post harvest is the time that begins with separation of commodity from growing of production medium like fruits/ vegetables from the plants, fishes from ponds etc. The term post harvest technology means the use of science and technology for the management of food commodities including pre-cooling, grading, sorting, packaging, transportation, storage, processing, and development of improved product, preservation and marketing etc. The use of the information generated by the techniques of Post Harvest Technology can ensure greater availability of a wide variety of food to the consumers with reduced post harvest losses. Based on the importance of the post harvest technology in agriculture and allied sciences, it is presumed that all people engaged in agriculture should know about the various terminologies being used in post harvest technology. But to understand the various concepts, some sort of documentation is needed. A manuscript having such information would be the strongest link between the industry, academia and the consumers.

The students, teachers, and researchers often need a direct reference which is complete on the subject for teaching undergraduate or postgraduate students. So the authors made an attempt to meet out this need. The book entitled "Post Harvest Management and Technology: An Illustrative Guide" contains important glossary of all related terms of post harvest technology for all alphabets from A to Z. This book is covering almost every important terminology pertaining to post harvest technology including other interrelated disciplines like physiology, processing, food science, food technology, food engineering, food packaging, food biochemistry and applied nutrition, industrial microbiology, health foods, and quality management etc. The terminology in each alphabet has been well illustrated with examples and photographs for better understanding. At present, there is no book available which gives complete information related to Post Harvest Technology. The book will be highly beneficial to both UG and PG students undergoing courses in Postharvest Technology, entrance examination for PhD in Post Harvest Technology at IARI, various examinations pertaining to JRF and SRF as well as NET.

We acknowledge from the core of our heart for the valuable guidance and support received from numerous publications of many committed and dedicated members of scientific community who have contributed tirelessly in the field of post harvest technology. Their assistance for the information obtained from literature and websites that were consulted during the preparation of the manuscript is thankfully acknowledged. Words are inadequate to express our gratitude towards our most venerable parents and family members whose constant inspiration and innumerable sacrifices have encouraged us to complete this goal.

It is hoped that the book will be welcomed and its benefits will be availed by an increased number of food professionals like students, researchers, teachers and all those who have interest in the subject. Although, every care has been taken by the authors while writing this text book even then spelling mistakes, overlapping; repetition of brief portion here and there could not be helped. Needless to say, errors and omissions are solely our. Constructive criticisms and suggestion by the readers are welcome for improving the quality of manuscript. Reader's indulgence in this regard is highly solicited.

Authors

Contents

A

Abnormal milk: It is milk that is not smooth and homogenous but is flaky, ropy, stringy, slimy, watery, discoloured or bad flavoured (Plate 1 & 2).

Abrassive peeler: It is equipment comprising a rotating drum lined with carborandum (an abrasive material) in which certain vegetables like potato and ginger are tumbled and exposed to the abrasive action of the carborandum lining, thereby facilitating separation of peel from the vegetables (Plate 3).

Absinthe: It is green liqueur which is flavored with wormwood, anise, and other aromatic herbs (Plate 4). Its commercial production is banned in ma ny countries for health concerns.

Plate 1: Abnormal ropy milk

Plate 2: Abnormal Slimy milk

Plate 3: Abrassive peeler

Plate 4: Absinthe

Absolute alcohol: It is water free alcohol. It contains atleast 99 per cent pure alcohol by weight and obtained by digesting rectified spirit over quicklime and redistilling the same.

Absolute density: It is the ratio of weight of certain volume of a substance at 1°C to that of weight of same volume of water at 4°C.

Absolute humidity: Absolute humidity is defined as kg of water vapour per kg of dry air at a given temperature and pressure. Absolute humidity is the mass of water vapour divided by the mass of dry air in a certain volume of air at a specific temperature. The warmer air absorbs more water than cooler one. Absolute humidity is the measure of water vapour or moisture in the air, regardless of temperature.

Absolute pressure: It is equal to atmospheric pressure minus vacuum pressure (Plate 5).

Absolute zero: It is a theoretical temperature characterized by complete absence of heat and motion and equivalent to exactly -273.15°C or -459.67°F (Plate 6).

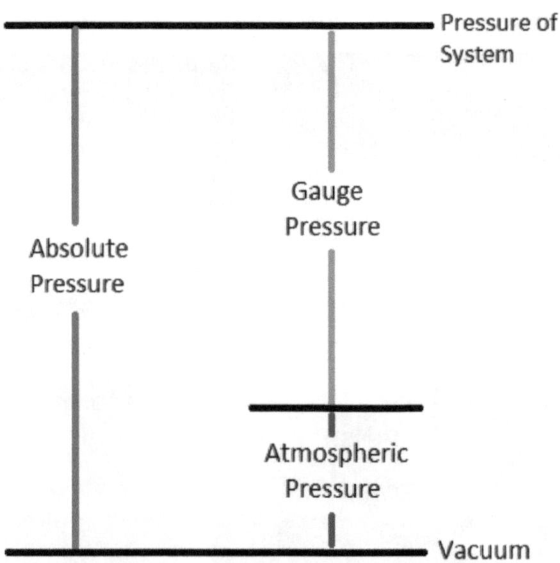

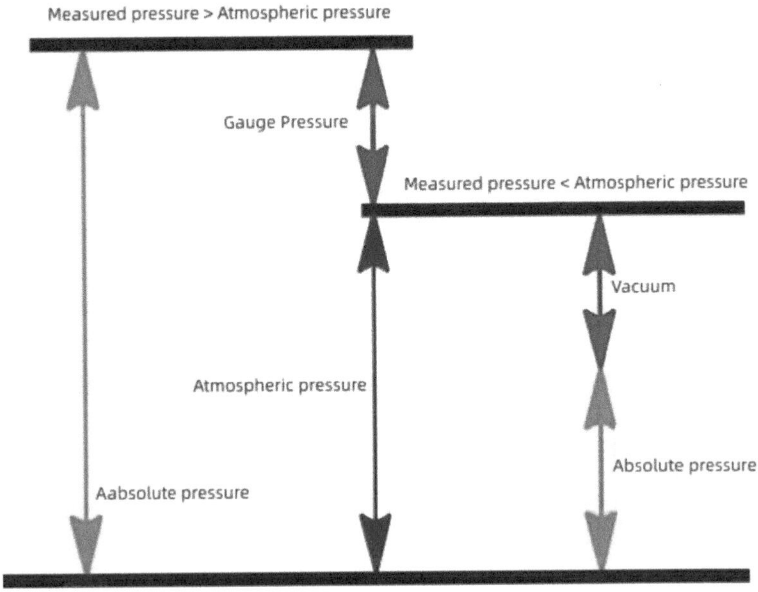

Plate 5: Absolute pressure

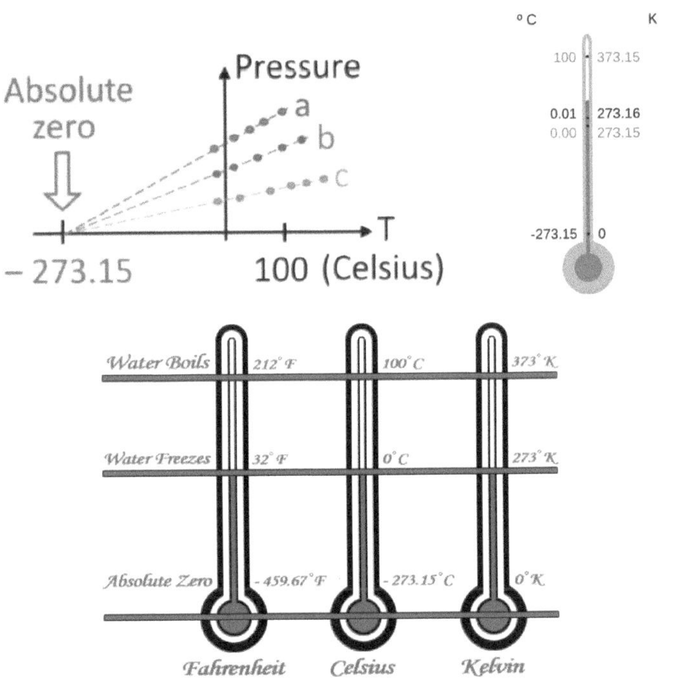

Plate 6: Absolute zero

Absorbance: It is the ability of a layer of a substance to absorb radiation expressed mathematically as the negative common logarithm of transmittance. It is equal to log (100/%T), T is the transmittance **(Plate 7).**

Absorptiometer: It is an instrument used to measure the absorption of monochromatic radiation in the visible range.

Absorption spectrum: An electromagnetic spectrum in which there is a decrease in intensity of radiation at specific wavelengths is known as absorption spectrum **(Plate 8)**. The decrease in the intensity of the radiation is due to absorption of radiation in the absorbing substances and is especially characterized by a pattern of dark lines or bands.

Absorptivity (α): The ratio of the absorbed energy to the incident energy is called absorptivity.

Abstraction licence: A licence issued by the water authority for the abstraction of water from the water source for domestic or commercial use.

Accessory Fruit: A fruit (apple) in which a conspicuous part consists of tissue other than that of the ripened ovary is known as accessory fruit **(Plate 9)**.

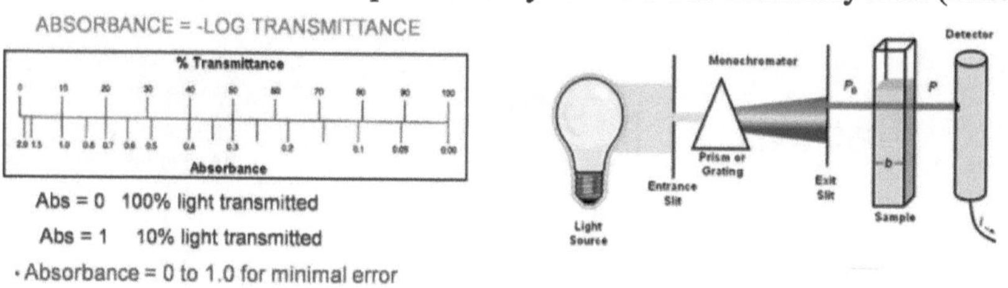

Plate 7: Illustration of Absorbance mechanism

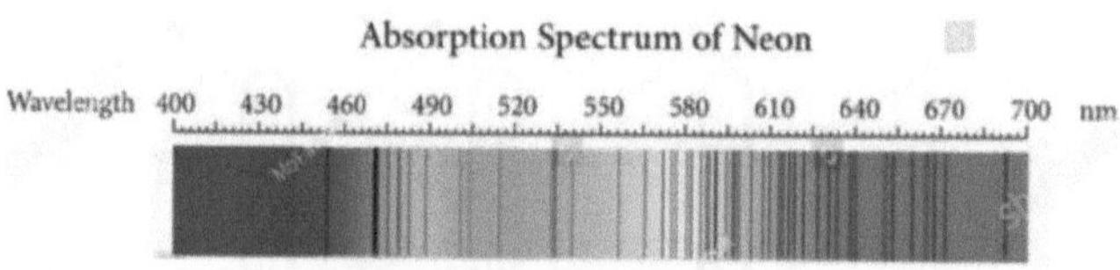

Plate 8: Absorption spectrum

Plate 9: Accessory Fruit example

Accessory pigments: The pigments such as carotenoids and phycobili – proteins which obtain light energy and transfer it to the reaction centre for photosynthesis.

Accredited milk: The milk sold by a producer to retailer from a herd accredited as free from Brucellosis.

Acetaldehyde: It is a colourless volatile water-soluble liquid aldehyde C_2H_4O used chiefly in organic synthesis.

Acetic Acid: It is a colorless pungent liquid acid $C_2H_4O_2$ that is the chief acid of vinegar and is used especially in pickle making and manufacture of plastics.

Acetic acid fermentation: It is type of fermentation where an aerobic respiration takes place in the presence of oxygen; ethyl alcohol is converted into acetic acid and water by Acetobacter (**Plate 10**).

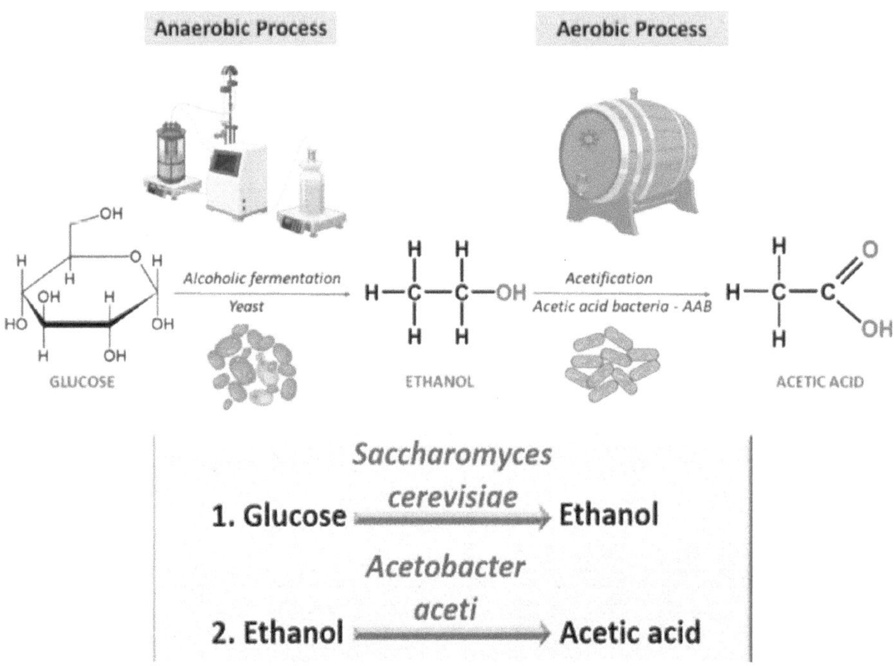

Plate 10: Illustration of Acetic acid fermentation

Achrodextrin: The dextrins which are produced when the hydrolysis of starch has carried out to the stage that it impart no colour with iodine.

Acid foods: Foods having pH 4.5 to 3.7 which are usually spoiled by non-spore forming aciduric, butyric anaerobes etc. e.g. Product of tomato, pear, fig, pineapple etc (**Plate 11**).

Acid resistant (AR) cans: These are golden coloured R enamel and cans are coated with R enamel. Fruits having water soluble pigments are canned in such lacquered cans (Plate 12).

Plate 11: Foods of pH 4.5 to 3.7

Plate 12: AR cans

Acidity: The state or degree of a food to which a food is acidic is known as acidity.

Acidulation: The process to make a food acidic or slightly acidic is called acidulation.

Acidulent: A food or a substance which is somewhat acidic or harsh in taste is known as acidulent. It is also called as acidulous.

Acidulous: A food or a substance which is somewhat acidic or harsh in taste is known as acidulous.

Ackee fruit: The fruit of an African tree (*Blighia sapida*) of the soapberry family grown in the Caribbean area, Florida, and Hawaii for its white or yellowish fleshy aril that is edible when ripe but is poisonous when immature or overripe and that has a toxic pink raphe attaching the aril to the seed **(Plate 13)**. Immature ackee fruit contain –toxin hypoglycine. Fully mature fruit of ackee contain little or no hypoglycial but it splits and results in very short life.

Acoustic and vibration test: These tests are used to judge the maturity of melon and pineapple **(Plate 14)**.

Plate 13. Ackee fruit *Blighia sapida*

Plate 14: Acoustic and vibration test

Acridity: Sense of irritation in tasting different aroids due to the presence of needle like crystals of calcium oxalate in the different plant parts of the aroids (Taro, elephant foot yam etc.).

Acrolein: It is a colorless, highly irritant and volatile pungent liquid aldehyde C_3H_4O formed when glycerol is heated to the point where two molecules of water split from it. It is used chiefly in organic synthesis.

Acrylamide: It is an amide C_3H_5NO that is derived from acrylic acid. Potatoes chips on frying produces acrylamide content which is carcinogenic.

Actinometer: It is one of the instruments for measuring the intensity of incident radiation. It is the instrument which is used for measuring the intensity of radiation by the speed of a photochemical reaction (**Plate 15**).

Plate 15. Actinometer

Action spectrum: Effect of different wavelengths of light on the rate of photosynthesis and photo-physiological processes.

Activation energy: The minimum amount of energy required to convert a normal stable molecule into a reactive molecule is called activation energy (Plate 16).

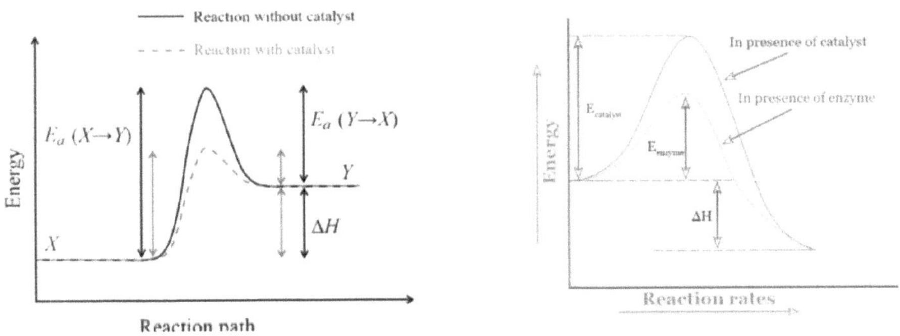

Plate 16: Activation energy

Active Packaging: The incorporation of certain additives into the packaging film or within packaging containers to modify head space atmosphere and to extend product shelf life (Plate 17). Active packaging refers to interactive or smart packaging which is intended to sense internal or external environmental changes and responded by changing its own properties and thus internal package environment. Active packaging results inclusion of a desiccant or O_2 absorber within or as part of the packaging material. Active packaging products are ageless and iron. Ascorbic acid based sachets (also generate CO_2) and catecol based sachets are also used as O_2 absorbers. Ascorbic acid based sachets used for O_2 absorption are marketed as Ageless G or Toppan C or Vitabon GMA (when combined with iron). Cathecol based sachets is called Tomotsu. Mineral powders are incorporated into some films for fresh produce particularly in Japan.

Active Packaging Film: It is defined as a type of packaging that changes the condition of packaging to extend shelf life or improve safety or sensory properties while maintaining the quality of the food. It involved oxygen; carbon dioxide and ethylene scavengers. Propylene films are used for active packaging of meat, cooked food and fresh foods **(Plate 18)**.

Plate 17: Active packaging of food

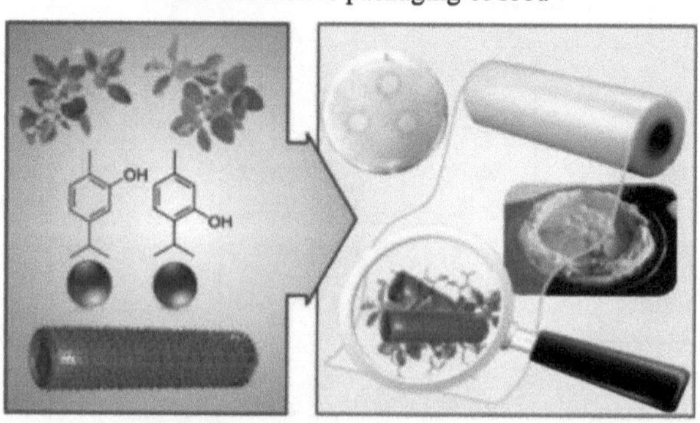

Plate 18. Active Packaging Film

Active scrubbing: It is the process of removal of the undesirable gases from the store or packaging material is called active scrubbing. In CA storage, when the level of carbon dioxide is increased, then the atmosphere is passed through a chemical which remove CO_2.

Active transport: It is the movement of a chemical substance by the expenditure of energy against a concentration gradient.

Adams consistometer: It is an instrument used to measure the viscosity of tomato puree, apple sauce, and fruit pulp (Plate 19).

Adaptive enzyme: An enzyme which is able to synthesize only in the presence of the substrate is called adaptive enzyme.

Add back process: It is a process in which cooked potatoes are partially dried by adding back enough previously dried granules. Add back process also includes addition of recovered oil / flavours etc into the processed juice to compensate the loss during processing operations (Plate 20).

Additives: These are food additives which are added in the food intentionally or un- intentionally in very small amount and possess or not possess any nutritional value but added to increase the sensory and nutrition qualities along with stability (Plate 21).

Addled egg: The fertile egg whose embryo died between 7th and 14th day of the incubation. It is also called rotten egg (Plate 22).

Plate 19: Adams consistometer

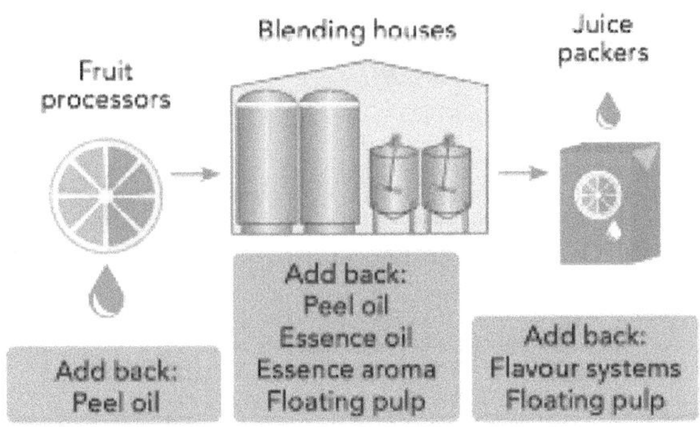

Plate 20: Illustration of Add Back Process

Plate 21: Additives

Plate 22: Addled egg

Adiabatic saturation temperature: It is the temperature at which the air can be brought to a saturation state adiabatically by the evaporation of water into the flowing air. It is also called thermodynamic wet bulb temperature.

Adiabatic: It is a process which is occurring without loss or gain of heat to the surrounding (do not involve heat transfer) is known as adiabatic. Or The expansion of a gas or a material without loss or gain of heat is called adiabatic. Adiabatic changes in temperature occur due to change in the pressure of gas while not adding or subtracting any heat. Adiabatic heating occurs when the pressure of a gas is increased while in adiabatic cooling pressure of a gas is decreased. Such temperature change can be quantified by using ideal gas law. In fact, no process is truely adiabatic because there is always some heat loss as no perfect insulator exist in the nature.

Adiabatic dryer: These are also called direct dryer or contact type dryers. The transfer of heat takes place by direct contact of the products with the hot gases. The gas transfers sensible heat into food to provide the heat of vaporization of moisture present in the solid. Adiabatic drying (drum type rotary dryer) is preferred due to high rate of heat transfer, short residence time and uniform drying (**Plate 23**).

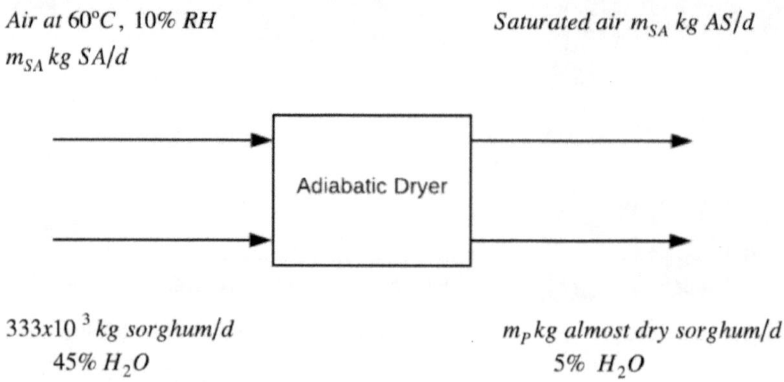

Plate 23 . Mechnism of Adiabatic dryer

Adjunct: Something that is added to another thing (beer) but not essentially a part of it is called adjunct. These are cereals like rice, corn etc which provide supplementary CHO, principally starch, for conversion to sugar for subsequent fermentation in beer making is known as Adjuncts.

Adulteration: Mixing of a food item with material of inferior quality or of an undesirable nature is known as adulteration (**Plate 24**).

Advertise: It is a process to make publicity by announcing publicly especially by a printed notice / broadcast by emphasizing desirable qualities of the products so as to arouse a desire to buy or patronize.

Advertising Agency: An agency that enables prospective buyers to know the quality of the product and decide about the purchase of commodities is known as advertising agency e.g. newspapers, radio, television, cinema slides.

Aerated water: Fruit juice when preserved with CO_2 under high pressure is known as aerated water **(Plate 25)**.

Aeration tank: It is a part of effluent treatment plant. Aeration tank is provided for the biological treatment of the organic matter, which is easily biodegradable. The essential features of this process are development of microbial floccules held in suspension in an aeration chamber. Prior to the biological treatment, nutrients like DAP & Urea are added to over flow of the primary clarifier. After the retention time, the water from aeration tank is sent to a secondary clarifier **(Plate 26)**.

Plate 24: Adulteration of food

Plate 25: Aerated water

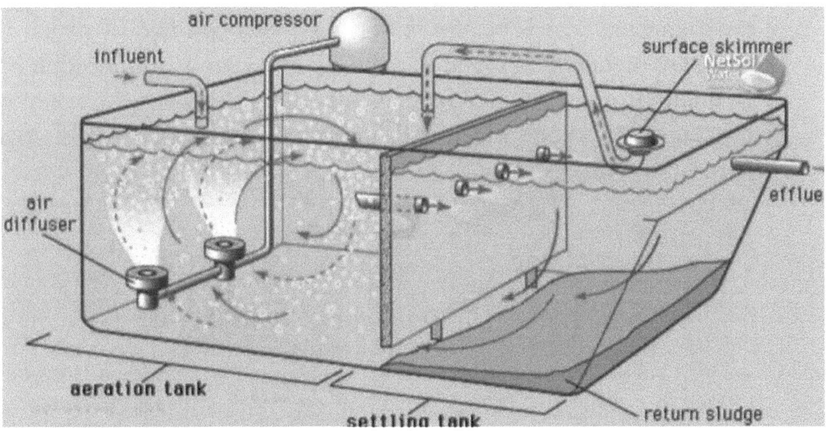

Plate 26: Aeration tank

Aerobe: An organism (bacterium) that lives only in the presence of oxygen is known as aerobe.

Aerobic respiration: It is the process of cellular respiration that takes place in presence of oxygen to produce energy from food (**Plate 27**). This type of respiration is common in most of plants and animals, birds, humans, and other mammals. Water and carbon dioxide are produced as end products in this process.

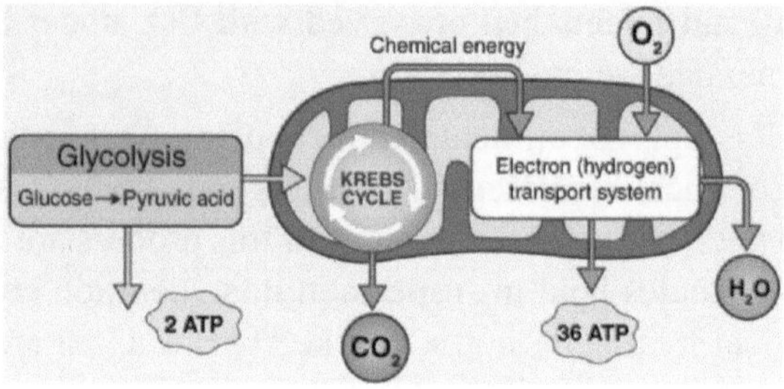

Plate 27: Aerobic respiration

Aerodynamics: It is a branch of dynamics that deals with the motion of air and other gaseous fluids and with the forces acting on bodies in motion relative to such fluids.

Aerometer: An instrument used for ascertaining the weight or density of air or other gases is known as aerometer (Plate 28).

Aerosol cans: These are two or three piece cans fitted with an aerosol valve (Plate 29).

Aflatoxins: It is the most important group of food borne mycotoxin. It is produced by *Aspergillus flavus* and few other moulds growing on food stuffs. There are 6 Aflatoxins and out of these four (β_1, β_2, G_1 & G_2) are of great concern which occurs in various foods and two of which (M_1 and M_2) are metabolites found in the milk of lactating animals that have eaten aflatoxin contaminated feed. Aflatoxin β_1 is found most commonly and occurs in ground nuts and in grain crops particularly maize (Plate 30, 31, 32, 33).

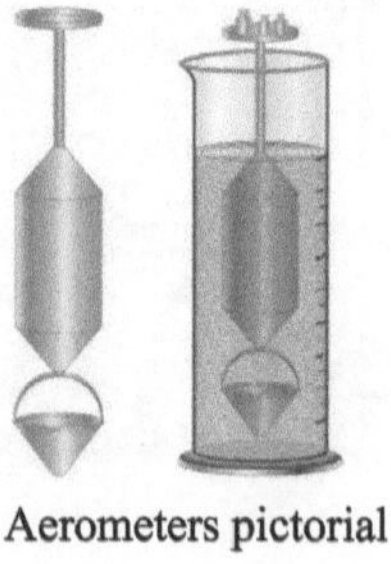

Plate 28: Aerometers pictorial diagram

Plate 29: Aerosol cans

Plate 30: Aflatoxin in Groundnut

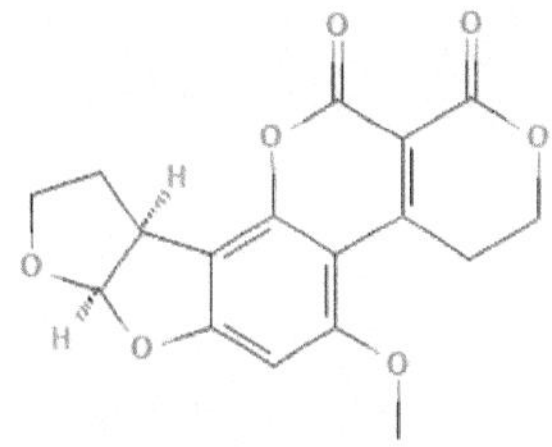

Plate 31: Aflatoxin β2

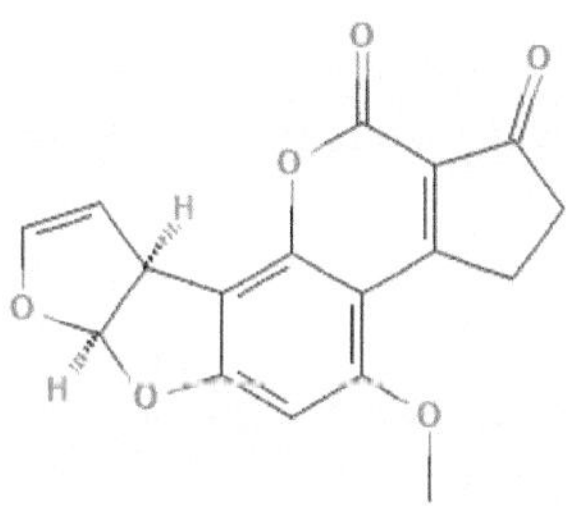

Plate 32: Aflatoxin β1

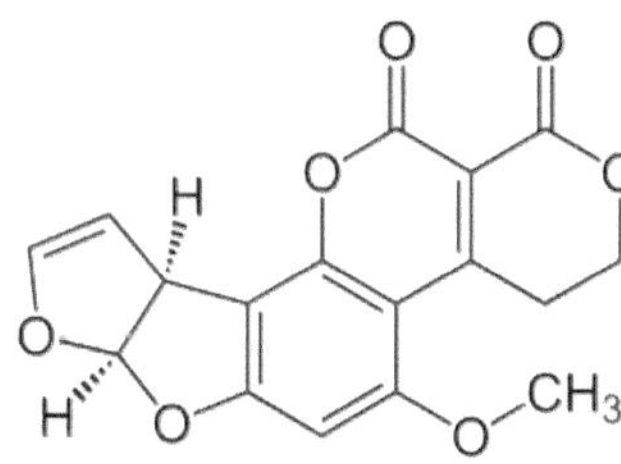

Plate 33: Aflatoxin G1

After cooking discolouration (ACD): ACD develops in cooked potato products after exposure to air, mostly in boiled and steamed potatoes (Plate 34).

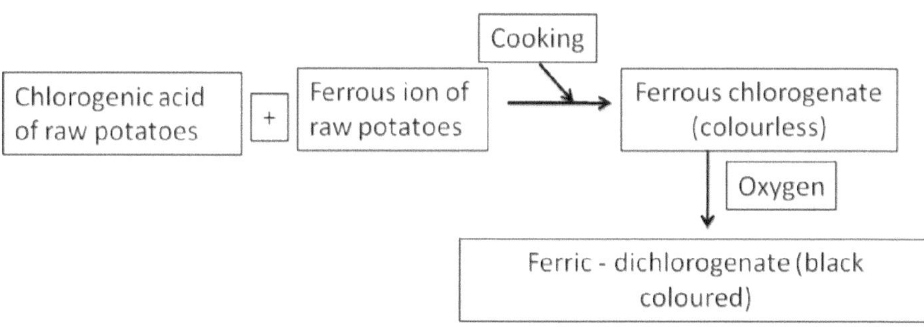

Plate 34: Mechanisms of after – cooking blackening in potatoes

After ripening: It is the physiological change that takes place with in a seed (or bud) during dormancy, such as maturation of embryo which enables germination to take place.

After taste: It is also called Come back taste. It is development of some bitterness on the tongue due to delayed reactions.

Agar: A gelatinous and complex polysaccharides obtained from sea weed (Gracilaria sp, Gelidium sp. etc.) widely used as a substrate in aseptic cultures and in making jelly, ice cream, soap, pastries as emulsifier. Agar is a dried polysaccharide extract of red algae. Or Agar is a gelatinous colloidal extractive of a red alga (as of the genera *Gelidium*, *Gracilaria*, and *Eucheuma*) used esp. in culture media or as a gelling and stabilizing agent in foods (Plate 35, 36, 37).

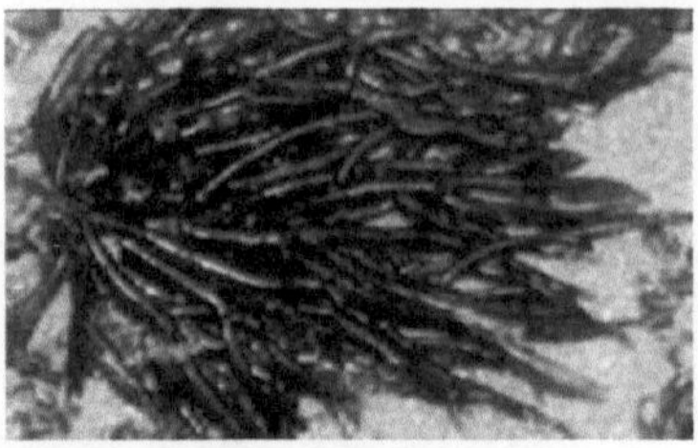

Plate 35: *Gracilaria* sp used for extraction of agar-agar

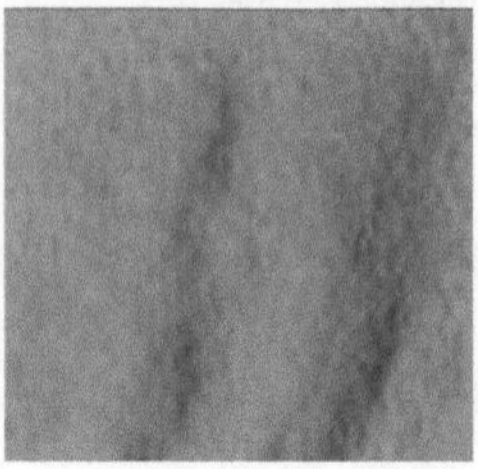

Plate 36: Agar-agar powder

Plate 37: Product prepared from Agar-agar

Agarose: It is a polysaccharide obtained from agar and used as a supporting medium in gel electrophoresis.

Ageing: Ageing is any increment of time which may or may not be accompanied by physiological change.

Ageing of citrus: A type of stem end rind break down that occurs in long storage (8-10 weeks for orange), usually accompanied by a stale or off-flavour in fruits.

Ageless G: This is ascorbic acid based sachets used for O_2 absorption and are marketed as Ageless G (when combined with iron).

Ageless Scavenger: It is ferrous based oxygen scavenging plastic packaging materials **(Plate 38)**.

Agglomeration: It is a process of collecting in a mass. Agglomeration is used to create large clumps from small particles of dry milk, chocolate or coffee, which rehydrate very rapidly **(Plate 39)**.

Plate 38. Ageless Scavenger

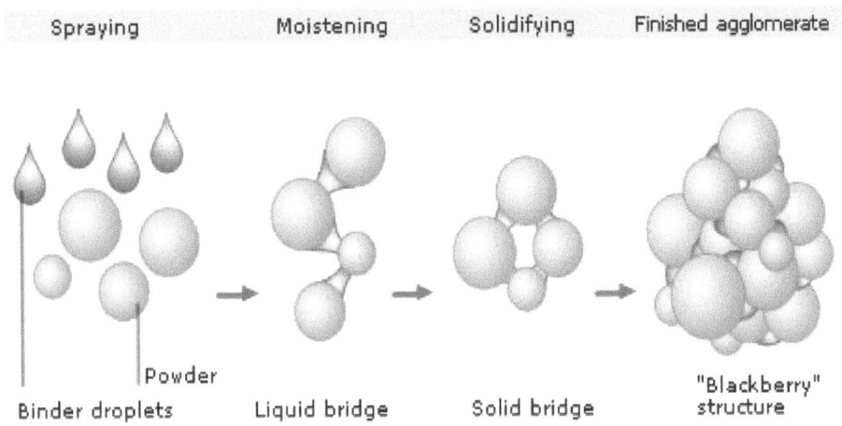

Plate 39. Agglomeration

Aging or maturation: Aging or maturation is the process which the wine undergoes after the completion of fermentation.

Aging vats: Aging vats are also known as casks. These are barrel shaped vessels of staves, headings, and hoops usually for holding wines /liquids (Plate 40).

Plate 40: Aging vats used for holding wines

AGMARK: It stands for agricultural produce grading and marketing act, 1937. This act is voluntary in nature. Different commodities like pulses, roasted channa, whole spices and powdered spices, vanaspati oils (mustard, groundnut, til, sunflower, coconut and mixed oil), flours, maida, sooji, milk products (ghee, buuter, cream) etc are covered under this act.The logo of AGMARK is presented in Plate 42.

Agrexco: It stands for Agricultural Export Company for Israel (1957). They sell their produce by its brand level "Carmel" (**Plate 42**).

Plate 41: AGMARK logo

Plate 42. Agrexco logo

Agricultural marketing: It includes all the services or functions which are involved in bringing produce from a farmer to the final consumer.

Agrochemical / Agrichemical: The agricultural chemical such as herbicide or insecticide is known as agrochemical / agrichemical.

Air blast freezing: Food products are frozen by vigorous circulation of cold air. It is mainly used for freezing fishery products like shrimps, fish pellets etc (**Plate 43**).

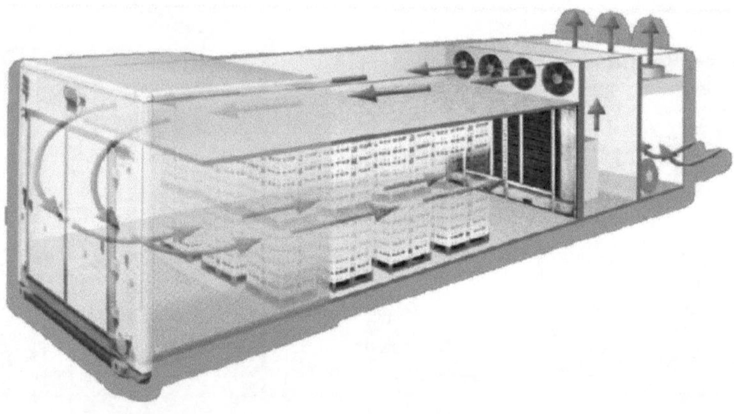

Plate 43. Air blast freezer

Air dry: A substance dried to such a degree that no further moisture is given up on exposure to air. Here the moisture is in equilibrium with the surrounding air.

Air lock: A stoppage of flow caused by air being in a part where liquid ought to circulate.

Air pump: A pump for exhausting air from a closed space or for compressing air or forcing it through other apparatus.

Air storage: Storage in the room at normal air temperature providing proper air flow and ventilation which facilitates the removal of ethylene produced by fresh fruits and vegetables and maintains the CO_2 and O_2 level about 0.03% & 21.0%, respectively.

Air tight: Any package which is impermeable to air is called air tight.

Alanine: Alanine is a simple nonessential crystalline amino acid $C_3H_7NO_2$.

Alanyl: The amino acid radical or residue of alanine is known as alanyl.

Alar: It is a growth regulator which is used for red colour development in apple. Its synonyms are B9 or B 995 (N- dimethyl amino succinamic acid).

Alarm water level: It is the minimum water content in the particular food which prevents mould growth or spoilage during storage. An increase in even 1% beyond this level favours the mold growth.

Albedo: It is the ratio of the amount of the radiation reflection to the amount of the coming radiation of the surface.

Albinism: It is a condition or disorder in which the normal pigment is not produced. It is a physiological disorder of strawberry which produces colourless fruits and is due to high K: Ca and N: Ca ratios (Plate 44).

Plate 44. Albinism of strawberry

Albumin: It is one of the simple heat coagulable water soluble proteins that occur in blood plasma or serum, muscle, the whites of eggs, milk, and other animal substances and in many plant tissues and fluids.

Albuminoids: These are simple proteins which are insoluble in all neutral solvents and in dilute acids and alkali's but are coagulated by the heat.

Alcohol: A mixture of ethanol and water that is usually 95 percent ethanol is called alcohol.

Alcohol meter: It is a hydrometer used for measuring the degree proof of the alcohol content in the wine, beer, whishkey or brandy etc (Plate 45&46).

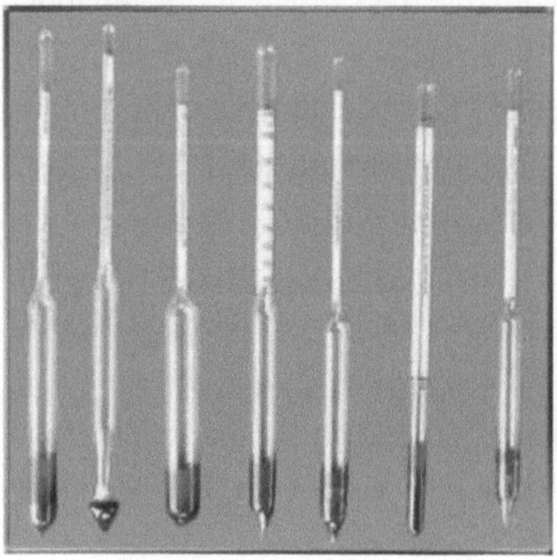

Plate 45: Alcohol meters

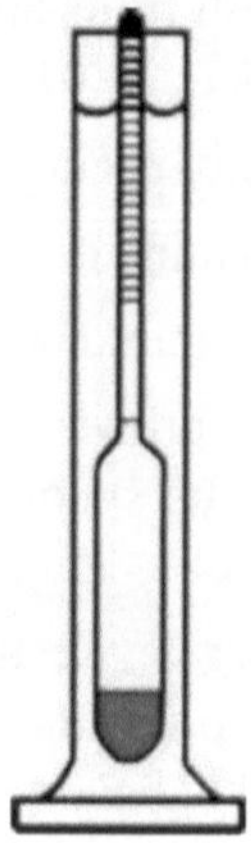

Plate 46: Dipping of alcohol meter in alcohol

Alcohol insoluble solids: It is measure of the texture rather than index of maturity. It is used as maturity indices for pea and beans.

Alcoholic fermentation: The production of the ethanol by fermentation of carbohydrate or a fermentable sugar or a polysaccharide that can be hydrolysed to a fermentable sugar.

Aldente: It is cooking of a food just enough to retain a somewhat firm texture.

Aldolase: A crystalline enzyme that occurs widely in living systems and catalyzes reversibly the cleavage of a phosphorylated fructose into triose sugars.

Aldose: A sugar containing in its acyclic form one aldehyde group per molecule is known as aldose.

Ale: Ale is an alcoholic beverage brewed especially by rapid fermentation from an infusion of malt with the addition of hops. It is a type of beer made from malt. It is old fashioned beer (Plate 47).

Ale house: It is a place where ale is sold to be drunk on the premises

Aleurone: It is a protein in the form of minute granules in grains / seeds occurring in endosperm or in special peripheral layer (**Plate 48).**

Aleurone layer: It is the layer of large cells between the bran coat and endosperm of cereal grains. It is rich source of protein, minerals and vitamins (Plate 48).

Plate 47: American ale

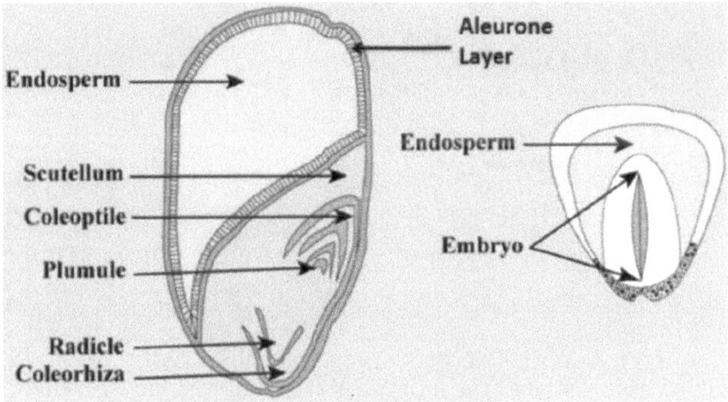

Plate 48. Aleurone layer

Alevin: It is a young fish especially a newly hatched salmon when still attached to the yolk sac.

Algin: The algin is a colloidal substance (alginate or alginic acid) derived from marine brown algae and used as emulsifiers or thickeners.

Alginic Acid: It is an insoluble colloidal acid $(C_6H_8O_6)_n$ that in the form of its salts is a constituent of the cell walls of brown algae.

Alitame: It is an artificial sweetener made from amino acids and is 2000 times sweeter than sugar (Plate 49).

Alkaline phosphatase test: This enzyme has ability to liberate phenol from the phenol phosphoric acid compound. The liberated phenol gives blue colour with certain organic compounds. If the incubated / pasteurized milk does not give blue colour with the indicator it means the pasteurization is complete. The test is so sensitive that presence of 0.1% raw milk added and if the pasteurization temperature is less by 1-degree F to pasteurize milk, can be detected (Plate 50).

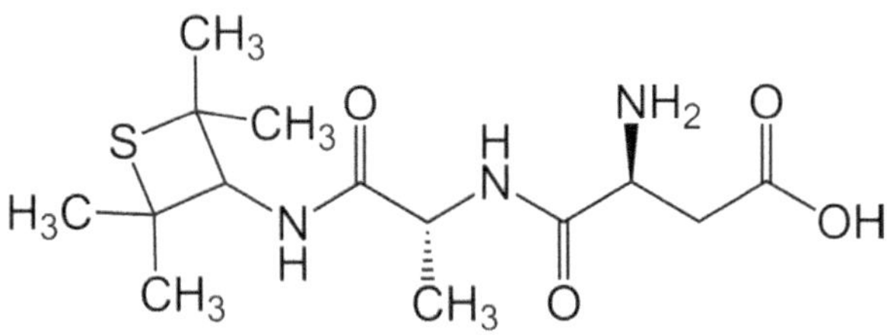

Plate 49: Formula of sweetener Alitame

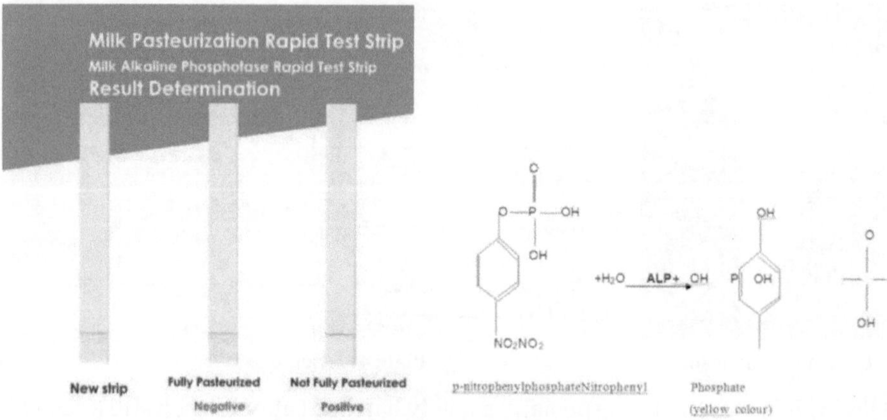

Plate 50: Alkaline phosphatase test

Alkaline phosphatase: Alkaline phosphatase is used as indicator of proper pasteurization treatment in milk.

Allergens: Some food components can cause an allergic or food intolerance response in sensitive individual and these reactions can range from mild to extremely serious. Major allergens of concern include peanuts (groundnuts), tree nuts (walnuts, hazelnuts, pecans, and cashews), eggs, milk products, shell fish. Other minor allergen includes soybean, wheat gluten, certain colourings, sulphites.

Alpha lipoic acid: It is a vitamin like substances that has very good antioxidant properties and plays an important role in supporting longevity. It plays an important role in conversion of glucose into energy (Plate 51).

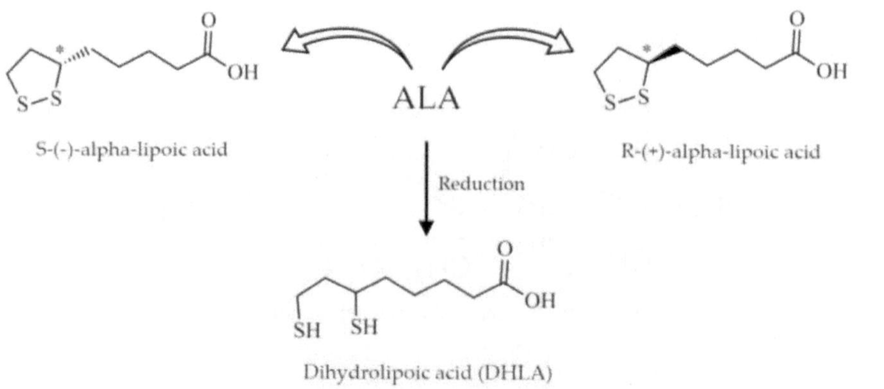

Plate 51: Alpha lipoic acid

Alpha tocopherol: It is a tocopherol ($C_{29}H_{50}O_2$) with high vitamin E potency. It is also called vitamin E.

Alveograph curve: It is related to the bread. It analyzes the physical properties of the dough. The Alveograph (Chopin) test measures gluten strength, using a piece of dough that is inflated until it breaks. Dough is prepared and developed, and dough disks are cut and rested, after which each disk is inflated with air and the resistance to expansion is measured (Plate 52).

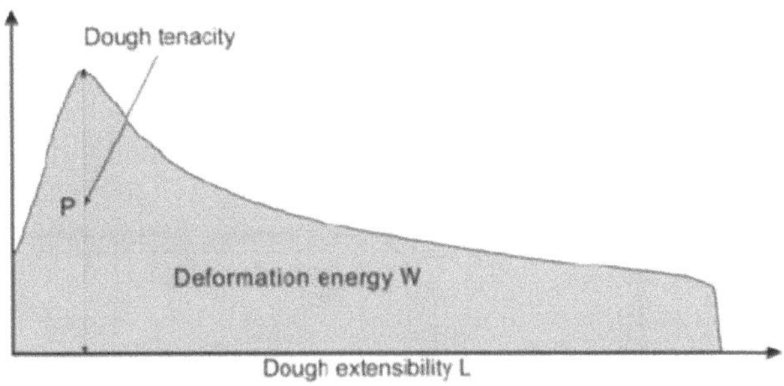

Plate 52: Alveograph curve

Aluminium foil: It is produced by cold reduction process in which pure aluminium (>99.4%) is passed through rollers to reduce thickness to lesser than 0.152mm and annealing to give dead folding.

Aluminium phosphate: A chemical which act as a fumigant for grain. It is available in tablet forms and grains are exposed for 4-5 days.

Amagat's law: Amagat's law is for partial volume; V = Va + Vb + -------+Vn.

Ambient temperature: It is the temperature of the surrounding atmosphere air.

Amchur: The dried raw mango slices or their powder is known as amchur. It is used in North India as curry powder.

Amelioration: The process of adjusting the sugar content of grape juice for wine making is known as amelioration.

Amino acid: An amphoteric organic acid containing the amino group NH_2 or amino acids having the amino group in the alpha position and are the chief components of proteins and are synthesized by living cells or are obtained as essential components of the diet.

Ammoneotelic animals: When the nitrogen is excreted as ammonia, then these animals are known as ammoneotelic animals e.g. aquatic vertebrates, fishes etc. all belongs to this class.

Ammonia injury: Refrigerated fruit/vegetable may get damaged by escaped ammonia gas (refrigerant) which causes brown to greenish black discoloration of outer tissue and also softening of the tissues.

Ammonium acetate: It is a white deliquescent solid used as meat preservative.

Ammonium bicarbonate: It is white crystalline solid and used as baking powder.

Amorphous: It is form of a solid lacking an organized crystalline structure.

Ampaper: It is the mango leather. It is also known as mango slab.

Amphoteric: It is a compound that has at least one group that can act as an acid and one group that can act as a base. It is related to proteins and amino acids which carry both negative and positive charges in the molecules. Such substances or proteins get precipitated at such a pH at which these are electrically neutral **(Plate 53).**

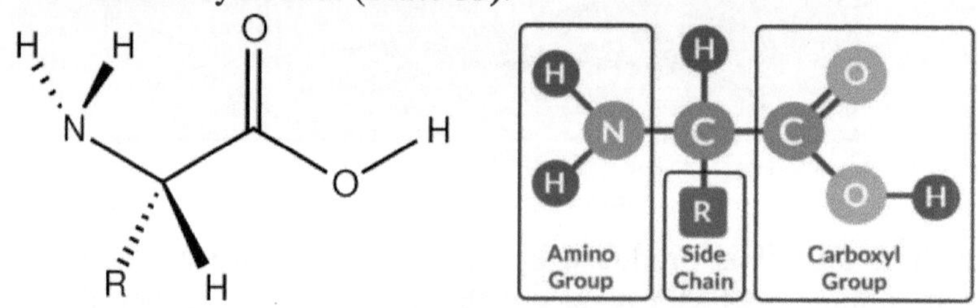

Amino acid structure

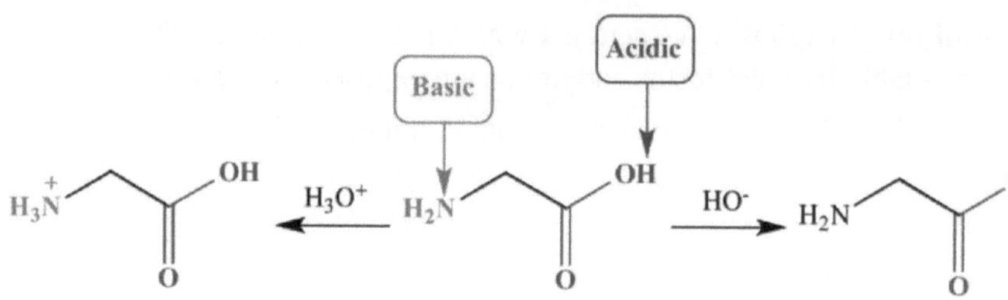

Plate 53. Amphoteric substance

Amygdalin: A white crystalline cyanogenetic glucoside $C_{20}H_{27}NO_{11}$ found mostly in the seeds of the apricot, peach, and bitter almond.

Amylase: A group of enzymes (amylopsin) that catalyze the hydrolysis of starch and glycogen or their intermediate hydrolysis products like maltose, ptyalin, and amylopsin.

Amylograph: It is a device designed to control the temperature of a starch paste and to measure its viscosity.

Amylolytic: It is an enzyme that is capable to split starch into soluble products is known as amylolytic enzyme.

Amylopectin: It is a component of starch that has a high molecular weight and branched structure and does not tend to gel in aqueous solutions (Plate 54).

Amyloplast: It is a colorless plastid that forms and stores starch.

Amylose: A component of starch characterized by its straight chains of glucose units is known as amylose (**Plate 55**).

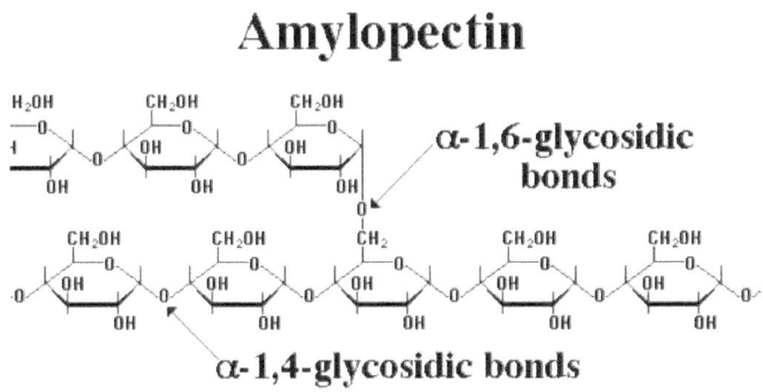

Plate 54. Amylopectin

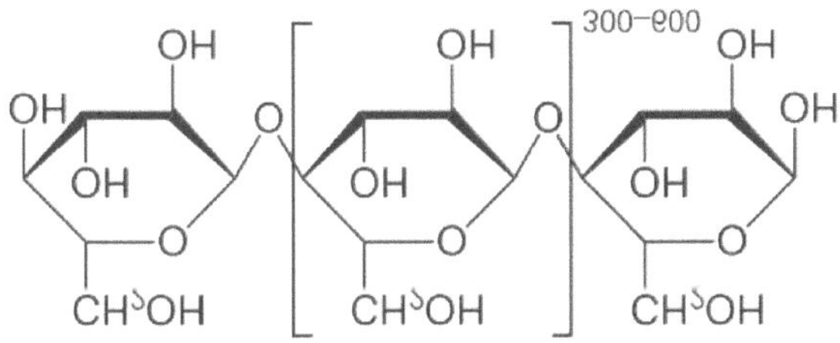

Plate 55. Amylose

Anabolism: The constructive phase of metabolism wherein complex organic substances are produced from simple materials. In this process the organisms gain in potential form which is consumed for future use e.g. Photosynthesis. It is constructive part of metabolism concerned with macromolecular synthesis.

Anaerobe: An organism that survives in the absence of air or molecular oxygen is known as anaerobe.

Anaerobic: These are the microbes or organisms which require the absence of the oxygen for growth or can exist in the absence of free oxygen.

Anaerobic respiration: Respiration which involves incomplete oxidation of foods with the release of energy in the absence of atmospheric oxygen is known as anaerobic respiration (Plate 56).

Anaeroid barometer: Instrument used for measuring atmosphere pressure is known as anaeroid barometer (Plate 57).

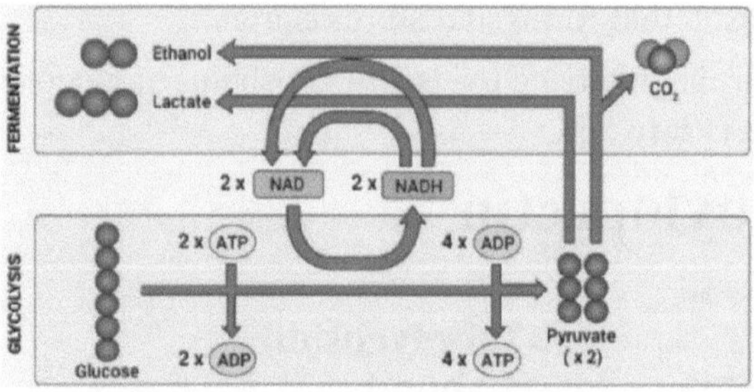

Plate 56: Anaerobic respiration

Plate 57: Barometer

Analysis: The identification or separation of ingredients of a substance from the constituents of a mixture is known as analysis.

Analysis of Variance: Analysis of variation is an experimental outcome and statistical variance in order to determine the contributions of given factors or variables to the variance. It is used to accept or reject the null hypothesis.

Analytical chemistry: It is a branch of chemistry which deals with detection and estimation of the elements and compounds.

Animal fats: The fats derived from animals and characterized by a high proportion of saturated fatty acid.

Annatto: A highly soluble colouring matter of vegetable origin derived from the pericarp of the seed of Bixa orellana (Annatto plant), used as a colouring matter of cheese and table butter. Annatto is a yellowish red dyestuff made from the pulp around the seeds of a tropical American tree (*Bixa orellana*) (**Plate 58**).

Plate 58. Seed of Bixa orellana (Annatto)

Annealling: The process of heating and then cooling (steel or glass) usually for softening and making less brittle is called annealing.

Antagonism: It is interaction of two or more substances / microorganisms such that the action of any one of them has ill effect on living cells or tissues of other.

Antemortem: It is stage preceeding death of the animals.

Anthocyanidin: It is an anthocyanins type pigment that lacks sugar in its structure **(Plate 59)**.

Anthocyanin: Flavonoid compound (C_6-C_3-C_6) mainly present as glycosides which are responsible for red, blue and purple pigmentation of plant parts. The colour of anthocyanin results from the structure of anthocyanidin which is combined with monosaccharide and occasionally a pentose sugar **(Plate 60)**.

Anthoxanthin: These are phenolic compounds contributing white to yellow colour and some flavour to the plants / food. These are creamy yellow, creamy white or yellow white water soluble compounds occurring in the cell saps of some fruits and vegetables e.g. potato and cauliflower **(Plate 61)**.

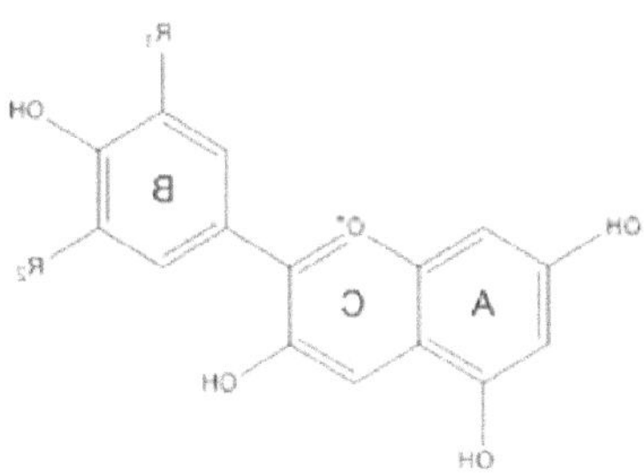

Plate 59. Anthocyanidin

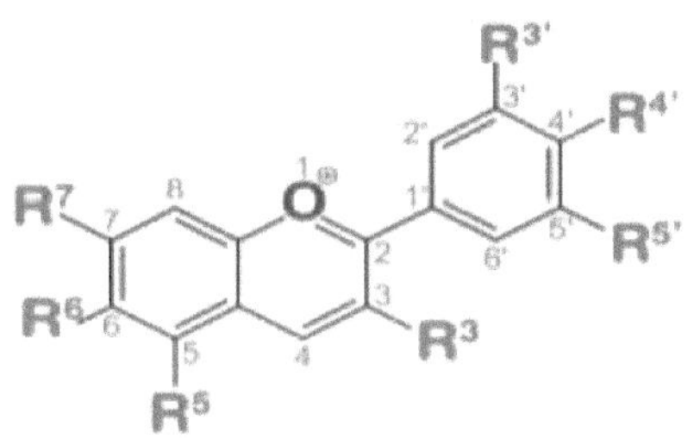

Plate 60. Anthocyanin

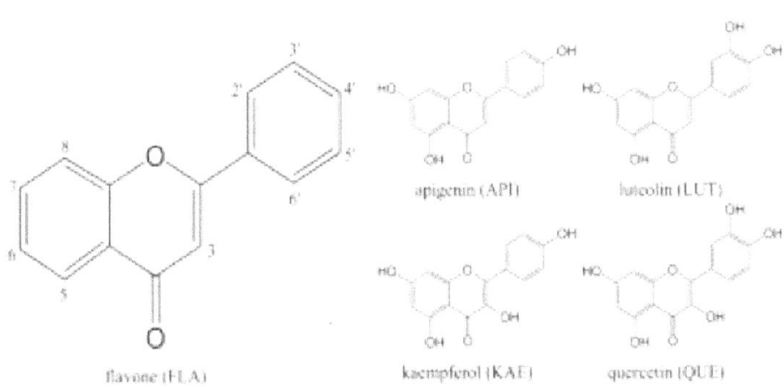

Plate 61. Anthoxanthin

Antibiosis: It is an antagonistic association between organisms and is detrimental to one of them or antagonistic association between one organism and a metabolic product of another organism.

Antibiotic: It is a substance (semi synthetic) produced by or a derived from a microorganism and able in dilute solution to inhibit or kill another microorganism. Antibiotics are chemotherapeutic agents (subtilin, nisin, penicillin, terramycine, myprozine, ampicillin and Streptomycin) obtained from metabolism of living organisms that have inhibitory/ germicidal effect against other living organism in vitro or in vivo, used to control the diseases and to preserve fruits, vegetables, meat, fish etc. are known as antibiotics.

Antibody: A large number of proteins of high molecular weight that are produced normally by specialized B cells after stimulation by an antigen and act specifically against the antigen in an immune response, that are produced abnormally by some cancer cells, and are typically consist of four subunits (two heavy chains and two light chains) and is also known as immunoglobulin.

Anticaking agents: Anticaking agents prevent food particles from adhering to each other and becoming solid lump in damp weather. These absorb excess moisture by coating food particles to make them water repellent. Calcium silicate is used to prevent caking in baking powder (up to 5%) and in table salts (up to 2%). It is a useful anticaking agent in complex powdered mixtures of certain spices that contains essential oils. Calcium stearate is often added to powdered food to prevent agglomeration during processing (Plate 62 and 63).

Plate 62: Calcium stearate

Plate 63: Free flowing property due to use of anti-caking agent

Anticancerous: These are the substances which are used against or tending to arrest /prevent cancer.

Antidote: It is a remedy to counteract the effects of poison.

Antidotic: A chemical substance which is produced by microorganism and has the capacity to inactivate the injurious effect of the bacterial toxins.

Anti-egg white injure factor: The vitamin biotin is known as Anti-egg white injure factor. The other synonyms are vitamin H, vitamin 7, Bios II and

Coenzyme R. RDA is 100 to 300 µg /day. It plays role in gluconeogenesis and fatty acid synthesis. List of biotin rich foods given in Plate 64

Plate 64: Boitin Rich foods

Antifreeze: A substance added to a liquid (as the water in an automobile engine) to lower its freezing point or substances (proteins or alcohols) that are found in some living organisms (as certain fish and insects) and serve to lower the freezing point of body fluids by limiting ice crystal growth. Or Any substance that prevents the formation of ice crystals is known as antifreezing agent.

Antifungal: A substance that is used to destroy fungi or inhibit their growth.

Antigen: A substance (immunogen or hapten) present in the body that evokes an immune response either alone or after forming a complex with a larger molecule (protein) and that is capable of binding with a product (antibody) of the immune response.

Anti-nutrient: Any substance that destroys, inactivate or in other way render unavailable an essential dietary constituent is known as antinutrient.

Anti-nutritional factor: The factors which are responsible for making the nutrients unavailable are known as anti-nutritional factors (ANF) (Plate 65).

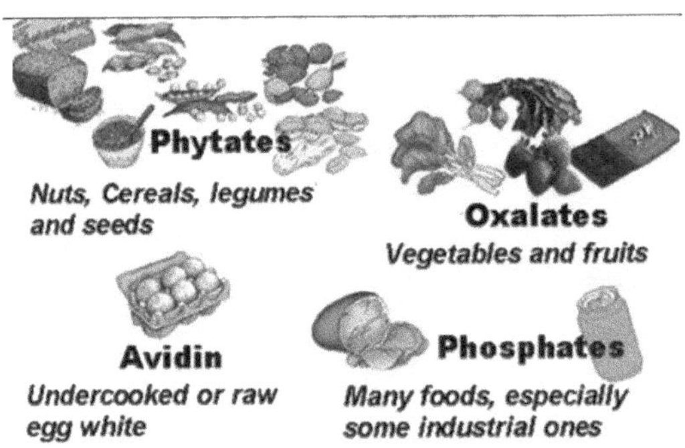

Plate 65: Anti-nutritional factor

Antioxidant: Any substance that reduces oxidative damage by scavenging free radicals/electrons. Antioxidants terminate these chain reactions by removing free radical intermediates and inhibit other oxidation reactions by being oxidized themselves. Or Antioxidant is a substance which prevents oxidation of foods caused by oxygen, thus prolongs their shelf-life, wholesomeness & palatability. or Any micronutrient that neutralizes the action of free radicals is called anti-oxidants. A number of phenols (Hydroquinone, pyrogallal) and naturally occurring substances like ascorbic acid, glutathione, tocopherol (Vit E) are used as antioxidants. In the production of dehydrated potato product, BHT (Butylated Hydroxy Toluene), BHA (Butylated hydroxyl anisole), propyl gallate are used as antioxidents and KMS (Potassium metabisulphite) is used as antioxidant in preserving fruit juices.

Antipollution Peeling Process: It is Dry Caustic Peeling System with Infrared heat. This industrial process is known as USDA – Magnusson Infrared Antipollution Peeling Process and is used for potatoes.

Antistaling: The tasteless or unpalatable form of bread is known as antistaling.

Antitoxin: It is a type of the antibody that acts against a poison or toxin that has entered the body.

Anti-transparent: The chemicals which either partially close the stomata or form a thin film over the produce and check transpiration rendering to additional benefit in moisture conservation inside the plant e.g. Mobileaf, Kaolinite, Phenyl mercuric acetate (PMA) etc.

Aperitif: An alcoholic drink taken before a meal as an appetizer is called aperitif (**Plate 66**).

Apiin: A glycoside presents in celery (**Plate 67**).

Plate 66. Aperitif

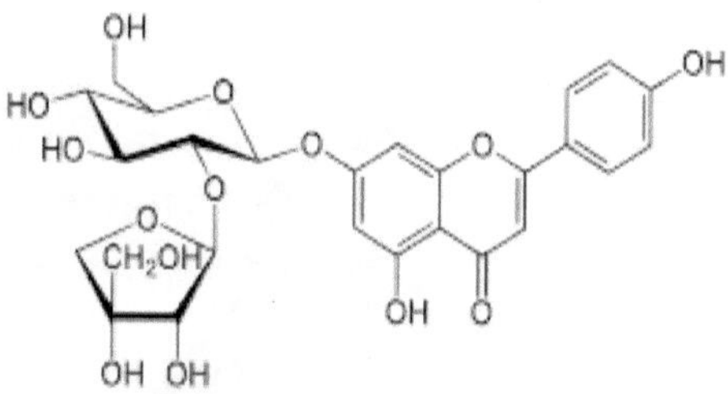

Plate 67. Apiin

Apo enzyme: A protein that forms an active enzyme system by combination with a coenzyme and determines the specificity of this system for a substrate. So, the protein component without co-factor is termed as Apo enzyme.

Appert process: Appert process involve 4 step (i) enclosing the food to be preserved in bottles (ii) Corking the bottles carefully, (iii) submerging the bottles of food in boiling water for a time depending upon the nature of food (iv) Removal of bottles and cooling them. Appert process has been given by Nicolas Appert, also known as father of canning (**Plate 68**).

Apple brandy: Apple brandy is also called fermented apple juice. It is also called applejack.

Applejack: Distilled apple cider is called applejack.

Apple scald: Accumulation of toxic levels of volatiles in the skin tissues leading to superficial death and browning of affected tissues, a storage disorder where immature fruit and the greener side of the fruit is most affected.

Aquavit: Clear Scandinavian liquor flavored with caraway seeds is called aquavit (**Plate 69**).

Plate 68. Nicolas Appert:father of canning.

Plate 69. Aquavit

Aqueous solution: A solution made from water is called aqueous solution.

Arabinose: A white crystalline aldose sugar $C_5H_{10}O_5$ occurring in vegetable gums.

Arabinoside: A glycoside that yields arabinose on hydrolysis.

Arachidonic Acid: A liquid unsaturated fatty acid $C_{20}H_{32}O_2$ that occurs in most animal fats, is a precursor of prostaglandins, and is considered essential in animal nutrition. In the body it is synthesized from the essential fatty acid, linoleic acid.

Arecotine: An alkaloid found in betel nut (Arecanut).

Arginine: A crystalline basic amino acid $C_6H_{14}N_4O_2$ derived from guanidine. Arginine is known as non essential amino acid because it can be synthesized from other amino acids in the body.

Argol: A winery bi-product of grapes which settle on walls and base of wine tanks and are purified as "Rochelle Salt". It is a crust of crude cream of tartar (potassium acid tartarate) that forms as hard crystal in stored grape juice and wine vats. These are removed by the extended storage of the juice at refrigerated temperature.

Argon ionization detectors (AID): The cross section can be made much more efficient if argon is used as carrier gas. The β particles produce Ar atoms, which then transfer their energy very efficiently to sample molecules producing ions and electrons and a very strong signal 10-100 times more sensitive than a H_2 flame ionization detectors. Material that can be ionized by 11.6 eV can be detected, which means that most materials (except noble gases and some inorganic gases) can be detected **(Plate 70)**.

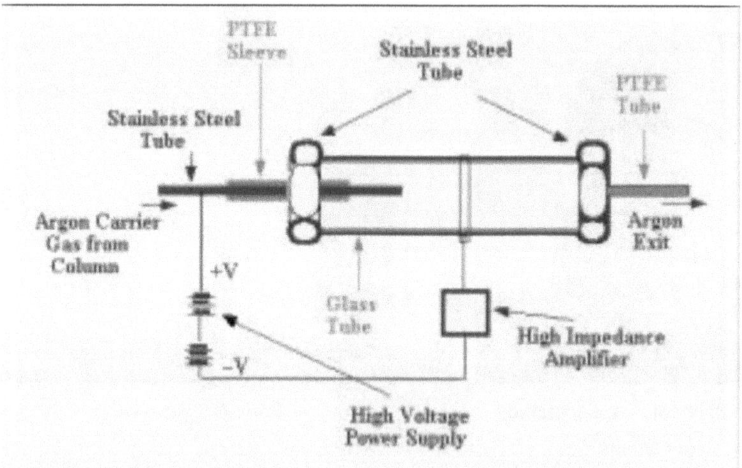

Plate 70. Working of Argon ionization detectors

Aroma: A pleasant, distinctive smell or fragrance of any substances which can be detected by nose only, plays an important role in the development of optimal eating quality in fruit due to synthesis of many volatile compounds (esters, alcohols, acids, aldehydes, ketones etc.) during the ripening phase of fruits. Aroma is due to stimulation of olfactory senses with volatile organic compounds. Aroma may be fragrant, acidic, burnt, pungent, enzymatic, spoilage.

Aroma recovery: In this process the volatile aroma compounds are transferred from fruit juices to another phase. This second phase can be a gas (Evaporative gas stripping), a liquid (solvent extraction) or a solid (adsorption). In the second step of the process the separated aroma is concentrated to aroma rich solution which can be added back to the concentrated juice.

Aromatase: An enzyme or complex of enzymes that promotes the conversion of an androgen into estrogen is known as aromatase **(Plate 71)**.

Aromatic compounds: These compounds contain atleast one benzene ring or a ring of six carbon carbon atom with three double bonds in the alternate position e.g. phenol, benzoic acid, nepthalene etc.

Arrack: A commercial product obtained by distillation of fermented toddy.

Arrowroot: It is starchy substance obtained from the root of arrowroot plant, almost pure starch **(Plate 72)**.

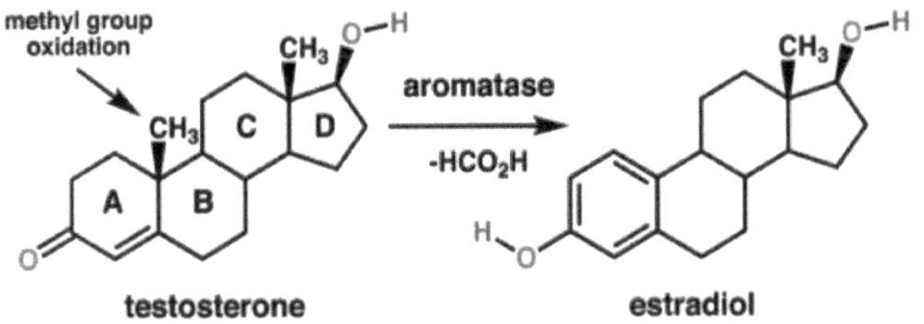

Plate 71. Aromatase

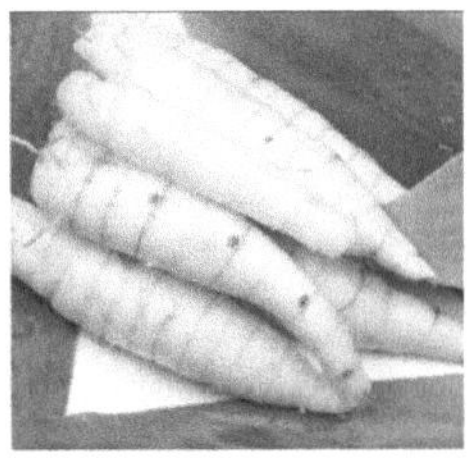

Plate 72. Arrowroot

Artificial vinegar: It is prepared by diluting synthetic acetic acid or glacial acetic acid to a legal standard of 4% and coloured with caramel. This is also known as non-brewed vinegar.

Ascorbic acid: Ascorbic acid is one of the most common and effective anti-oxidant, being beneficial in protecting both the colour and flavor of the fruit product. Vitamin C or ascorbic acid is an enediol isomer of 2 – keto-L gluconolactone with a configuration similar to L- glucose. Oxidation of ascorbic acid gives rise to dehydro – ascorbic acid and both forms are physiologically active. Ascorbic acid helps in tooth formation healing of broken bones, healing of wounds. Deficiency cause scurvy, spongy gums, loose teeth, swollen joints, hemorrhages and impaired healing of wounds. It is also called vitamin C, hexauronic acid and has formula of $C_6H_6O_8$ and is water soluble.

Ascorbic acid oxidase: The enzyme which is involved in the oxidation of ascorbic acid into dehydro-ascorbic acid. Cu^{++} acts as a cofactor for this enzyme. Heat treatment of substance containing ascorbic acid helps for better retention of vitamin C by destroying ascorbic acid oxidase.

Asepsis: Absence / exclusion of the microorganisms that cause disease or infection is called asepsis.

Aseptic canning: The ultra high temperature (UHT) processing followed by aseptic filling is called aseptic canning. It is a technique in which food is sterilized outside the can and then aseptically placed in previously sterilized cans which are subsequently sealed in the aseptic environment **(Plate 73).**

Aseptic preservation: Preservation of sterile products with sterile packages under sterile environment in a continuous and closed system without refrigeration with the retention of high nutritional and sensory qualities. Here high temperature for short time (HTST) or ultra high temperature (UHT) is used to sterilize the product (Plate 73).

Aseptic processing and packaging: It refers to a technique in which the food is sterilized out side the package and then aseptically placed in a previously sterilized package, which are subsequently sealed in a microbe free aseptic environment. Aseptic package used for aseptic processing have different layers (Plate 74).

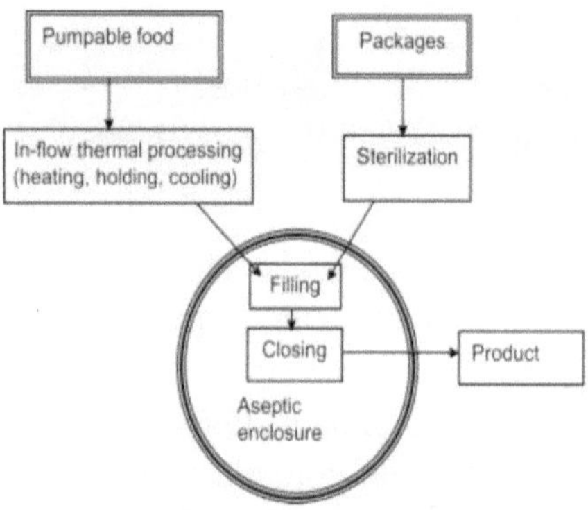

Plate 73. Aseptic canning/Preservation

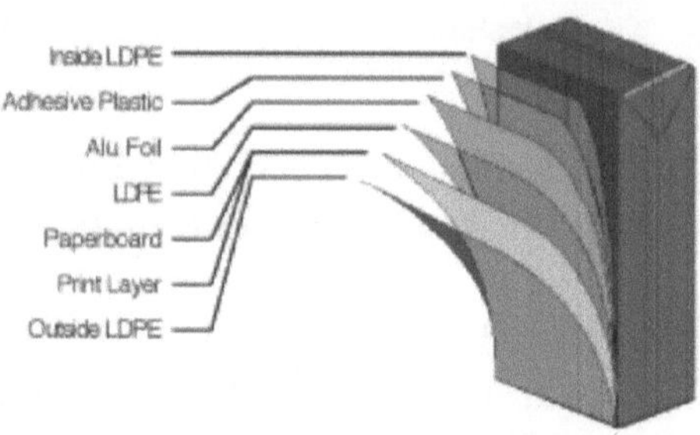

Plate 74: Aseptic package

Aseptic processing: It is method in which the product and package are sterilized prior to filling the product in to the sterile package and filling and closing of the package is in a sterile environment.

Aseptic/Ultra high temperature (UHT) packaging: Packaging material for aseptic processing / packaging must be sterilizable, suitable for aseptic filling and permit the maintenance of sterility of the product during storage and handling. Glass containers offer advantages similar to metal containers but have the additional disadvantage of fragility and high density. Plastic materials offer versatility, but cannot be used alone as they do not possess all of the required stiffness and barrier properties.

Consequently, they are laminated to other materials to give packages with the appropriate requirements.

Ash: Ash content of the food stuff represents inorganic residue remaining after destruction of organic matter.

Asparagine: A non-essential amino acid $C_4H_8N_2O_3$ that is an amide of aspartic acid is called asparagine.

Aspartame: It is a sweet crystalline compound $C_{14}H_{18}N_2O_5$ that is a diamide (dipeptide), synthesized from phenylalanine and aspartic acid and is used as a non nutritive or low-calorie sweetener. It is used in diet beverages. It is used as high intensity sweeter. It has no undesirable after taster but is unstable in water.

Aspartic Acid: A crystalline amino acid $C_4H_7NO_4$ found in plants.

Assimilatory Quotient: It is the ratio of carbon dioxide intake to the oxygen output.

Atmometer: It is an instrument used for measuring the evaporation. It is also called atmidometer or evaporimeter. An atmometer or evaporimeter is a scientific instrument used for measuring the rate of water evaporation from a wet surface to the atmosphere. Atmometers are mainly used by farmers and growers to measure evapotranspiration (ET) rates of crops at any field location (Plate 75).

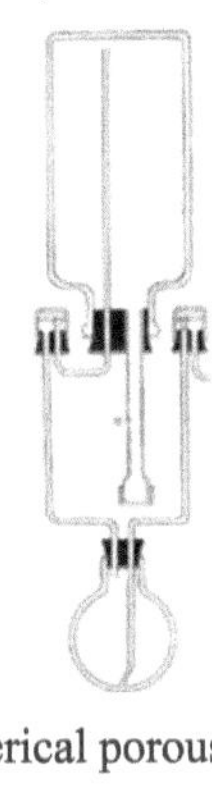

Plate 75: Spherical porous cup atmometer

Atmosphere generator: These are used to remove O_2 by N_2 or burning O_2 in air with propane.

Atomization: Atomization is the most important singe step in spray drying. Atomization means to reduce large particles to minute size /atom by a fine spray (**Plate 76**).

Auto recorder: It is a milking machine comprising a container on a spring balance and is used to record the weight of milk from each cow.

Autoclave: A steam pressure vessel that is sterilized by steam under pressure at temperature above 100°C. It is a strong sealed vessel in which super heated steam under pressure is used for sterilizing or cooking(Plate 77).

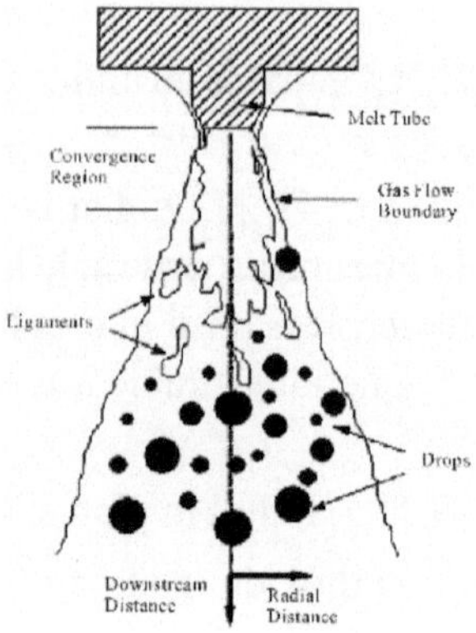

Plate 76. Atomization

Plate 77: Autoclave

Autoclaving: The most practical and dependable process for sterilization of a medium where heat in the form of saturated steam under pressure is the main killing agent which is carried out in a autoclave at 121^{0}C (lb/inch^2 pressure) for atleast 20 mint.

Autointoxication: It means self-poisoning which occurs due to the absorption of the uneliminated toxins produced within the body.

Autolysis: It is the process of self digestion of the cell or tissues by the enzymes present in the tissue (self produced enzymes) which leads to deterioration of quality of fresh foods especially (Plate 78).

Automation: The automation is the technique of making an apparatus, a process, or a system which operate automatically under controlled conditions by mechanical or electronic devices and take the place of human labor (Plate 79).

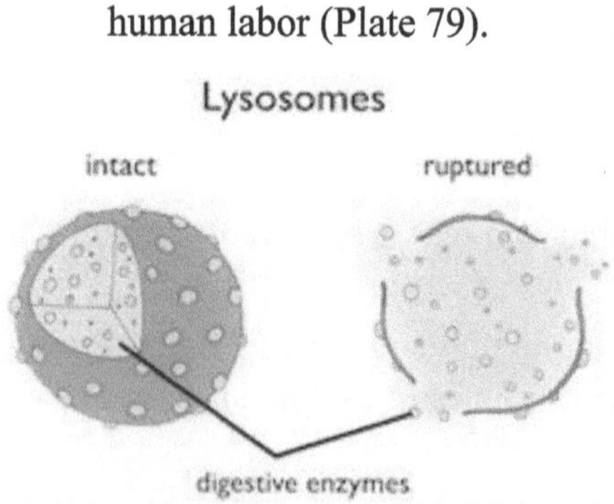

Plate 78: Autolysis by lysosome

Plate 79: Automation

Auto-sterilization: Killing of microorganisms by their own metabolic end products commonly found in case of micro-organisms causing flat sour spoilage in canned foods.

Autotrophic: Organisms capable of manufacturing complex organic nutritive compounds from simple inorganic sources e.g., green plants.

Autotrophs: Organisms capable of manufacturing complex organic nutritive compounds from simple inorganic sources e.g. green plants (Plate 80).

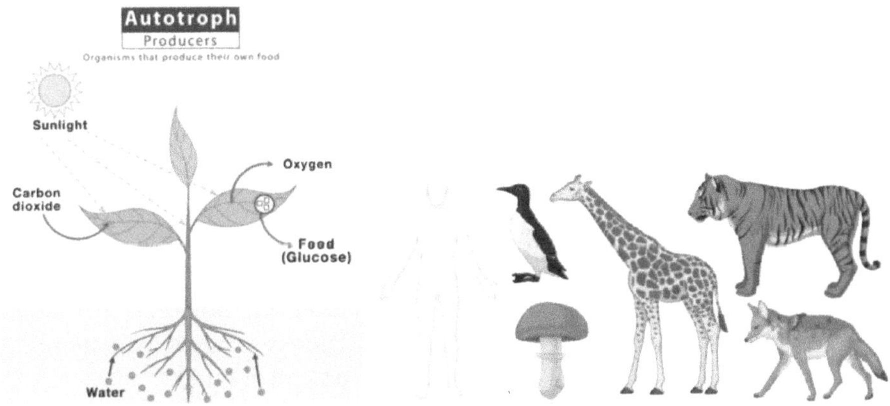

Plate 80: Autotrophs vs Heterotrophs

Autoxidation: The process of oxidation of foods by direct combination with oxygen (as in air) at ordinary temperatures.

Auxin: A category of plant growth regulators that can control rate of cell enlargement, retard abscission of young or mature fruit, stimulate ethylene synthesis in fruits, rooting in stem cutting, induce cambial division, inhibit lateral bud development and may also act as herbicide e.g. Indole acetic acid (IAA) which is found in plants and synthetic compounds Naphthalein acetic acid (NAA), NAAm (Naphthaleneacetamide), B-nepthoxyacetic acid (BNOA), IBA, 2,4-D; 2,4,5-tri iodobenzoic acid (TIBA), 2,4,5-T; 2,4,5 –Trichlorophenoxy propoinic acid (2,4,5 –TP).

Auxin synergists: Refers to a number of non-auxinic chemicals particularly phenolic compounds which synergise the action of exogenously applied auxins in the regeneration of roots e.g. parahydyoxybenzoic acid (PHB), ferulic acid etc.

Auxotrophic: These microorganisms are requiring a specific growth substance beyond the minimum required for normal metabolism and reproduction by the parental or wild-type strain / mutant strains of bacteria.

Available water: It is the ratio of the water vapour of the solution to the water vapour of saturated air and is measured in term of water activity. $A_w = P/P_o$.

Avidin: Avidin is an egg white protein (0.05%) which binds with biotin and makes it unavailable (Plate 81).

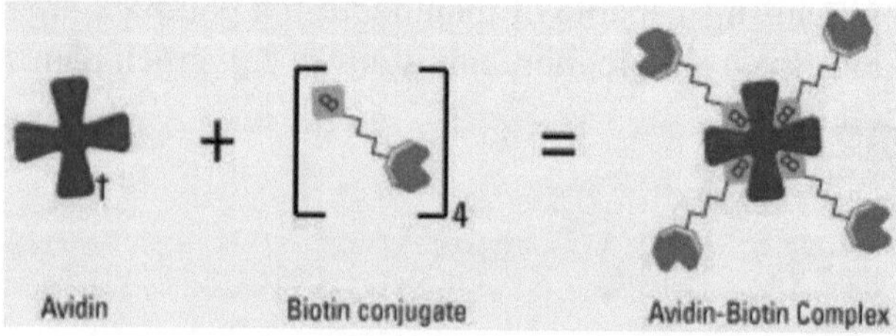

Plate 81: Avidin Biotin Complex

Avitaminosis: It refers to the deficiency of vitamins.

Avogadro's number: These are number of molecules in 1g mole of gas i.e. 6.023 x 10^{23}. Avogadro's in 1811 introduced the principle that equal volume of gases at the same temperature and pressure contain the same number of molecules. It is also *called* Avogadro number.

Azeotrope: It is a liquid mixture that is characterized by a constant minimum or maximum boiling point which is lower or higher than that of any of the components. or It is a stable mixture of refrigerant whose vapour and liquid phase retain identical composition over a wide range of temperature. Each azeotrope has a characteristic boiling point. The boiling point of an azeotrope is either less than the boiling point temperatures of any of its constituents (a positive azeotrope), or greater than the boiling point of any of its constituents (a negative azeotrope). For both positive and negative azeotropes, it is not possible to separate the components by fractional distillation and azeotropic distillation is usually used instead. The term *azeotrope* is derived from the Greek words with the prefix "a" mean "no", "zeo" mean "boil" and trope mean "turning / change" to give the overall meaning, "no change on boiling". The term was coined in 1911 by English chemist John Wade and Richard William Merriman. Due to this, azeotropes are also called *constant boiling point mixtures,* because their composition is unchanged by distillation,.

Azeotropic drying: This process use solvent to form mixture and this mixture decrease the boiling point of water and thus help in faster drying.

B

B.T.U.: It stands for British thermal unit. The heat required to raise the temperature of one pond of the water through 1°F is called BTU (Plate 82).

Baby back ribs: Meaty pork ribs cut from the lower back rib section.

Bacitracin: A polypeptide antibiotic isolated from a bacillus (*Bacillus subtilis or B. licheniformis*) and usually used against gram-positive bacteria.

Back Bacon / Canadian bacon: Bacon cut from the loin that has little fat and is cut into round or oblong slices.

Backward feed evaporator: In this evaporator, fresh feed enters the last and the coldest effect and continues on until the concentrated product leaves the first effect (Plate 83).

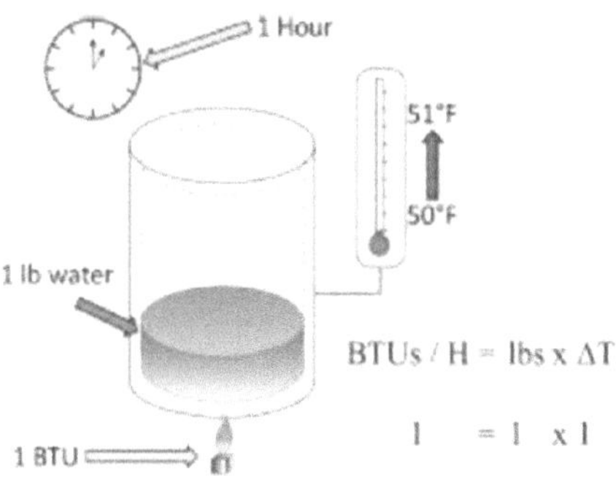

Plate 82: BTU

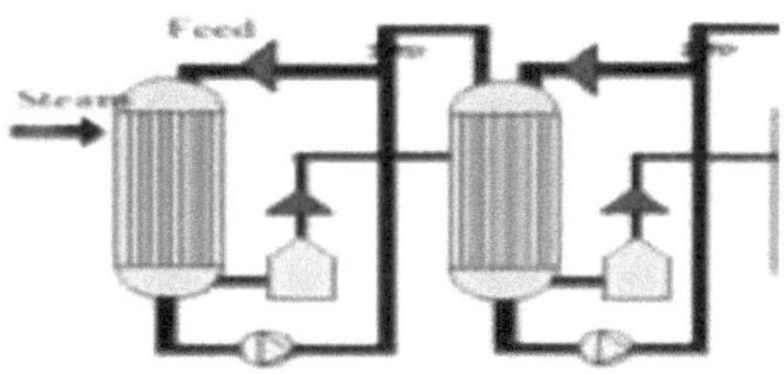

Plate 83: Backward feed evaporator

Bacon: Bacon is the meat cut from belly portion of hog carcass (a domestic swine when weighing more than 120 pounds or 54 kilograms) and has high percentage of fat. It is cured and smoked.

Bacon side: A side of a pig cured and smoked is called bacon.

Bacteria: These are unicellular (one celled microorganisms), microscopic organism lacking in chlorophyll (shizomycetes), may cause several diseases in plants and spoil the preserved products (Plate 84). Bacteria are broadly divided into two types i.e. Gram +ve and Gram –ve, depending upon a simple colour reaction produced by gram staining (Plate 85). As a general rule gram –ve bacteria tend to exert their effect through invasion of the host (food borne infection) whereas the effects of Gram +ve bacteria are usually mediated via preformed toxins (food poisoning). Mostly bacteria exhibit three types of shapes viz. cocci (spherical), bacilli (rods) and spirilla (twisted rods) (Plate 86).

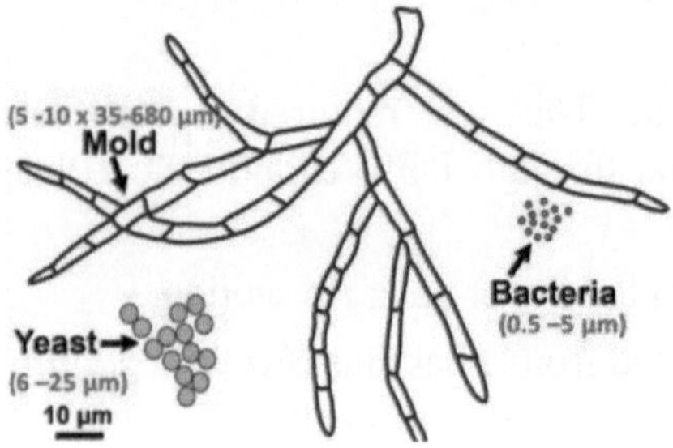

Plate 84: Size of Bacteria, yeast and mold

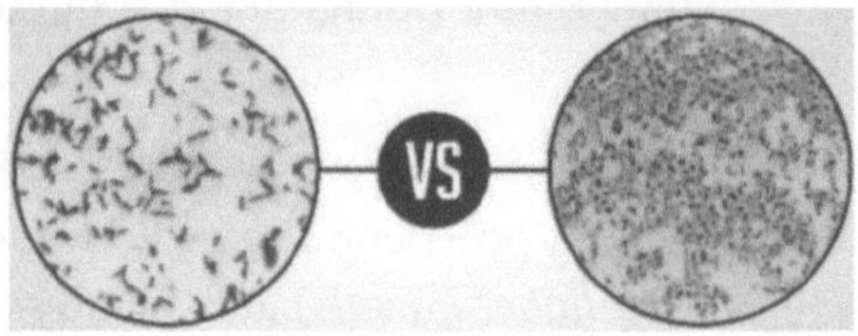

Plate 85: Gram positive vs Gram Negative bacteria after Gram staining

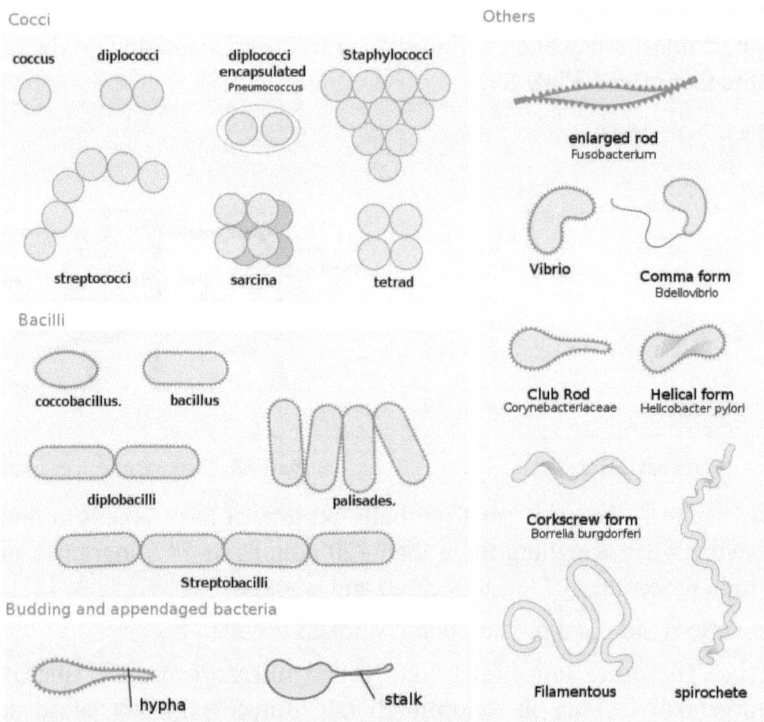

Plate 86: Shapes of different bacteria

Bacterial food infection: Food poisoning due to the entrance of bacteria into the body through ingestion of contaminated foods and the reaction of the body to the presence to their metabolites.

Bacterial food intoxications: Food poisoning due to the consumption of foods contaminated with toxin formed in food due to bacterial invasion prior to eating.

Bacteriocidal method: Method in which most of the micro-organisms are killed by preserving through heating, pasteurization or sterilization.

Bacteriocin: It is an essential biologically active protein i.e. bacteriocins are small bacterial proteins moiety possessing a bacteriocidal mode of action. The bacteriocins are small bacterial proteins / peptide with bactericidal properties. These are used as biopreservative against bacteria to arrest their growth in the food material.

Bacteriostasis: Bacteriostasis is the inhibition of the growth of bacteria without destruction

Bacteriostatic method: Method in which micro-organisms unable to grow in food/food product. These methods are dehydration, salting, smoking or freezing.

Bactofugation: It is the process of removal of the bacteria from milk by centrifugation. Bactofugation is a process in which specially designed centrifuge called Bactofuge is used to separate micro- organisms from milk (Plate 87).

Bactofuge: A bactofuge is a high-speed centrifuge specially designed to remove bacterial spores from milk at high temperatures.

Bactotherm Process: Clarified and standardized milk is pumped into a plate heat exchanger, where it is heated to a temperature of 60-75°C prior to being fed to the bactofuge. The centrifugal acceleration is increased to 10,000g.

Bag – In Box: In a bag in Box concept, a pre-made box having a bag inside is loaded in a machine. The machine automatically opens the box, closes the bottom, opens the bag and fills the product. Spice, Tea and Custard powder are major uses of such machines. Tin cans, drums, large bottles, jerry cans etc are no longer the preferred choice for sensitive liquids, pulps and concetratrates. These products have varying pH values, Brix values and other properties which have to be evaluated and catered to when considering storage and packaging conditions. Bag in box packaging is indeed an innovative packaging system to meet variety of needs for retail, intermediate and bulk packaging (Plate 88).

Bagasse: The mill residue obtained from sugarcane industry is called bagasse. It consists of crushed stalks from which the juice has been extracted (Plate 89).

Bagel: It is a firm doughnut shaped roll traditionally made by boiling and then baking (Plate 90).

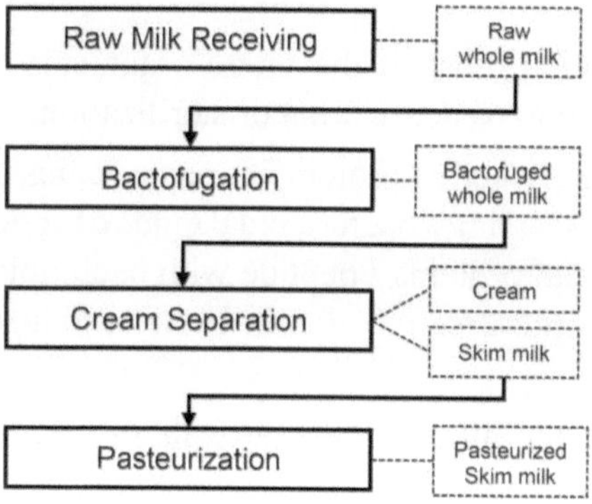

Plate 87: Bactofugation

Plate 88: Bag – In Box

Plate 89: Bagasse

Plate 90: Bagel

Bail: It is a device used for transporting animals.

Bail milking: The milking of the cows in a portable shed is called bail milking.

Baker's yeast: Yeast (*Saccharomyces cerevisiae*) used or suitable for use as a leavening agent is called baker's yeast.

Baking: Baking is the process to cook or dry or harden (as food) by dry heat especially in an oven. Baked products include bread, cookies, pies, and pastries. Ingredients used in baking include flour, water, leavening agents (baker's yeast, baking soda, baking powder), shortening (fats, oils), eggs, milk, and sugars. These are mixed together to create dough or batter, which is then transferred to a pan or sheet and heated. Leavening agents produce gas that becomes trapped in the dough, causing it to rise. Shortening makes dough more easily workable and the final product more tender. Egg whites are used to produce a light, airy texture, and yolks contribute colour, flavour, and texture. Milk is used for moistening and flavouring and sugars for sweetening and aiding fermentation.

Baking powder: A powder used as a leavening agent in making baked goods (as quick breads) that typically consists of sodium bicarbonate.

Balance diet: The food taken by the human being is called diet. A diet which contain adequate amount of all essential nutrients like carbohydrate, fats, proteins, minerals and vitamins for normal growth is called a balance diet.

Balling: The measurement of sugar percentage in the sample syrup at a fixed temperature is called balling. It is also called brix.

Balling Scale: Balling Scale is a measurement (expressed as °**Balling**) of the concentration of dissolved solids (mainly sugars) in a brewery wort (Plate 91).

Banana flour: The meal prepared from fully mature unripe bananas in called banana flour or banana meal.

Bar code: A code consisting of a group of printed and variously patterned bars and spaces and sometimes numerals that is designed to be scanned and read into computer memory and that contains information (as identification) about the object it labels. It is called bar code (Plate 92).

Plate 91: Balling Scale

Plate 93: Bar code used on packages

Bar code symbology: The automatic identification of the product with the help of computer by optical character recognition, voice data entry, machine vision, radiofrequency, data communication, magnetic strips and smart cards.

BAR: It stands for Brix to Acid ratio.

Barbecue / barbeque: It is a place where large animals like steer, meat and fish are roasted whole or split over an open fire or fire pit or portable fireplace. Or It is a place where barbecued food is eaten in a social gathering in the open air (Palte 93).

Plate 93: Barbecue

Barbecueing: It is a process to roast or broil on a rack or revolving spit over or before a source of heat (as hot coals). The cooking in a highly seasoned vinegar sauce is also known as barbecueing.

Barfoed's Test: It is a biochemical test which is used for testing monosaccharides content. Barfoed's agent is a solution of copper acetate and acetic acid. It is added to the sample. The positive results indicate development of red precipitates and the negative results indicates no precipitation. In this test cupric acetate on reduction converted to red insoluble cuprous oxide (Plate 94).

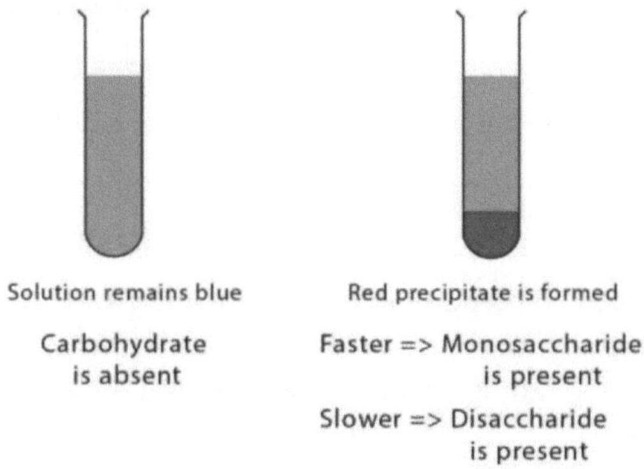

Plate 94: Barfoed's Test

Barley water: It is essentially the squash of a fruit in which a part of added water is replaced by an extract of dehusked polished pearl barley. It is a fruit beverage containing minimum of 25 % juice, 30% TSS, 0.25% barley starch and 1% acid (Plate 95).

Barm: The froth of the fermented liquor is called barm. Yeast formed on fermenting malt liquors is also called barm (Plate 96).

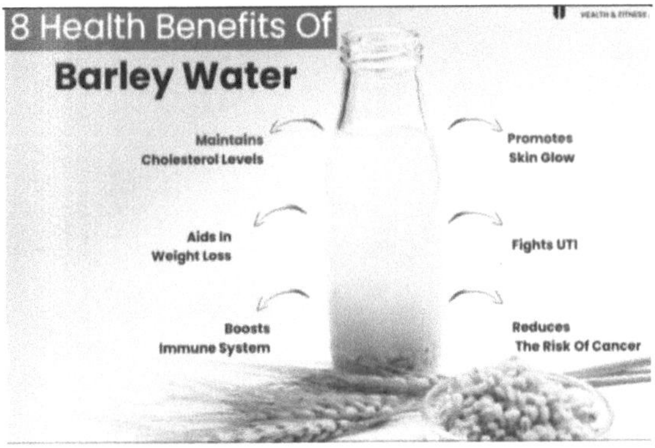

Plate 95: Barley water

Plate 96: Barm

Basal metabolic rate: The rate at which heat is given off by an organism at complete rest is called basal metabolic rate.

Bating: It is one of the steps used in the tanning of the leather (Plate 97).

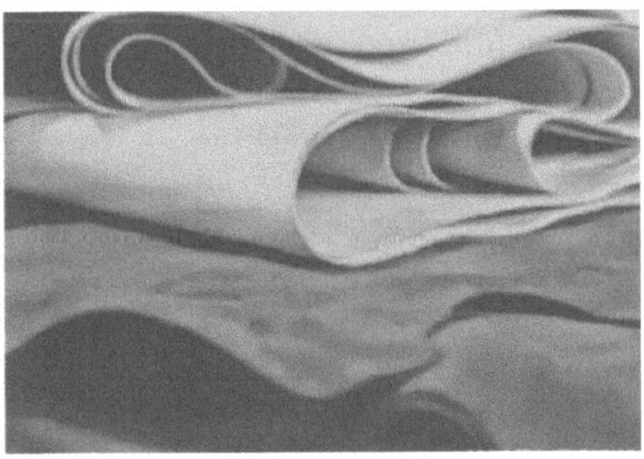

Plate 97: Tanning of Leather

Bath tub gin: Preparation of gin by illegal methods is called bath tub gin. The gin is prepared when essential oils may be added to the diluted alcohol in which no further distillation is required.

Batter: It is a mixture consisting chiefly of flour, egg, and milk or water and being thin enough to pour or drop from a spoon **or** a mixture (as of flour and egg) used as a coating for food that is to be fried or to coat (food) with batter for frying.

Battering: Battering is also known as enrobing. It is blending of batters i.e. flour, leavening (rising agents) and seasoning. The process involves dipping of food products in semi liquid.

Baume hydrometer: It is hydrometer used for measuring the brix or total soluble solids of the food. Liquid samples are required for measurement (Plate 98, 99 and 100).

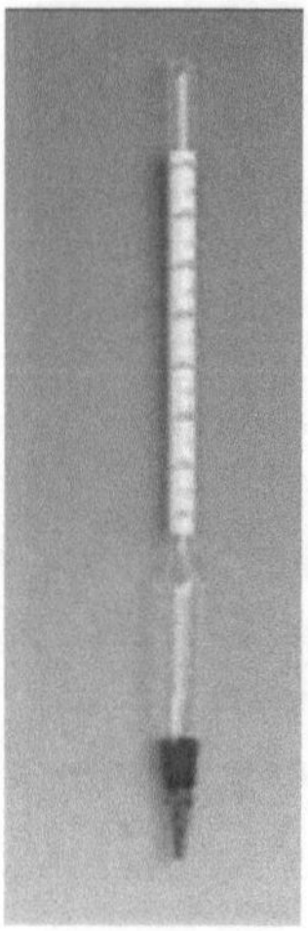

Plate 98: Baume hydrometer

Plate 99: Baume hydrometer

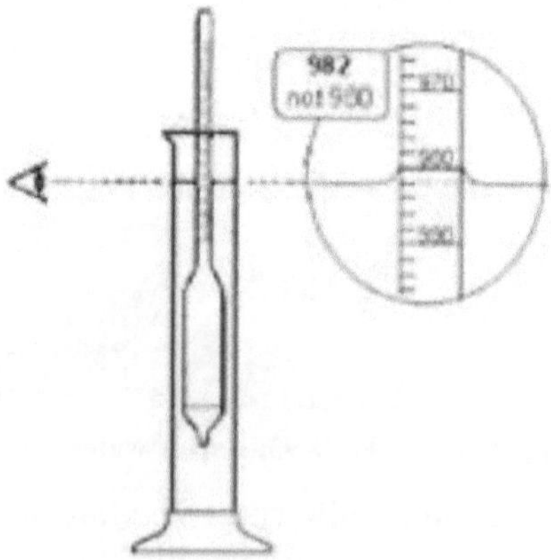

Plate 100: Baume hydrometer Working

BCP Test: It is a test which is used to determine the presence of antibiotic in the milk. It involves the addition of the bacterial culture to the milk together with the colour indicator Bromo Cresol Purple (BCP). If no antibiotic is present then culture grows and produce lactic acid and colour changes from green to yellow. If antibiotic is present, then the culture do not grow and no LA formed and no colour change occur.

Beamage: It is the loss of the weight of the animal carcase by evaporation.

Beating: It is the process of moving a device back and forth to blend ingredients together to achieve a smooth texture is called beating. It is practiced in making an omelette.

Becquerel (Bq): It is that quantity of the radioactive substances in which one unit atom disintegration per second. Earlier the unit was curie.

Beef: It is the term applied to meat of cattle over one year. or The flesh of an adult domestic bovine (as a steer or cow) used as food.

Beefsteak tomato: It is a very large globe shaped red tomato with dense flesh.

Beer: A class of alcoholic beverage brewed from malt or malt substitute with the addition of hops to give a bitter taste. Or an alcoholic beverage made usually from malted barley, flavoured with hops, and brewed by slow fermentation (Plate 23). It is produced by employing either a bottom fermenting yeast, which falls to the bottom of the container when fermentation is completed or top-fermenting yeast, which rises to the surface. Lager beers are bottom-fermented and stored at a low temperature for several months; mostly light in colour, with high carbonation, medium hop flavor, and alcohol content of 3–5% by volume. Top-fermented beers include ale, stout, and porter; characterized by a prominent released carbon dioxide, a sharper and more strongly hopped flavour than lagers, and an alcohol content of 4–6.5% by volume. Lager yeasts typically settle at the bottom of the fermentor kept at 8-14^{0}C while ale fermentation processed at 16-12^{0}C and yeast rises to the top (Plate 101).

Plate 101: Beer in glass

Benedict's test: This test is used for testing of the sugars. This test gives positive results if large quantity of sugar is present otherwise Fehling test can be used.

Bentonite: Absorptive and colloidal clay used as a sealing agent or suspending agent.

Benzoic acid: It is a preservative which is effective against mold and bacteria. It is for preservation of Jam, jellies and other fruit juice beverages @ 50-750 ppm. Cranberries contain upto 5% of benzoic acid. It is white crystalline acid C_6H_5COOH found naturally (benzoin or cranberries) or made synthetically and used as a preservative of foods, in medicine and in organic synthesis.

Beriberi: It is a deficiency disease caused by a lack of or inability to assimilate thiamine and marked by inflammatory or degenerative changes of the nerves (polyneuritis). Edema, emaciation (becoming lean), digestive system, and cardiac disturbances (enlargement of heart and very rapid heartbeat) are the characteristic symptoms of beriberi (Plate 102).

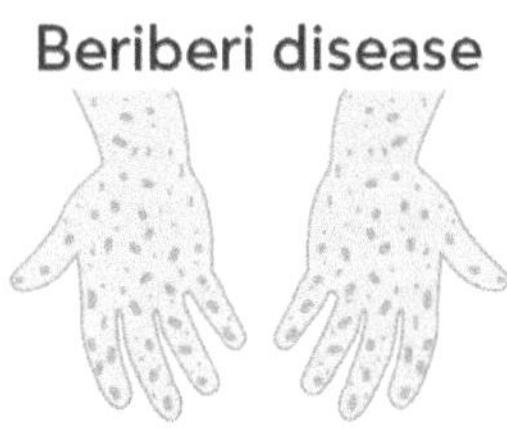

Plate 102: Beriberi disease

Berry wine: The wine prepared from berries like cran berry, blackberry and rasp berry is known as berry wine.

Beta ray: The radiant energy that is longer than the gamma ray and can penetrate food. It can not penetrate aluminium.

Beta-carotene: It is a fat soluble plant pigment and is a precursor of vitamin A. It is an isomer of carotene found in dark green and dark yellow vegetables and fruits. The name beta-carotene is derived from the Latin name for carrot. It gives yellow and orange fruits and vegetables their rich hues. Beta-carotene with the molecular formula $C_{40}H_{56}$ is also used to color foods such as margarine. It is a member of the carotenes, which are terpenoids, synthesized biochemically from eight isoprene units and thus having 40 carbons (Plate 103).

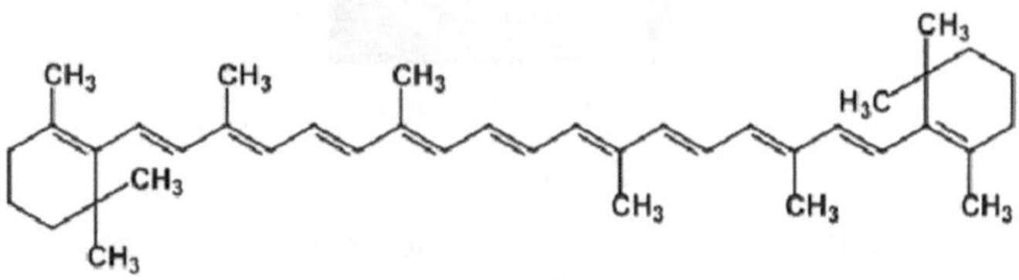

Plate 103: Beta- carotene

Betalins: These are group of red and yellow pigments that are found in red beet. These two groups of the pigments contribute to the anthocyanins like colour to beets but differ from anthocyanins.

Beverage: A drink or fruit juice drink which contains less than 100% fruit juice is called beverage. It is a drinkable liquid fermented or unfermented, sweeten or unsweetened.

BHA / butylated hydroxyanisole: A phenolic antioxidant $C_{11}H_{16}O_2$ used to preserve fats and oils individually or in food.

BHT: A phenolic antioxidant $C_{15}H_{24}O$ used to preserve fats and oils individually or in food in food.

Big neck: A goiter of animals which occurs due to deficiency of iodine.

Bingham plastic: The plastics which resist a small shear stress but flows easily under larger shear stresses are known as Bingham plastics e.g. tooth paste, jellies etc.

Biocatalyst: A biological entity exhibiting enzyme activity is called biocatalyst.

Biochemical Oxygen Demand (BOD): It is the quantity of oxygen used in the biochemical oxidation of organic matter in a specified times under specified condition, usually as mg of the oxygen used per litre of the effluent at 20°C in a five day period. The oxygen used in meeting the metabolic needs of aerobic microorganisms in water (water polluted with sewage) which is rich in organic matter. BOD is a measure of the amount of the oxygen consumed by microorganisms by biological process (aerobically) for breakdown the organic matter. It is the quantity of oxygen utilized by bacteria in degradation of organic substance is known as Biological Oxygen Demand (BOD). Aerobic bacteria are involved in BOD. BOD is used as an index of organic pollution in water i.e. swages water. More the oxidisable organic matter present in water more is the amount of oxygen required degrading it biologically, hence more is the BOD. BOD is evaluated by measured concentration of oxygen in sample before and after incubation in dark at 20°C for 5 days. BOD is directly proportional to organic waste in water. BOD is used as a measure of waste strength and also an indicator of degree of pollution.

Bio-chemistry: A science that deals with the composition, structure, and properties of substances and with the transformations that they undergo in a living system is known as bio-chemistry.

Biocompost: A mixture of organic matter, as from leaves and manure, that has decayed or has been digested by organisms, used to improve soil structure and provide nutrients (Plate 104).

Biocomposting: The process of conversion of the waste material with the help of living organisms into compost is called biocomposting. Bio-composting is the aerobic bio- degradation of organic materials under controlled conditions, resulting in a rich humus-like material.

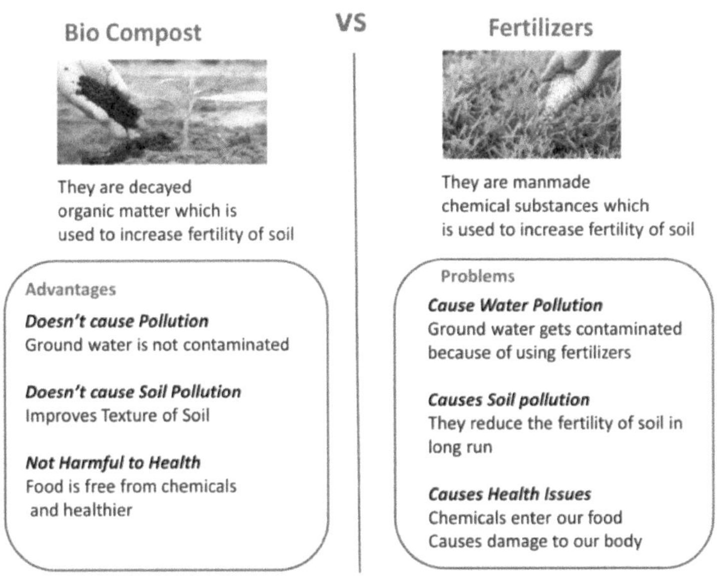

Plate 104: Biocompost vs Fertilizers

Bio-degradable plastics: Bio-degradable plastics are those which can be decomposed into CO_2 and H_2O by the action of naturally occurring microorganisms e.g. Starch polysaccharide based plastics.

Bioflavonoids: It is also known as vitamin P and is a group of water soluble substance. These possess better free radical scavenging power. It is one of the biologically active flavonoids (hesperidin and quercetin) derived from plants, and found in citrus fruits and vegetables (Plate 105 & 106).

Plate 105: Citrus bioflavonoides

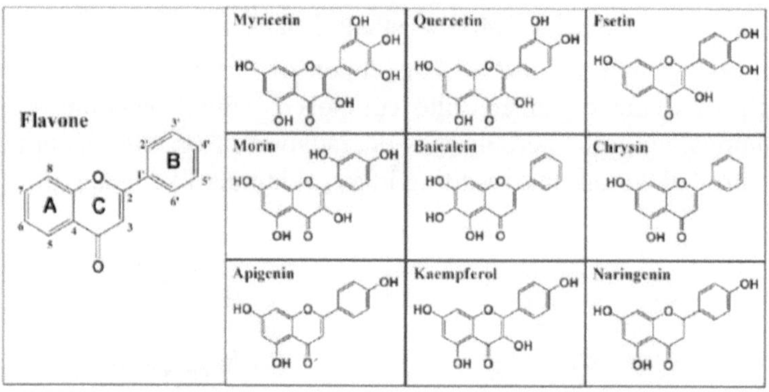

Plate 106: Chemical structure of different Flavonoides

Biological hazards: Danger posed to food safety by the contamination of food with pathogenic micro-organisms or naturally occurring toxins. Biological hazards usually cause greatest and broadest danger to consumers. Major biological hazard in foods is the microbiological contamination caused by various organisms like bacteria, moulds, insects & their fragments.

Biological indicator units (BIUs) for sterilization: Plastic rods called biological indicator units (BIUs) containing bacterial spores. These are introduced into the container along the central vertical axis to the cold point. During the sterilization the microbes present in the plastic rod act as indicator. In the test some of the containers are filled with BIUs and some are filled with thermocouples. A biological indicator is made up of a carrier material, on which bacterial spores with a defined resistance to the sterilization process have been applied. The carrier material is enclosed within a glassine envelope or a vial (Plate 107 and 108).

Plate 107: Warnig sign of Biological hazards

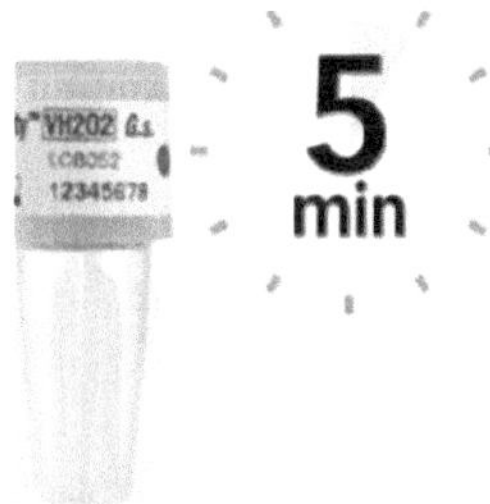

Plate 108: Biological indicator units for sterilization (BIUs)

Biological magnification: It is the process of gradual increase in the concentration of the harmful chemicals in the body of higher animals through food chain.

Biological score: It is a chemical method of defining the nutritive value of proteins as proposed by Block and Mitchell (1946). Chemical score numerically equals to protein score. It is also known as Chemical score.

Biological value: It is the quantitative measurement of the nutritional value of proteins in food i.e. it is the amount of protein retained in the body for repairement or body growth. Or Biological value (BV) is a measure of the proportion of absorbed protein from a food which becomes incorporated into the proteins of the organism's body (Plate 109).

Biological Value of Proteins	
Egg	94
Milk	90
Rice	83
Quinoa	83
Fish	76
Beef	74
Soybeans	73
Whole wheat	64
Beans	58

Plate 109: Biological Value of different foods

Biomass: The amount of living matter / plant materials/ animal waste produced per unit area or volume of habitat and used as a source of fuel / energy is called biomass. Biomass is the total amount of the living material stored in the body per unit area.

Biopol: Biopol is poly hydroxybutyrate and is thermoplastic material and marketed as biopol by ICI. It can be easily broken down by bacteria withstand 100^{0}C. It is biodegradable plastic as well as thermoplastic material consists of polyhydroxybutyrate (Plate 110). Biopol PHB, can be injection moulded and vacuum formed. It has a range of uses such as, packaging, shampoo bottles, disposable razors, disposable cups, surgical stitches, surgical pins, disposable knives and forks, woven medical patches and nappy linings.

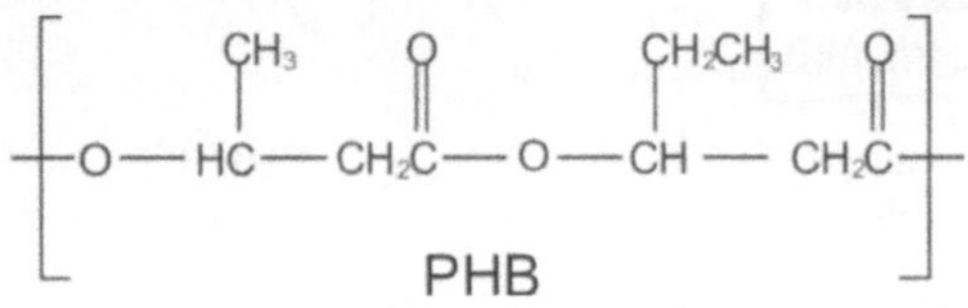

Plate 110: Biopol

Biosynthesis: The synthesis of complex compound from simple substances in a living body is called biosynthesis.

Biotechnology: The manipulation (as by genetic engineering) of living organisms or their components to produce useful commercial products (as pest resistant crops, new bacterial strains, or novel pharmaceuticals).

Biotin: A colorless crystalline growth vitamin $C_{10}H_{16}N_2O_3S$ of the vitamin B complex found mostly in yeast, liver, and egg yolk. It is also called as Vitamin H.

BIS: It stands for Bureau of Indian Standards (1986). It is mandatory in nature and establishes standards for weight and measures (Plate 111).

Biscuit: It is hard / crisp dry baked products. A small quick bread made from dough that has been rolled out and cut or dropped from a spoon is called biscuit.

BITOT's spots: These are foam like irregular plaques on the surfaces of the eyes due to deficiency of vitamin A (Plate 112).

Bitter pit: Bitter pit is a physiological disorder of apple due to Ca deficiency (Plate 113).

Black body: A black body in one whose absorptivity is cent (hundred) percent. Black body absorbs all the incidence radiations. Perfect black body does not exist (Plate 114).

Black neck: The formation of black ring in the neck of bottles containing ketchups.

Black tip: It is a disorder of mango caused due to Brick-Kiln fumes and is characterized by blackening and hardening of the distil end of the fruits which ripe prematurely. or Blacktip is the etiolation (yellowing) of distal end of the fruit, the monocarp and seed being unaffected. Brick kiln fumes like sulphur dioxide, ethylene and carbon monoxide result blacktip (Plate 115).

Blanching: Heats treatment of vegetable in boiling water (Plate 116) or steam for 2-5 mint to inactivate enzymes and to destroy peroxides. or It is a unit operation in which food is heated rapidly to a predetermined level to inactivate enzymes and to destroy peroxides. It is also known as scalding, par boiling or pre-cooking. The blanching operation usually takes place at 85 to 90 °C but below 100°C. The produce after blanching is cooled immediately.

Plate 111. BIS logo

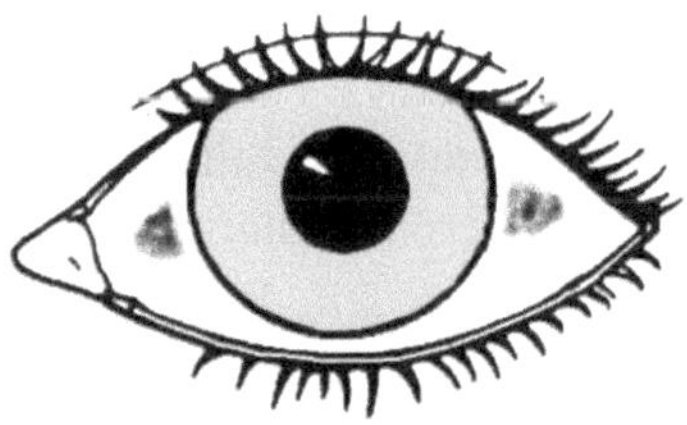

Plate 112. BITOT's spots

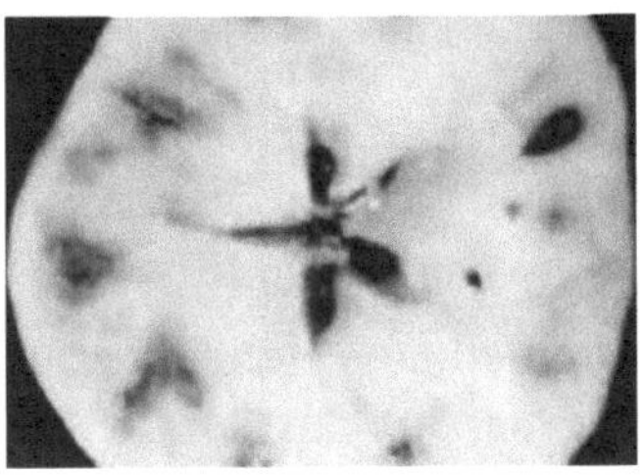

Plate 113. Bitter Oit

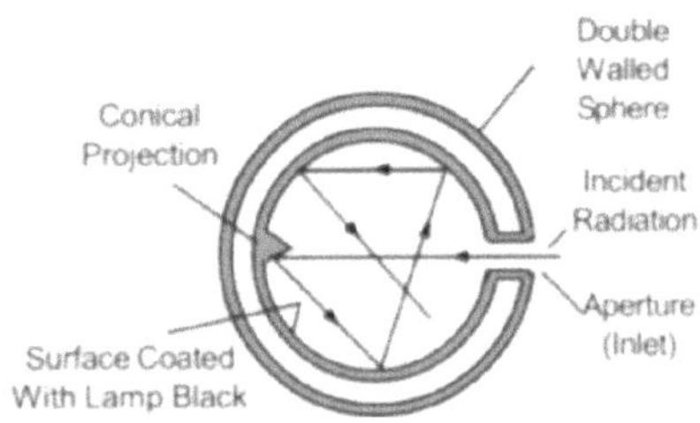

Plate 114. Black body

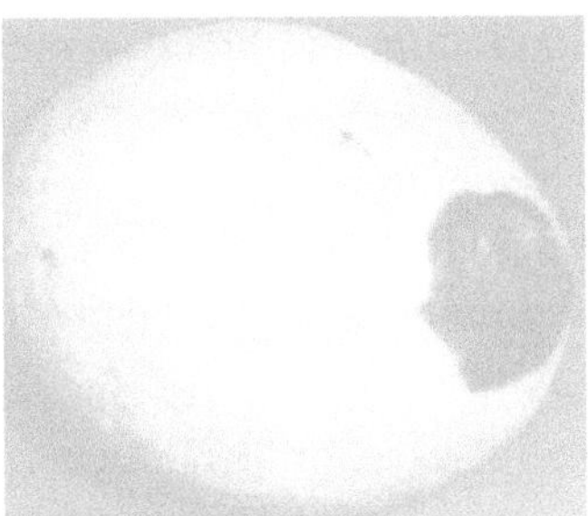

Plate 115: Black tip of mango

Plate 116. Blanching operation

Blanching indicator: Peroxidase and catalase enzymes are used as indicator of sufficient blanching. These are most heat resist enzyme of plant origin.

Blancmange: The sweetened and flavored dessert made from gelatinous (starch) ingredients and milk is called Blancmange (Plate 117).

Plate : Blancmange

Bland diet: It is a food that contains a minimum of crude fibre or roughage and is there fore non-irritating and soothing to the intestine.

Bland oil: Oil often used as inert, soothing ingredients in medicines e.g. olive oil etc.

Blast freezing: The blast freezing (-30 to -40°C) is the freezing in which cold air is blown across the food. It is freezing in air blast freezer with rigid air circulation at high velocity. It causes rapid freezing and small ice crystal formation. The Blast Freeze process reduces temperature from +70°C to –18°C in no more than 240 minutes. Current food legislation requires that all food must be cooled as quickly as possible.

Bleaching Powder: A white powder consisting chiefly of calcium hydroxide, calcium chloride, and calcium hypochlorite and used as a bleach, disinfectant, or deodorant

Bleeding: The heavy / uncontrollable escape of blood from vessels or the hemorrhage, a copious discharge of blood from the blood vessels is known as bleeding.

Blends: Blend means something produced by blending or a product prepared by blending. After blending individual component cannot be seen and the product become stabilized. Blender is used for blending of the products.

Blending: The process of mixing two or more different products together to obtain the desired final production. Blender is used for blending purpose (Plate 118).

Plate 118: Blender (electrically operated)

Blister pack: It is a kind of plastic flexible film packaging where individual pieces or entities of food are enclosed in a blister formed of the packaging film (Plate 119). Or It is a package holding and displaying merchandise in a clear plastic case sealed to a sheet of cardboard. Blister packaging, Blister Packs, or simply blisters, are terms used to describe several types of pre-formed clear plastic cavities typically used for securely packaging small consumer goods. Often these packages will hang on a merchandising wall.

Blood meal: The residue from the slaughterhouse, rich in protein, low in minerals content and used as feedstuff for poultry.

Bloom gelometer: It is the instrument used for measuring the tenderness of the gels. It is the modification of the penetrometer (Plate 120).

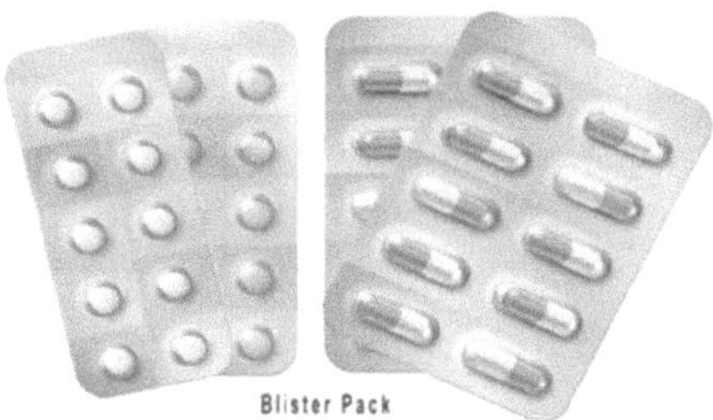

Plate 119: Blister Pack

Plate 120: Bloom gelometer

Blossom end rot of the grapes: It is a disorder of grapes due to Ca deficiency and characterized by black sunken spot developed at the blossom end of the berry which later on spread with water soaked area.

Blower: A can with tightly bulged ends due to formation of CO_2 and other gases inside the can as result of microbial decomposition of foods is called blower (Plate 121). This food has an offensive and sour odour and is generally discoloured and unfit for consumption. It is also called Swell.

Plate 121: Bulged Can / Blower can

Blue value index: It is primarily used to measure the quantity of straight chain amylase and not of the branched amylopectin, which is present is larger quantity than amylose. The intensity of the resultant starch iodine blue colour is measured colourimetrically and is the index of the free starch.

Bock beer: A special lager beer is called bock beer. It is made during spring and prepared from very concentrated wort with little hops. It contains considerable amount of unfermented sugars and shelf life is very short. It is generally used as spring tonic.

Boil in bag Packs: Boil in bag packs are used for the packaging of prepared meals such as frozen vegetable, biryanis, and shrimps etc which require only reheating by immersion in boiling water before being opened over a serving dish or plate. The essential requirement of a Boiling Bag Pack is that it should withstand the temperature and stresses involved in the reheating operation (Plate 122).

Boiler: It is a vessel used for boiling **or** the part of a steam generator in which water is converted into steam and which consists usually of metal shells and tubes (Plate 123).

Plate 122: Boil in bag

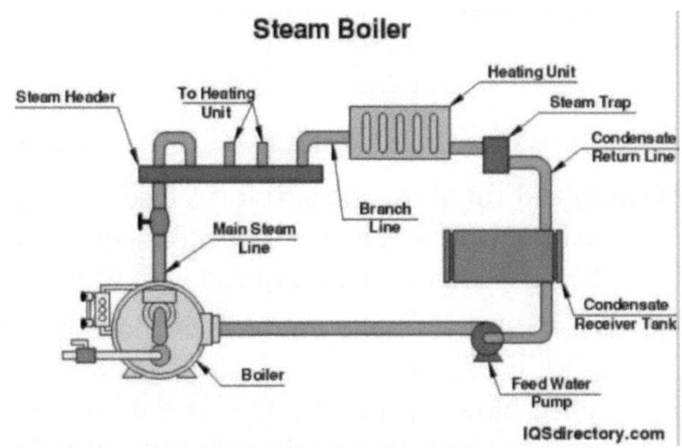

Plate 123: Steam boiler

Boiler corrosion: The decay or disintegration of the material of steam boiler due to chemical or electrochemical reactions is called boiler corrosion.

Boiling point: It is the temperature at which any liquid boils at a given pressure and the material is in equilibrium both as liquid and gas. It is the temperature at which the pressure on the surface of the liquid is equal to the standard atmospheric pressure i.e. 760 mm Hg. Boiling boiling of a substance is constant at a particular pressure.

Boiling wine: The palm sap (from spathe) is allowed to ferment naturally for 24 hrs before sale. The bottles have constant foaming/bubbling due to the fermentation process called as herbiendo or boiling wine. In 30 kg sap about 4.5 kg sugar is added. Or Boiling wine removes much of the alcohol in that wine, affecting its taste. Non-alcoholic wine brands that use heat to remove the alcohol in their wines do so very carefully, making sure not to cook the wine.

Bomb calorimeter: It is an apparatus for measuring quantities of absorbed / emitted heat or for determining specific heats of the food (Plate 124, 125).

Bombe dessert: A frozen dessert containing ice cream and formed in layer in a mold is called bombe. It is a French ice cream *dessert* frozen in a spherical mould so as to resemble a cannonball, hence the name ice cream *bomb* (Plate 126).

Bonbon: A candy with chocolate or fondant coating and fondant centre that some times contains fruits and nuts is called bonbon / banban. A *bonbon*, sometimes *bon-bon*, is a small chocolate confection (Plate 127).

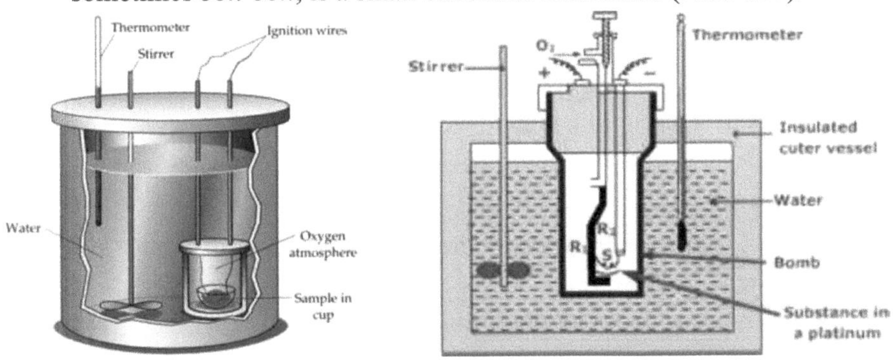

Plate 124: Typical sketch diagram of bomb calorimeter

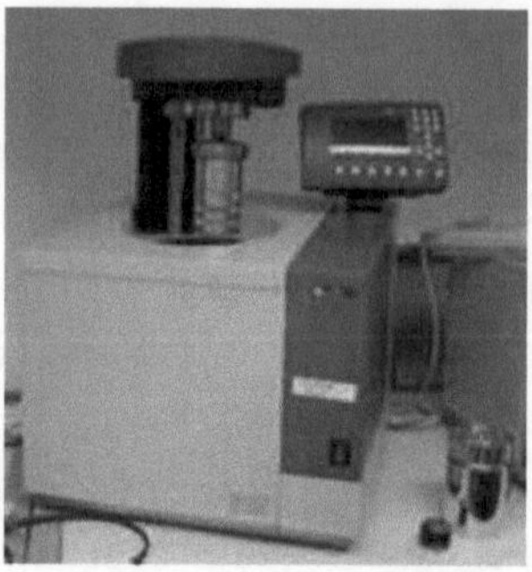

Plate 125: Bomb calorimeter

Plate 126: Bombe dessert

Plate 127: Bonbon

Bonded whiskey: A 100-degree proof whiskey aged for at least four years under government supervision before being bottled is called bonded whiskey.

Bone dry food: A food which is completely dried is called bone dry food.

Bone fish: It is a slender silvery small scaled fish (*Albula vulpes*) that is a food fish of warm seas.

Bone matrix: It is the protein ground work in which the minerals are deposited.

Bone meal: The crushed or ground bone used as fertilizer or feed is called bone meal.

Bordeaux: A white or red wine of the Bordeaux region of France is called Bordeaux.

Boro-silicate glass: It is the glass made up of SiO_2 (85%), B_2O_3 (12.9%), Na_2O (3.8%), K_2O (0.4%) and Al_2O_3 (2.2%). It is called called Pyrex.

Bostwick consistometer: Bostwick consistometer is used to measure the viscosity of tomato ketchup and sauce (Plate 128).

Bottle fermentation method: Bottle fermentation method is called as champanoise method.

Bottle method or holding method of pasteurization: This method of pasteurization is commonly used for preservation of fruit juices at home. The extracted juice is strained or clarified and filled in bottles (sterilized and not necessarily hot); living sufficient head space for the expansion of the juice during heating i.e. juice in bottles is not filled up to brim. The bottles are then sealed air tight and pasteurized.

Bottle neck in ketchup: A defect of ketchup characterized by the formation of black ring at the surface of the ketchup in the neck of the bottle, caused due to oxidation of ferrous tannate which is formed by the reaction between tannin of spices and iron from boiling equipment or metal of the cap through the action of acetic acid.

Bottle sterilizer: Bottle sterilizer is a form of oven having insulated chamber and a tray chamber. Oven is used for bottle sterilization and electricity is used for heating and air circulation purposes (Plate 129).

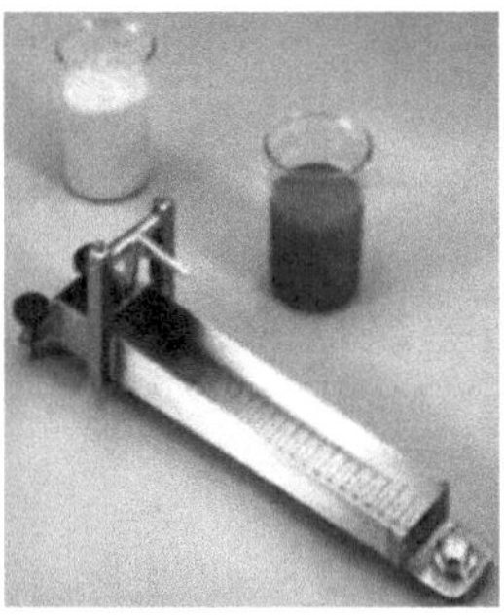

Plate 128: Bostwick consistometer

Plate 129. Hot air oven

Bottle washing machine: This machine is used for washing of the bottles. Bottles are washed with brushes driven by motor and washed bottles are collected on the trays where the water is drained off.

Botulinum Toxin: It is a neurotoxin formed by botulinum that causes botulism and that is injected in a purified form for therapeutic and cosmetic purposes (as to treat blepharospasm and reduce wrinkles).

Botulism: Botulism is an acute paralytic disease caused by botulinum toxin especially in food contaminated by *C. botulinum*. A food poisoning caused by the toxin produced by *Clostridium botulinum* which affects the peripheral nervous systems and typical symptoms appear after 10-12 hrs of ingestion of contaminated foods. The effects are acute disturbance in digression following by vomiting, haves, diarrhea with headache, constipation etc. and in fatal cases, death may occur within 3-6 days after consumption

Boukha: A wine made from figs is called Boukha.

Bound moisture /water: It is the liquid held by a material, which exerts a vapour pressure less than that of pure liquid at the same temperature.

Bouquet: A distinctive and characteristic fragrance of wine.

Bourbon whiskey: A whiskey distilled from a mash made up of not less than 51 percent corn plus malt and rye. It is popular in USA.

Bouyancy: The hydrometer works on the principles of the buoyancy. The buoyancy of the hydrometer is directly proportional to the density of the solution. The degree upto which hydrometer sinks into the juice can be calibrated to a Brix scale on the neck of the hydrometer. Dissolved gases affects this buoyancy, so deaeration of the juice aids in the accurate degree Brix determination by the hydrometer.

Brabender Farinograph: It is used to measure the plasticity of wheat dough for preparing bread products (Plate 130).

Bragg's law: This law relates the wavelength of the electromagnetic radiation to the diffraction angle and lattice spacing in the crystalline sample (Plate 131).

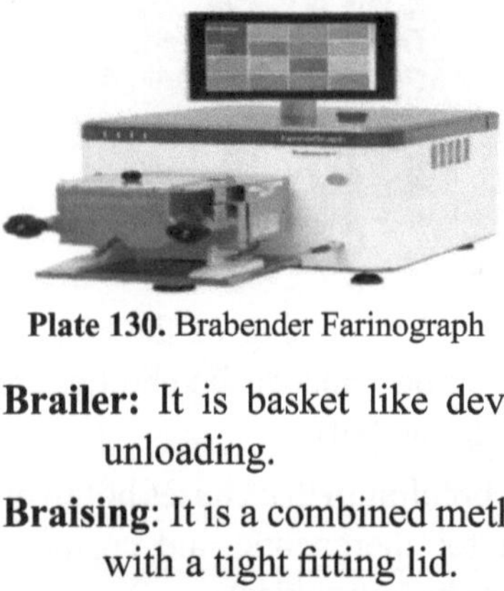

Plate 130. Brabender Farinograph

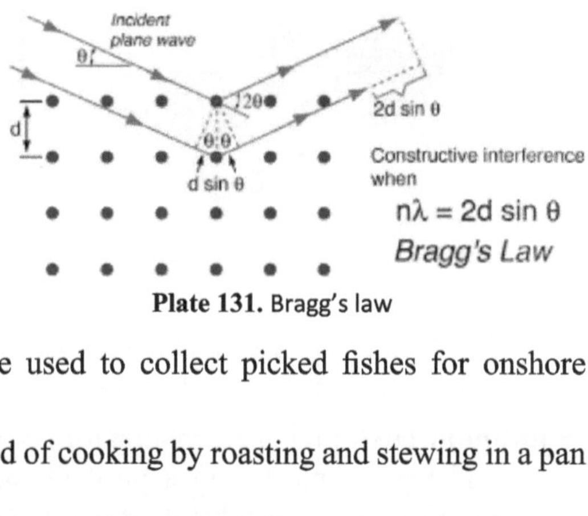

Plate 131. Bragg's law

Brailer: It is basket like device used to collect picked fishes for onshore unloading.

Braising: It is a combined method of cooking by roasting and stewing in a pan with a tight fitting lid.

Brand Name: An arbitrarily adopted name that is given by a manufacturer or merchant to an article or service to distinguish it as produced or sold by that manufacturer or merchant and that may be used and protected as a trademark or one having a well-known and highly regarded or marketable name.

Brandy: An alcohol produced obtained by distillation of almost any fermented fruit juice or an alcoholic beverage distilled from wine or fermented fruit juice. Brandi is an alcoholic distillate from the fermented juice mash or wine. Brandy from grapes.

Brandying: The processes to flavor, blend, or preserve with brandy is known as brandying.

Braxy mutton: The meat of the sheep that died due to braxy (bacterial disease) or accident.

Bread: Baked food product made of flour that is moistened into dough, kneaded, and usually leavened with yeast (Plate 132). Flat unleavened bread is still eaten in the Middle East, Asia, and Africa. The principal grains used in such breads are wheat, barley, millet, buckwheat, rye, and corn. Raised bread is common in Europe and the U.S., is usually made of wheat or rye. Both contain the elastic protein substance gluten, which traps gas produced by fermentation during leavening, helping the bread to rise. While the simplest breads contain only flour, water, and yeast, other common ingredients are milk, shortening (fats, butter, oils), salt, eggs, and sugar. Bread is a source of complex carbohydrates and B vitamins; whole-wheat bread contains more protein, vitamins, minerals, and fibre than white-flour bread.

Breadcorn: The corn, from the flour of which the bread is made is called cornbread.

Breading: It is a three tier process in which the product is first dredged in flour, and then dipped in egg mixture liquids and finallyconverted into crumbs.

Break even crop value ratio (R): It helps the growers to estimate the price for mechanical harvesting. R = 1-(H-M) / G where, H = hand operated cost, M = mechanical harvesting cost and G = Total income. It helps growers to estimate the price for mechanical harvesting.

Breather: A can with a very minute leak that permits air to move in or out but does not necessarily allow microorganisms to enter into the can. Damage of food inside the can usually occurs due to rusting of can.

Brew: A brewed beverage (as beer) or the process of brewing.

Brewed vinegar: It is made from fruits, starchy materials (potato) and sugar containing substances (molasses and honey) by alcoholic and subsequent acetic acid fermentation.

Brewery: A plant where malt liquors are produced is known as brewery.

Brewing: Process of boiling wort along with hops in brew kettle for about 2.5 hr is known as brewing. Hot water extraction of plant material is also known as brewing (tea, coffee and beer). Brewing is a process of boiling to inactivate microorganisms, denature protein and inactivate enzyme activity.

Brie: Brie is a soft surface ripened cheese with a whitish rind and a pale yellow interior. It is also called brie cheese (Plate 133).

Brine floatation: It is used for judging maturity and grading of peas for canning.

Brine: 1-2% solution of common salt which is generally used as covering liquid in caning vegetable. Or A solution of the salt in water is called brine.

Brining: The process of preservation of vegetables in a salt solution of suitable concentration for certain period of time is called brining. Hot brine of 1-3% concentration is used for covering vegetables and is filled at 79 – 82°C, leaving a head space of 0.3 – 0.5 cm.

BRIT: It stands for Board of Radiation and Isotope Technology (Plate 134).

Brix: A measure of the total soluble solids in fruit/ vegetable juice, calibrated in terms of pure sucrose i.e. 1^0Brix = 1% sucrose measured with a refractometer.

Brix saccharometer: An instrument which is used for measuring the degree and gives reading of the total solids dissolved in the juice.

Brixometer: It is a special hydrometer which is calibrated to read off directly the quantity of the total solids in the sugarcane juice. It can also be measured by hand refractometer or digital brixometer. Other brixometer is abbe's refractometer (Plate 135, 136 and 137).

Plate 132: Bread

Plate 33. Brie cheese

Plate 134. BRIT logo

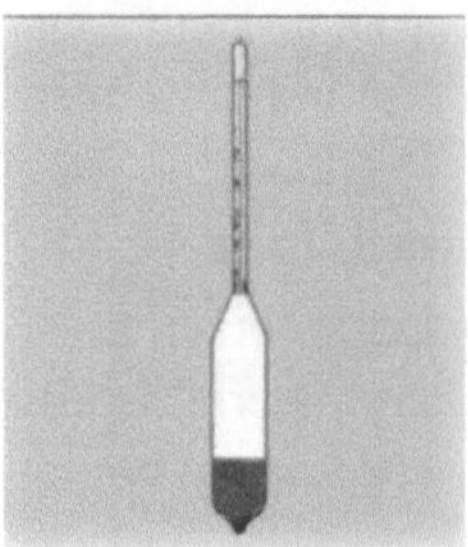

Plate 135: Hydrometer

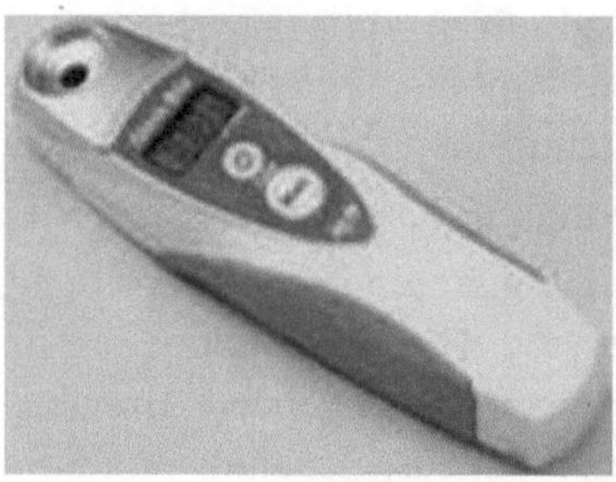

Plate 136: Digital brixometer

Plate 137: Working with brixometer

Broiler: A bird fit for broiling usually a chicken that is younger and smaller than a roaster.

Broiling: It is the process of cooking meat by direct exposure to radiant heat or oppressive heat.

Bromelain: A protease obtained from the pineapple. It is also called as bromelin.

Brominate vegetable oil (BVO): BVO is a vegetable oil whose density has been increased to that of water by bromination. Due to bromination, it does not separate out in the beverages and is extensively used in stabilization of flavouring oil in synthetic beverages. As bromination result cloudiness of the drink, it protect from light induced reaction. BVO are used @ 15 ppm (USA) but banned it India.

Brookfield viscometer: This instrument is used to measure the resistance to rotate (Plate 138).

Brown core: It is a low temperature disorder of apple where browning and necrosis of the flesh takes place around seed cavity (Plate 139).

Brown staining: The irregular discolouration of the peel of mandarins and grape fruit due to chilling injury is called brown staining.

Plate 138: Brookfield viscometer

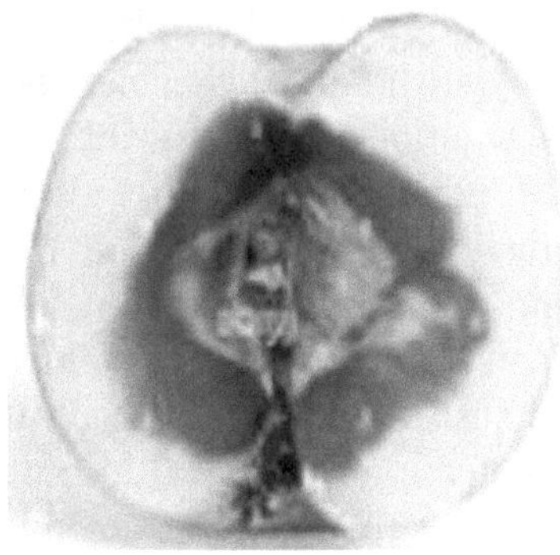

Plate 139: Brown core of apple

Buckling: It is bursting or distortion of cans due to improper cooling, swelling. The contents spoiled due to entry of the microorganisms.

Buffer: A substance capable in solution of neutralizing both acids and bases and thereby maintaining the original acidity or basicity of the solution is known as buffer. The solution containing such a substance is also known as buffer solution. Or **Buffer s**olution usually containing a weak acid and its conjugate weak base, or a salt, of such a composition in which the pH is held constant within a certain range. An example is a solution containing acetic acid (CH_3COOH) and the acetate ion (CH_3COO^-). The pH depends on their relative concentration and can be found with a simple formula involving their ratio. Relatively small additions of acid or base will change the concentration of the two species, but their ratio, and hence the pH, will not change much. Different buffers are useful in different pH ranges; they include phosphoric acid, citric acid, and boric acid, each with their salts. Biological fluids such as blood, tears, and semen have natural buffers to maintain them at the pH required for their proper function. These are the substances that resist change in the pH (alkalinity and acidity). Or It is solution containing both a weak acid and its conjugate weak base whose pH changes only slightly on addition of acid or alkali.

Bulk roughage: It is the mostly indigestible material in food that stimulates the intestine to peristalsis. It is also called Fiber.

Bunging process: This process is used during preparation / fermentation of beer. In this process, the fermenting beer is transferred to a fresh tank while it still contains about 2% fermentable extract. A bunging apparatus is attached to the tank to collect the CO_2 generated by yeast by fermenting the residual sugars.

Bureau of Indian Standards (BIS): BIS is the National Standards Organization established as Society in 1947 as Indian Standards Institution and subsequently made its statutory body as BIS under Bureau of Indian Standards Act 1986 (Plate 140). It revoked Indian Standards Institutions (Certification Marks) Act 1952 but incorporated all its provisions. The Bureau is a body corporate and responsible for laying down policy guidelines for BIS. It comprises of members Representing Industry, Consumer Organizations, Scientific and Research Institutions, Professional/technical institutes, Central Ministries; State Government and Members of Parliament. The act is mandatory for milk powders, sweetened condensed milk, infant formula etc. This act generally covers hygienic conditions of manufacture, raw material quality and safety. It also ensures the quality to the consumers by certification.

Burette: A graduated glass tube with a small aperture and stopcock for delivering measured quantities of liquid is known as burette (Plate 141).

Plate 140: BIS

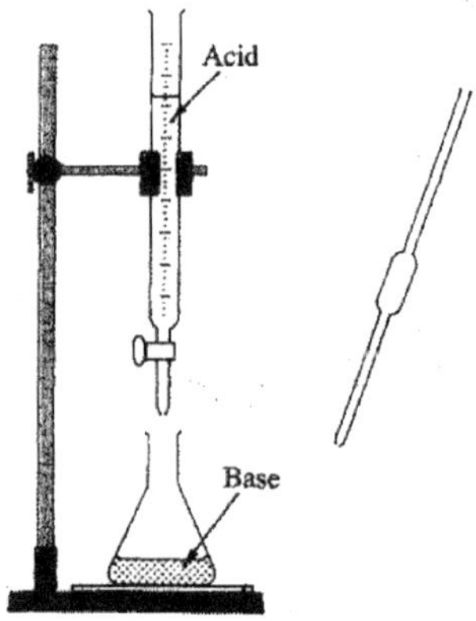

Plate 141: Set up of Burette on burette stand

Burger: A sandwich similar to a hamburger.

Burgundy: A red or white unblended wine from Burgundy / produced elsewhere is known as burgundy.

Bursting: It is caused by the excess pressure of gases produced by the fermentation of the food by microorganisms or by H_2 gas formed by the chemical action of the food acids on the tin plate.

Butter: A solid emulsion of fat globules, air, and water made by churning milk or cream and used as food. It is fat concentrate obtained by churning the cream or curd. It is solid emulsion of fat globules and water made by churning cream and is used as a food. Butter has long been used as a cooking fat and as a spread. It is a high-energy food, containing about 715 calories per 100 grams. It is high in butterfat (80–85%) and low in protein. Colouring is sometimes added to enhance its natural yellow colour (from carotene), and salt is often added. It is a unique example of emulsion technology.

Butter fat: The natural fat of milk and chief constituent of butter consisting essentially of a mixture of glycerides as those derived from butyric, capric, caproic, and caprylic acids.

Butter milk: The by-product left after churning is known as butter milk. Or It is the liquid left after butter has been churned from milk / cream or it is cultured milk made by the addition of suitable bacteria to sweet milk.

Butter scotch: A candy made from brown sugar, butter, corn syrup, and water.

Butternut squash: A smooth somewhat bottle shaped light yellow colored winter squash (Cucurbita moschata) with orange flesh.

Butyric acid: It is isomeric fatty acids $C_4H_8O_2$ especially a straight-chain acid of unpleasant odor normally found in perspiration and rancid butter.

Bypass valve: Splitting of fluid streams into two parallel streams, one of which is processed while passing through equipment while the other unites (Plate 142).

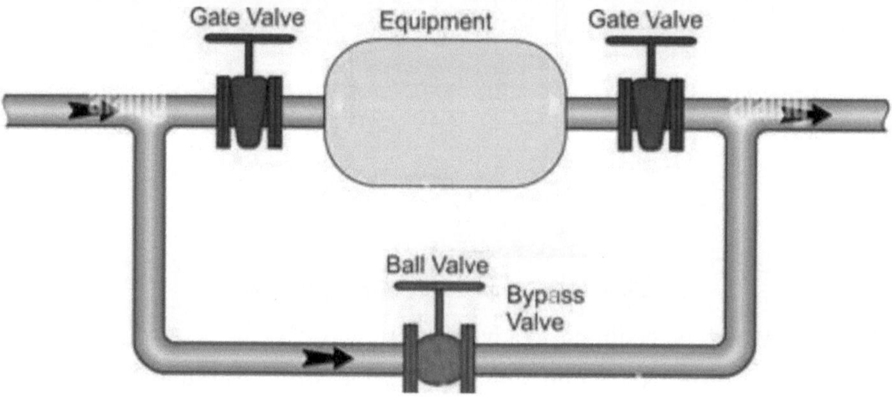

Plate 142: Bypass valve

C

C - enamel can: The sulphur resistant lacquer is golden coloured and the can coated with it are called C-enamel can or sulphure resistant can, used for packing non acid products (Plate 143).

CA storage: Storage of commodities under gaseous atmosphere like gas storage, where the composition of atmosphere is controlled accurately. MA and CA differ only in degree of control and CA is more exact. These storagc mcthods in combination with refrigeration markedly enhance storage life of fruit and vegetables (Plate 144).

Plate 143: SR cans or C enamel cans

Plate 144: CA storage unit

CAC: It stands for Codex Alimentarious commission (1962). It is a joint commission of FAO and WHO, comprising of 146 member countries. It ensures consumer food safety, establish fair practices in food trade and promote the development of international food standards. CAC make standards for food additives, veterinary drugs, pesticide residue and other miscellaneous substances that affect consumer safety. These are published in Codex Alimentarius.

Cacao: The dried partly fermented fatty seeds of *Theobroma cacao* of the family Sterculiaceae that are used in making cocoa, chocolate, and cocoa butter. It is also called cacao bean (Plate 145).

CAD: Computer-aided design (Plate 146). CAD design is used by architects, construction managers, and engineers and has replaced manual drafting. It helps users create designs in either 2D or 3D to visualize construction, and enables the development, modification, and optimization of the design process.

Plate 145: Cacao seeds

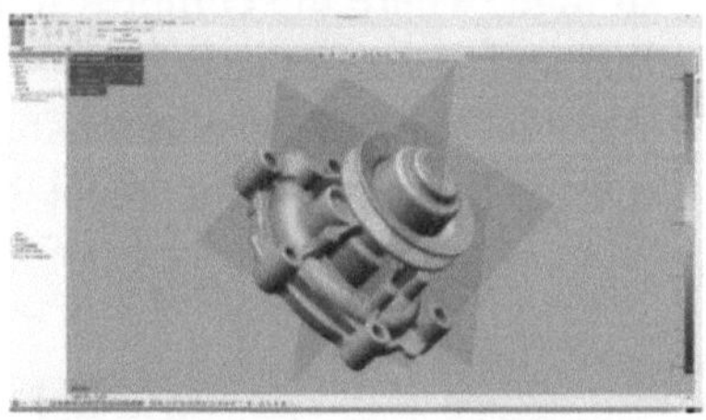

Plate 146: CAD

Caffeine: A bitter alkaloid $C_8H_{10}N_4O_2$ found in coffee, tea, cacao, and kola nuts and used medicinally as a stimulant and diuretic.

Cake: A breadlike food made from a dough or batter that is usually fried / baked in small flat shapes and is often unleavened or a sweet baked food made from a dough or thick batter containing flour and sugar and often shortening, eggs, and a raising agent (as baking powder) or a flat and round mass of food that is baked or fried is called cake.

Calamus: The aromatic peeled and dried rhizome of the sweet flag that is the source of a carcinogenic essential oil. It is prohibited food additive.

Calandria: It is the heat exchange section in the evaporator (Plate 147).

Calciferol: An alcohol $C_{28}H_{43}OH$ usually prepared by irradiation of ergosterol and used as a dietary supplement in nutrition and medicinally in the control of rickets and related disorders. It is also called vitamin $D_{2.}$

Calcification: It is the process by which an organic tissue or substance become hardened by a deposit of calcium salts.

Calcium carbide: It is dark gray crystalline compound CaC_2 used especially for the generation of acetylene for ripening purpose and for making calcium cyanamide. It it now banned (Plate 148).

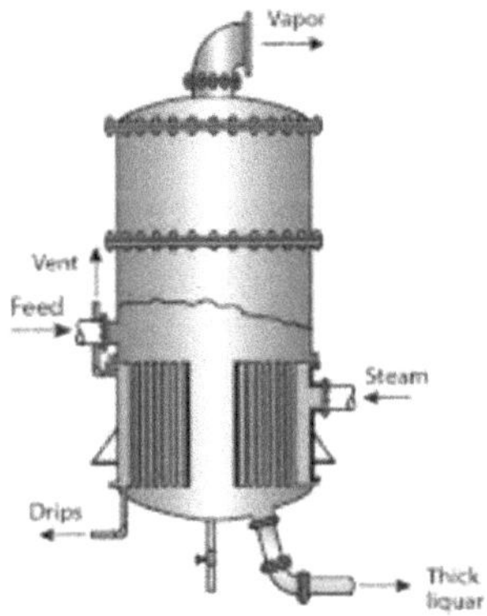

Plate 147: Calandria

Plate 148: Calcium carbide

Calcium chloride: A white deliquescent salt $CaCl_2$ used in its anhydrous state as a drying and dehumidifying agent. It is used as firming agent in canning industry.

Calcium hypochlorite: A white powder $CaCl_2O_2$ used as a bleaching agent and disinfectant.

Calcium phosphate: The phosphates of calcium $CaH_4P_2O_8$ used as baking powder. It is a sequestering agent and is used in ice cream mix.

Calcium silicate: It is an anticaking agent.

Calcium sulfate: A white salt $CaSO_4$ that occurs as anhydrite, gypsum, and plaster of paris and that in hydrated form is used as a building material and in anhydrous form is used as a drying agent. It is used for colouring vegetables.

Calorie: It is the amount of heat required (at a pressure of one atmosphere) to raise the temperature of one gram of water by 1°C (14.5° to 15.5°C) that is equal to about 4.19 joules (1 calorie = 4181 J). It is also known as cal / gram calorie / small calorie.

Calorimeter: It is an apparatus for measuring quantities of absorbed / emitted heat by oxidation of food. It is also used for determining specific heats.

Calvin cycle: It is the cycle of enzyme catalyzed dark reactions of photosynthesis that occurs in the chloroplasts of plants and in many bacteria and involves the fixation of carbon dioxide and the formation of a 6 carbon sugar.

Camembert cheese: A soft surface-ripened cheese with a thin grayish-white rind and a yellow interior.

Camerean point: It is point where there is partition of food from low acid food to high acid food i.e. pH of 4.7.

Can: It is a cylindrical vessel for holding liquids usually with an open top, often with a removable cover, and sometimes with a spout or side handles (as for holding milk or trash). It is a container (as of tinplate) in which products (as perishable foods) are hermetically sealed (airtight cans) for preservation until consumption (Plate 149).

Can coding: It is a systematic statement of can body or a system of symbols (as letters or numbers) used to represent assigned and often secret meanings in the can.

Can current air flow: Movement of hot air in the drier in the same direction as the movement of the product which the hottest air contacts the wettest product. Here air at outlet end becomes cool and the final product may not be sufficiently dry.

Can flanger: It is an apparatus used for making edges of the can. This operation is done just before seaming of the cans (Plate 150).

Can reformer: It is an apparatus used for giving the round shape of can. The flat cans are used in this operation (Plate 151).

Candied fruit: A fruit when impregnated with sugar free from syrup, drained and dried is called a candied fruit.

Plate 149: Can used for preservation

Plate 150: Can flanger

Plate 151: Can Reformer

Candling: It is the process to examine eggs by holding between the eye and a light. This test is done in eggs for cracks, staleness, blood clots, fertility, and growth. It is used for quality evaluation of eggs (Plate 152).

Candy: A fruit product prepared by gradually concentrating fruits in syrup by repeated boiling until the fruit is heavily impregnated with sugar and then fruits are dried to overcome stickiness. Or A crystallized sugar formed by boiling down sugar syrup or a confection made with sugar and often flavoring and filling is known as candy (Plate 153).

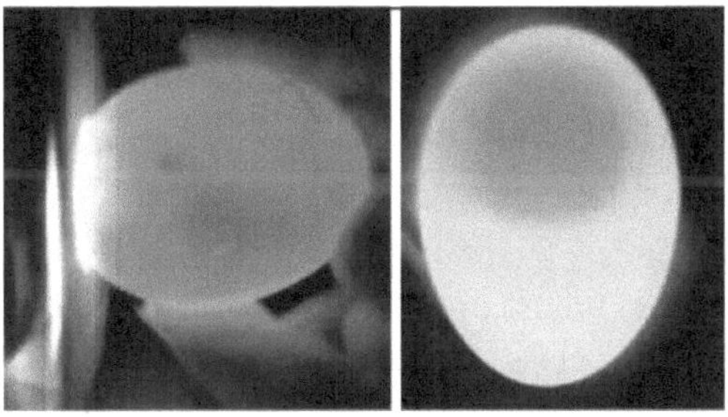

Plate 152: Candling of eggs to chech defects

Plate 153: Sand pear candy

Canister /Canister: It is often a cylindrical tin container for holding a specified object or substance.

Cannikin: It is a small can or drinking vessel.

Canning: Preservation of food in hermetically sealed containers and usually implies heat treatment for preservation. Or Method of preserving food by storing it in containers that are hermetically sealed and then sterilized by heat. The process was invented in 1809 by Nicolas Appert of France. In modern canning, food is passed under hot water or steam, transferred to a sterile container, sealed inside, and subjected to heat, sufficient to kill any remaining microorganisms. The process preserves most nutrients but often affects consistency and taste.

Caproic Acid: A liquid fatty acid $C_6H_{12}O_2$ that is found as a glycerol ester in fats and oils or made synthetically and used in pharmaceuticals and flavorings.

Caprylic Acid: A fatty acid $C_8H_{16}O_2$ of rancid odor occurring in fats and oils and used in the synthesis of esters for perfumes.

Capsaicin: A colorless irritant phenolic amide $C_{18}H_{27}NO_3$ found in various capsicums that gives hot peppers their hotness and that is used in topical creams for its analgesic properties.

Caramel: An amorphous brittle brown and somewhat bitter substance obtained by heating sugar and used as a coloring and flavoring agent.

Caramelization: It is the process of removal of water from sugar (such as sucrose or glucose) followed by isomerization and polymerization steps. Sugar in dry condition (or their syrup) when heated beyond their melting point decompose and form a brown mass known as caramel, which has a bitter taste, reaction occur/develop brown colour at low temperature while maillard with the help of suitable catalysts it is possible to carry out caramelization to provide either flavouring or colouring caramel for food use. For flavouring sucrose, syrup is caramelized. For the manufacture of caramel colours for use in beverages, glucose syrup is treated with dil H_2SO_4 and then partially neutralized with ammonia.

Caramelized sugars: It is dry sugar heated with constant stirring until melted and dark in colour, mostly used for colouring and flavouring of foods.

Carbohydrate: The neutral compounds of carbon, hydrogen, and oxygen (as sugars, starches, and celluloses) most of which are formed by green plants and which constitute a major class of animal foods.

Carbon dioxide (CO_2): It is a heavy, colorless, odourless gas that does not support combustion, dissolves in water to form carbonic acid and is formed in animal / human tissue by oxidation of carbon and eliminated as a result of respiration from lungs and in the decay or combustion of animal and vegetable matter. It is absorbed from the air by plants in photosynthesis, and used in the carbonation of beverages.

Carbon disulphide: It is a colourless liquid with slightly disagreeable odour, prepared from methane and sulphur; used as fumigant to check the losses caused by stored grain insect pests in godown.

Carbon dot: Carbon dots (CDots) are generally defined as small-carbon nanoparticles with surface organic functionalization and their classical synthesis is literally the functionalization of pre-existing carbon nanoparticles. The food packaging industry has higher requirements for carbon dots to extend product storage life and safety. The addition of carbon dots as active material enhances the physical, chemical, and mechanical properties of packaging without any effect on the remaining properties of the film and foods. Carbon dots are luminescent carbonaceous nanoparticles that can be endowed with chiral properties, making them particularly interesting for biomedical applications due to their low cytotoxicity and facile synthesis. The primary physiological functions of CDs are to promote seed germination and root growth, improve plant nutrient absorption, promote plant growth, increase biomass accumulation, enhance photosynthesis, and increase plant carbohydrate content, plant abiotic stress, and disease resistance parameters. Carbon dots (CDs) are an interesting kind of zero-dimensional (0D) carbon nanomaterial with lateral diameters <10 nm with lattice spacing of 0.21 and 0.34 nm, respectively.

Carbonated beverages: The beverages which are combined or infused with carbon dioxide are called carbonated beverages (Plate 154).

Carbonation: A method of fruit juice preservation with CO_2. It helps to make a mist atmosphere inside the bottled juice by displacing oxygen which ultimately prevents microbial spoilage and fermentation of the juice. It is addition of carbon dioxide gas to a beverage, imparting sparkle and a tangy taste and preventing spoilage. The liquid is chilled and cascaded down in an enclosure containing carbon dioxide (either as dry ice or a liquid) under pressure. Increasing pressure and lowering temperature maximize gas absorption. Carbonated beverages do not require pasteurization. or It is the process of dissolving or incorporating CO_2 in a beverage so that when served it gives off the gas in fine bubbles and has the characteristic pungent taste suitable to the carbonated beverages. Carbonation unit is used for carbonation purpose (Plate 155).

Carbonic maceration: Carbonic maceration is used for red wine production. It is the fermentation of intact (uncrushed) grapes, preserved under at atmosphere of CO_2 for some time to degrade sugars in the intact berries to produce 1.5-2.5 ethanol (v/v).

Carcass: A dead / dressed body of a meat animal is known as carcass (Plate 156).

Carnitine: The carnitine originally called as Vitamin $B_{T.}$ It is a quaternary ammonium compound $C_7H_{15}NO_3$ that is present in vertebrate muscle and in the levorotatory form. It is involved in the transfer of fatty acids across mitochondrial membranes.

Carnot refrigeration system: Reverse Carnot Cycle in a refrigeration system is known as Carnot refrigeration system. It is just a theoretical cycle but serve as an ideal cycle (Plate 157).

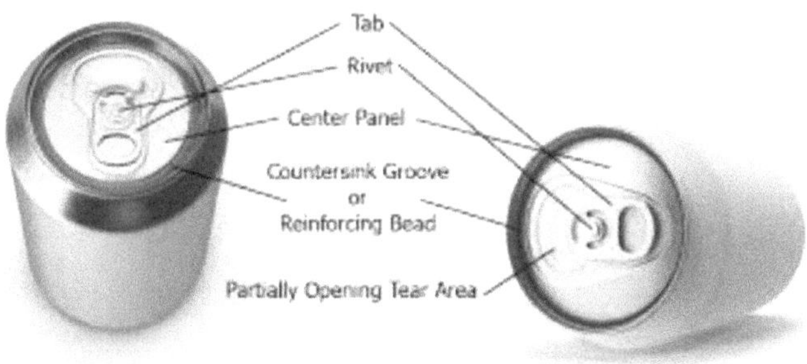

Plate 154: Cabonated beverage Aluminium can

Plate 155: Carbonation unit used for carbonation

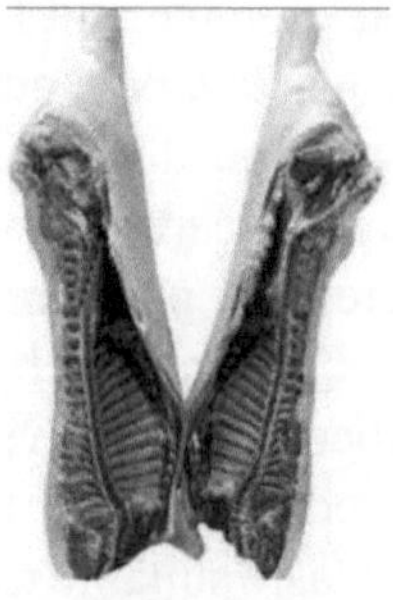

Plate 156: Carcass

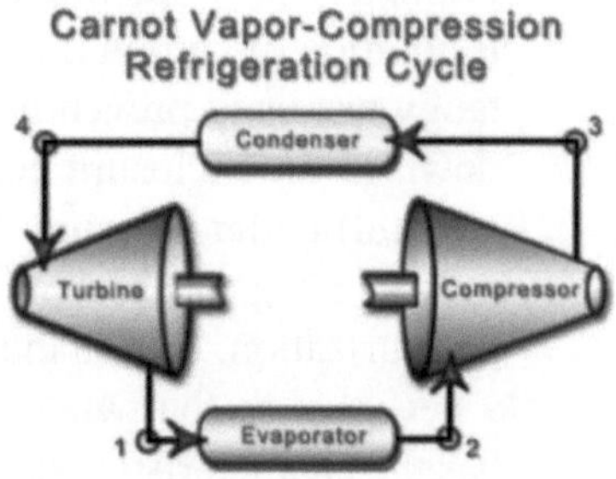

Plate 157: Carnot cycle refrigeration

Carotene: The orange red corotenoids pigment which is the precursor of vitamin A. One molecule of B-carotene is converted to two molecular of vitamin A by hydrolysis. They are pigments that gives orange, yellow, or sometimes red colours, for example, dandelions, apricots, carrots, sweet potatoes, butter, egg yolks, canary feathers, and lobster shells. Carotenes are converted in the body into vitamin A, but, unlike the vitamin, they are not toxic even at high doses. Carotene has an antioxidant effect and is therefore used in pharmaceuticals and as a food and feed additive, as well as to colour margarine and butter.

Carotenoids: A group of yellow, orange and orange red fat soluble pigment present in plant parts. These are either hydrocarbons or its derivatives and are composed of isoprene units. Carotenoids occur in different forms like carotene, lycopene, lutein, violaxanthin, neoxanthin, capsanthin, bixin, xanthophylls (Plate 158).

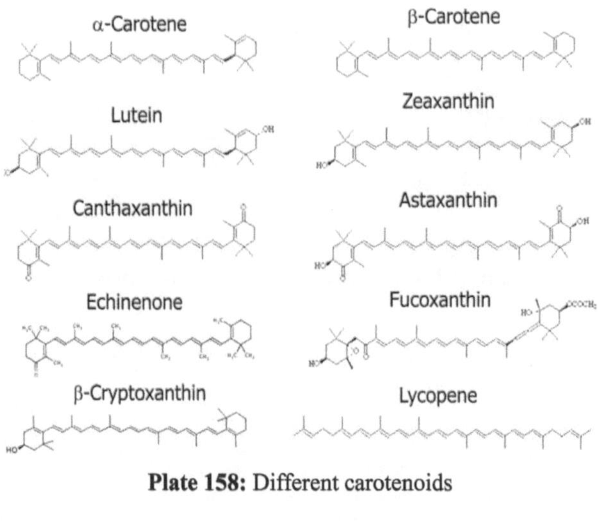

Plate 158: Different carotenoids

Carrageenan or **carrageenin:** It is a colloid extracted from various red algae (as Irish moss) and used as a stabilizing or thickening agent. Carrageenan trademarks are Gelcarin, seagel, lactarin, Novagel, Avicel and nutricol.

Carrier gas: Most common carrier gases are nitrogen, helium, argon and Co_2. The carrier gas must be chemically inert. The choice of carrier gas is often depends upon the type of detector used. The carrier gas system also contains a molecular sieve to remove water and other impurities.

Caryopsis: A small one-seeded dry indehiscent fruit (Indian corn or wheat) in which the fruit and seed fuse in a single grain is called caryopsis.

Cascade application: The dipping or spraying of the crop in a solution or suspension of the chemical, which may be in hot water to enhance control of the disease or insect infestation. The crop may be passed below a shower of the diluted chemical. This is called cascade application. Its application is in banana and citrus.

Case hardening: Case hardening is a special condition in dehydrated product related to shrinkage and sealing of the surface of food (pore clogging) with solute is known as case hardening. Case hardening is due to high surface temperature and unbalanced drying. Or During the dehydration of the food products a hard impermeable layer is formed at the surface resulting in a reduced drying rate. This phenomenon is known as case hardening. It is also called crust formation. This is formed if dehydration is carried out at high temperature.

Casein: It is the major phosphoprotein / protein of the milk. It is precipitated from milk by heating with an acid or by the action of lactic acid in souring. It is produced when milk is curdled by rennet and is the chief constituent of cheese. It is used in making edible films (Plate 159).

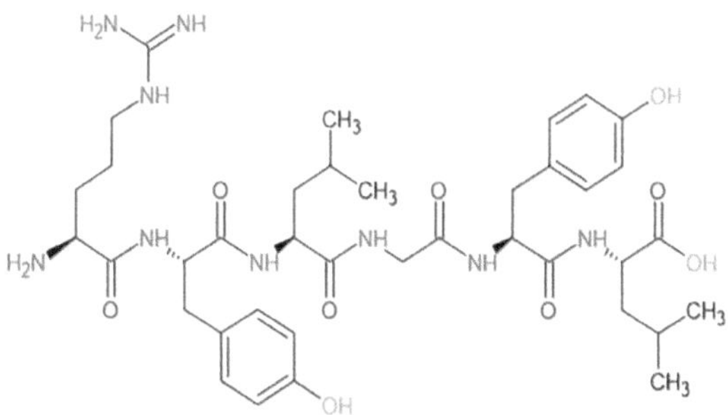

Plate 159: Casein structure

Cashew brandy: Cashew brandy is also called fenni (Feni or feon mean forth).

Casks: Aging vats are also known as Aging vats. These are barrel shaped vessels of staves, headings, and hoops usually for holding wines /liquids.

Cassava: It is also called tapioca. It is a tropical plant with edible starchy roots.

Catabolic process: It is a process in which big molecules are broken into smaller molecules e.g. respiration.

Catabolism: Degradative metabolism involving the release of energy and resulting in the break down of complex materials (as proteins or lipids) within the organism. It is a destructive phase of metabolic processes in which complex materials are converted into simple compounds e.g. Respiration. In this process potential energy stored during anabolism is converted into kinetic form (Plate 160 and 161).

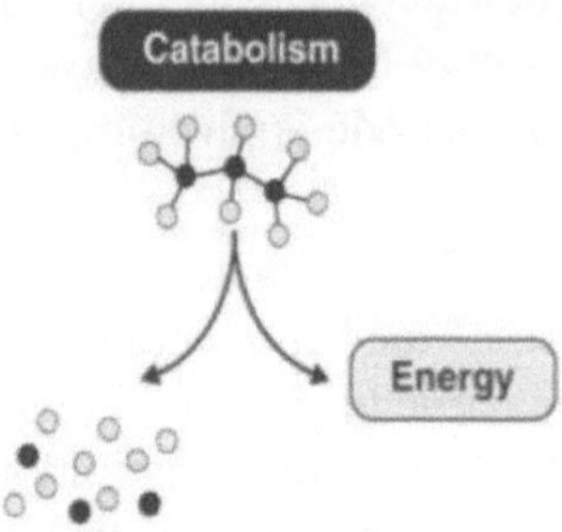

Plate 160: Catabolism

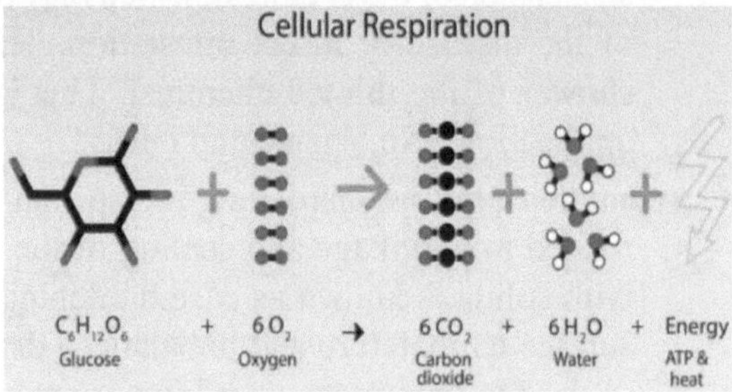

Plate 161: Respiration as catabolic process

Catalyst: A catalyst is an agent affecting the velocity of a chemical reaction without appearing among the final products of the reaction.

Catechol: A crystalline phenol $C_6H_6O_2$ obtained from various natural sources but usually made synthetically. It is synonym to catechin.

Caviar: It is processed salted roe (eggs of a fish) of large fish (sturgeon) (Plate 162).

CCP: It stands for critical control point. It is a point, stage, raw material, practice, formulation, process, step or procedure in a food production process at which we can apply control measures and can prevent / eliminate / reduce hazards to an acceptable level.

Cellar: It is a storage room below ground level in a house e.g. the stock of wine in a cellar.

Cellar master: A person who supervises the making of wine in a winery

Cellar storage: It is a sophisticated form of below ground storage used for storing of equal to the average annual air temperature. Good drainage a protection from rain is essential. The performance of cellar is improved by providing controlled ventilation opening for entrance of cold air and exit of warm air by conventional circulation when cooling is required.

The storage is usually not much below that of the outside air and is seldom lower than 15°C.

Cellarage: An underground space especially for storage is known as cellarage.

Cellarette or **cellaret:** A case or sideboard for holding bottles of wine or liquor (Plate 163).

Plate 162: Caviar

Plate 163: Cellaret

Cellophane Noodle: A translucent noodle made from mung beans is called cellophane noodle.

Cellulase: An enzyme that hydrolyzes cellulose is called cellulase.

Cellulolytic: An enzyme hydrolyzing or having the capacity to hydrolyze cellulose is known as cellulolytic.

Cellulose acetate: The compounds insoluble in water that are formed by the action of acetic acid, anhydride of acetic acid, and sulfuric acid on cellulose and are used for making textile fibers, packaging sheets, photographic films, and varnishes.

Cellulose: A polysaccharide $(C_6H_{10}O_5)_x$ of glucose units that constitutes the chief part of the cell walls of plants, occurs naturally in such fibrous products as cotton, and is the raw material of many manufactured goods (as paper, rayon, and cellophane). Cellulose is digested by non ruminant animals. The presence of cellulose in fruit and vegetables is much less (0.8% in date) as compare to and cotton (98%). Cellulose molecules are extremely stable and can be broken down only with stronger acid and enzyme into a simple molecule by celluloses. Cellulose is an aggregation of β-D-glucophyranose in which C1 linked to C4. Cellulose it major constituent of cell walls. It is an insoluble crystalline substance of natural form.

Celsius: It is a scale of temperature on which water freezes at 0° and boils at 100°. It is denoted by °C.

Certification: Certification is the procedure by which official certification bodies or officially recognized certification bodies provide written or equivalent assurance that food or food control systems conform to requirements. Certification of food may be based on a range of inspection activities which may include continuous on-line inspection, auditing of quality assurance systems and examination of the finished products.

Certification mark: **A** mark or device used to identify a product or service that has been certified to conform to a particular set of standards.

Certified milk: Milk produced in dairies that operate under the rules and regulations of an authorized medical milk commission.

CFU: It stands for colony forming units.

Champaigne: A sparkling, sweet wine produced chiefly in France from certain varieties of grapes where fermentation is allowed to process till its completion in bottle. Named for the site of its origin, the Champagne region of northeastern France, it is made from only three grapes viz. Pinot and Meunier (both black) and Chardonnay (white). The juice from these grapes is initially fermented in stainless-steel vats. A mixture of wine, sugar, and yeast is added, and it is then transferred to pressure tanks for a second fermentation that yields carbon dioxide and effervescence. It is chilled, sweetened, bottled, and left to mature. Depending on residual sugar in champagne; they are given different names like Brut if residual sugar is 2-10 g/l, Extra dry if residual sugar is 10-20 g/l, Dry if residual sugar is 20-40 g/l, Semi dry if residual sugar is 40-60 g/l, Sweet if residual sugar is 80-100 g/l.

Channa: It is also called as Paneer in certain part of the country in India. It comprises the milk solids obtained by acid coagulation of boiled hot milk and subsequent straining of whey.

Charmat process: Charmat process is the production of sparking wine by tank method.

Cheddar: It is a hard white, yellow, or orange smooth-textured cheese with a flavor that ranges from mild to strong as the cheese mature called cheddar cheese (Plate 164). It is pressed uncooked cheese. The name is originated from Cheddar, a village in Somersetshire, England. It contains not more than 39% moisture and not more than 50% of the solids consisting of milk fat.

Cheese: It is a food consisting of the coagulated, compressed (in a cake form), and usually ripened curd of milk separated from the whey or a product obtained by draining whey after coagulation of milk with harmless milk coagulating agents under influence of desirable bacterial culture is known as cheese (Plate 165). When milk sours, it forms both a protein-rich gel, or curd, and a lactose-rich fluid, or whey. Coagulation is often facilitated by adding rennin, an enzyme that acts on the milk's chief protein, casein. The resulting curd is then cut or broken to release most of the whey. Ripening and curing are affected by moisture content, acidity, presence of microorganisms, and other factors. Cheese is made from the milk of cows, goats, sheep, water buffalo, and other animals. In the West, cow's milk is most common. Products vary according to fat content of the milk, heating or pasteurization, and addition of enzymes or cultures of bacteria, molds, or yeasts. Cheese varieties include hard cheeses (e.g., cheddar, Edam, Emmental, Gouda, Provolone, Romano, Swiss), semisoft cheeses (Gorgonzola, Limburger, Muenster, Roquefort), and soft cheeses (Brie, Camembert, cottage, Neufchâtel, and ricotta). Cheese is a source of protein, fat, minerals (calcium, phosphorus, sulfur, iron), and vitamin A. Cheese may be defined as the concentration of all or part of the components of milk obtained through the coagulation of the major protein in milk.

Plate 164: Cheddar Cheese

Plate **165:** Unripened cheese prepared from milk

Cheese bitterness: Cheese bitterness / Protease bitterness in cheese caused by breakdown of protein.

Cheese cake: A dessert consisting of a creamy filling usually contains cheese baked in a pastry or pressed crumb shell.

Cheeseburger: A hamburger topped with a slice of cheese is known as cheese burger.

Cheesecloth: The cloth used in making of cheese is known as cheesecloth.

Cheesesteak: A sandwich consisting of thinly sliced beef topped with melted cheese and condiments (as fried onions or peppers)

Chelator: It is a binding agent that suppresses chemical activity by forming chelates.

Chemical additives: These are the substances other than food stuff incorporated in food either directly or indirectly as a result of any phase of production, processing, storage or packaging to improve or maintain the nutritional value, enhance quality and consumer acceptability. They may include antioxidants, preservatives, sequestrants, emulsifiers, colouring agents, stabilizers, flavouring agents, clarifying agents and other miscellaneous additives such as acids, antimicrobial agents etc.

Chemical hazards: Danger posed to food safety by the contamination of food by chemical substances such as pesticides, toxic metals, detergents etc. Chemical hazards include compounds that are present on the raw material and pose a threat to the consumer health. The effect of chemical contamination on consumer can be long term (chronic) such as for carcinogenic or accumulative chemicals (e.g. Mercury) which can build up in the body for many years or it can be short term (acute) such as the effect of allergenic foods. These hazards introduced deliberately or inadvertently which is potentially hazardous to health. They are insecticides, pesticides, fungicides, plant cleaning and sanitation compounds, lubricants, paints, aflatoxins and some heavy metals, food additives like colouring agents, preservatives, antioxidants, emulsifiers, stabilizers and environmental pollution.

Chemical Oxygen Demand (COD): The chemical oxygen demand (COD) is the amount of oxygen required by organic matter in the sample of water for its oxidation by a strong chemical oxidant and is expressed as ppm of oxygen taken from a solution of potassium dichromate in 2 hours. COD is not a perfect index of organic compounds present in water because many organic compounds are also oxidized and at the same time some organic compounds remain unaffected. COD is a measure of the oxygen consumed when organic matter is broken down chemically rather than by biological processes.

Chemical preservative: Food additives which are specially added to prevent the deterioration or decomposition of food by microorganisms, food enzymes or by purely chemical reactions. These include sodium benzoate, potassium metabisulphite, boric acid, propionic acid, lactic acid etc. but do not include salt, sugar, antioxidant, etc.

Chemical score: It is a chemical method of defining the nutritive value of proteins as proposed by Block and Mitchell (1946). Chemical score numerically equals to Biological score. It is also known as protein score.

Cherry candy: A candy prepared from cherry fruit is called cherry candy (Plate 166).

Plate 166: Cherry candy

Chhana: It is heat and acid precipitated / coagulated milk.

Chicken: It is the young of the domestic fowl. It is also a term is used for the flesh of the domestic fowl.

Chicory/chickory: A thick rooted blue flowered European perennial herb (*Cichorium intybus*) grown for its roots and as a salad plant. The dried ground roasted root of chicory used to flavor or adulterate coffee.

Chilli con carne: A spiced stew of ground beef and minced chillies or chilli powder along with beans.

Chill proof of beer: When beer is stored at lower temperature (5-8°C), it tends to become cloudy (hazy) due to the formation of colloidal insoluble matter by a reaction between tannins and proteins. This can be prevented either by removing tannins by treatment with polyvinyl pyrolidene or by breaking down the protein with an enzyme so that it can be no longer react with the tannins. The most commonly used enzyme for this purpose is papain.

Chill proofing agents: Chill proofing agents are proteolytic enzymes papain, bromelin, and ficin or tannic acid.

Chilled foods: Perishable foods which are maintained or stored at a low temperature (1°C to 8°C) to retain their shelf life, wholesomeness and safety.

Chilled juice: Flash pasteurized non frozen juice product made directly from blended fresh juices reconstituted frozen concentrate and held at or near 0°C.

Chilli sauce: A thick sauce of meat and chillies is called chilli sauce.

Chilli: A hot pepper of any of a group of cultivars (*Capsicum annuum* group *longum*) noted for their pungency. It is also called chilli pepper.

Chilling injury: A disorder in plant tissue, especially those of tropical or subtropical origin at temperature ranging from 0-15°C depending on crops causing the release of metabolites such as amino acid, sugars, mineral salts from cells, the degradation of cell structure and accumulation of toxic substance (Ethanol, acetaldehyde etc.). Symptoms includes pitting of skin, internal discolouration, uneven ripening development of off flavor, rotting etc. **or** It is a physical disorder occurred at low and non freezing temperature in certain food commodities like mangoes.

Chilling storage: Storage at temperature not far above freezing which usually involves cooling by ice or by mechanical refrigeration (-1^0C to 5^0C).

Chip: A small thin slice of food especially potato chip (Plate 167).

Plate 167: Potato chips

Chipped beef: Smoked dried and thin slice of beef is known as chipped beef.

Chloride of Lime: The bleaching powder is also known as chloride of lime.

Chlorinated lime: The bleaching powder is also known as chlorinated lime.

Chlorine test: This test is used to determine the mastitis. The normal chloride content of the milk is 0.08 to 0.14%. If in the test it detect more than this range then it show the presence of the mastitis as it results in increase in the chlorine level.

Chlorophyll: The green colouring matter of the chloroplast which trap the energy of sun light for photosynthesis. Chemical composition is C_{55} H_{72} O_5 N_4 Mg (Chlorophyll a), C_{55} H_{55} H_{70} O_6 N_4 Mg (Chlorophyll b). Chlorophylls are green coloured, fat soluble pigments involved in photosynthesis and present in many leafy vegetables. There are two types of chlorophyll a and b, which occur in the plants in the ratio of 3:1.

Chocolate: A beverage made by mixing chocolate with water or milk or a food prepared from ground roasted cacao beans or a small candy with a center (fondant) and a chocolate coating is known as chocolate.

Chocolatier: A person who is maker or seller of chocolate candy.

Cholecalciferol: It is a sterol $C_{27}H_{43}OH$ that is a natural form of vitamin D found in fish, egg yolks, and fish-liver oils and is formed in the skin on exposure to sunlight or ultraviolet rays. It is also called vitamin $D_{3.}$

Cholera: It is a food borne illness caused by *Vibrio cholerae.* It is one of the diseases of humans and domestic animals marked by severe gastrointestinal symptoms especially an acute diarrheal disease caused by an enterotoxin produced by a comma shaped gram-negative bacillus

(*Vibrio cholerae* syn. *V. comma*) when it is present in large numbers in the proximal part of the human small intestine.

Cholesterol: A steroid alcohol $C_{27}H_{45}OH$ that is present in animal cells and body fluids, regulates membrane fluidity, and functions as a precursor molecule in various metabolic pathways and as a constituent of low density lipoprotein (LDL) may cause arteriosclerosis. Cholesterol may be bad or good.

Cholesterol free food: Food having less than 2 mg of the cholesterol and 2 gm or less of the saturated fat per serving or per 100 gm.

Choline: A basic compound $C_5H_{15}NO_2$ that is found in various foods (egg yolks and legumes) or is synthesized in the liver and is a component of lecithin. It is a precursor of acetylcholine, and is essential to liver function.

Christal texturometer: Instrument used to measure the tenderness of meat.

Chromatography: Separation of discrete chemical compounds from a complex mixture by using solvents or gases.

Churn: It is a process of making butter in a vessel in which milk or cream is agitated in order to separate the oily globules from the watery medium.

Churning: Churning is a process that involves the agitation of cream at 10°C to bring a partial breakdown of oil-water emulsion. "Churning" is butter making technology results in the inversion of oil in water emulsion (Creams) to water in oil emulsion (butter) (Plate 168, 169, 170 and 171).

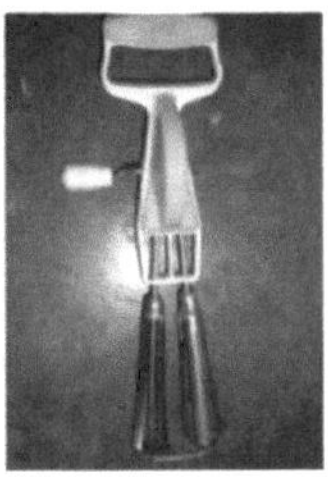

Plate 168: Hand operated churning machine (also used as egg beater)

Plate 169: Churning process to separate butter from milk / curd

Plate 170: Traditional churning in villages

Plate 171: Motor operated churner

Chutney: A preserved product where fully mature green fruits or mature but tender vegetable are peeled, boiled, crushed and mixed with sugar, salt coarsely powdered spices herbs and cooked to a reasonably thick consistency. Or A thick sauce of Indian origin that contains fruits, vinegar, sugar, and spices and is used as appetizer.

Cider: An alcoholic beverage made from fermented juice of apple. Cider is apple wine or fermented apple juice / unfermented apple juice. Soft cider contains 1-5% alcohol, hard cider contains 5-8% and apple wine contains >8%-14%.

CIE system: In this system, the light is reflected to the sensor and the sensor in turn reads the product characteristics and determines the value of L, a, b and give the colour of the products. Thus in the Hunter scale "L" measures lightness and varied from 100 for perfect white to 0 for black. Hunter scales 'a' measures redness when positive, gray when 0 and green when negative. Hunter scale 'b' measure yellowness when positive, gray when 0 and blue when negative,

CIP: CIP means cleaning in place. It is an automatic cleaning system used for cleaning industrial equipments and machines without dismantling the equipments (Plate 172). Caustic solution @ 2%, chlorine and iodophore are used for CIP.

Plate 172: CIP equipment

CIPC: CIPC also called chloropropham is applied as thermal fog. CIPC is a sprout suppressant in potatoes.

Citric Acid: A tricarboxylic acid $C_6H_8O_7$ occurring in cellular metabolism, obtained esp. from lemon and lime juices or by fermentation of sugars, and used chiefly as a flavouring and preservation

Clarification: The process by which a liquid can be made clear and transparent by sedimentation or removing the suspended material present in it.

Clarifying agent: Clarifying agents are used to remove the haziness or sediments produced by oxidative deterioration in fruit juice, wine, beer etc. Gelatin, bentonite and resins acts as clarifying agents for products in which foam formation takes place. Chill haze is a common problem in beer, wine and many fruit juices and oxidative deterioration are long standing problems. Protein and polyphenol complex along with pectic substances play a major role in this phenomenon. Bentonite (montmorillonite clay) is employed as clarifying agent. Gelatin is used commonly to clarify beverages.

Cleaver: It is a butcher's implement for cutting animal carcasses into joints or pieces. A cleaver is a large knife that varies in its shape but usually resembles a rectangular-bladed hatchet. It is largely used as a kitchen or butcher knife and is mostly intended for splitting up large pieces of soft bones and slashing through thick pieces of meat (Plate 173).

Climacteric Peak: The maximum point of respiration rate of the mature fruit (Plate 174).

Climacteric fruits: Fruits in which the respiration rate is minimum at maturity and remains rather constant ever after harvest which gradually increases at the beginning of ripening followed by sharp rise to a peak (climacteric peak) and then a slowly decline (Post climacteric stage) e.g. mango, banana, avocado, papaya, apple etc (Plate 175).

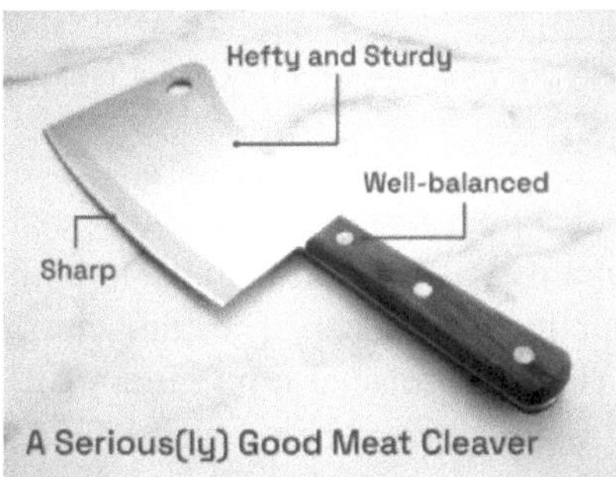

Plate 173: Cleaver

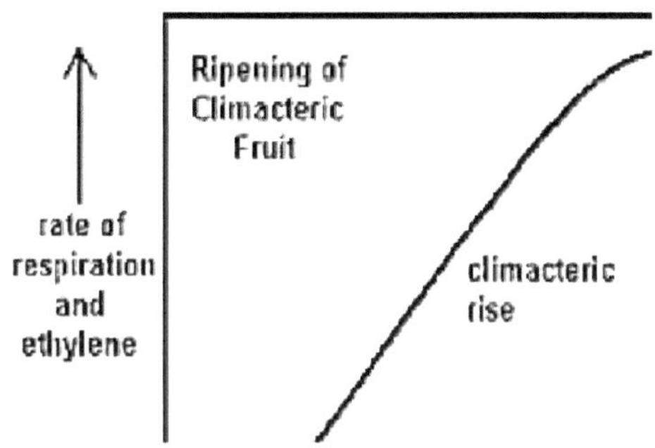

Plate 174: Climacteric Peak illustration

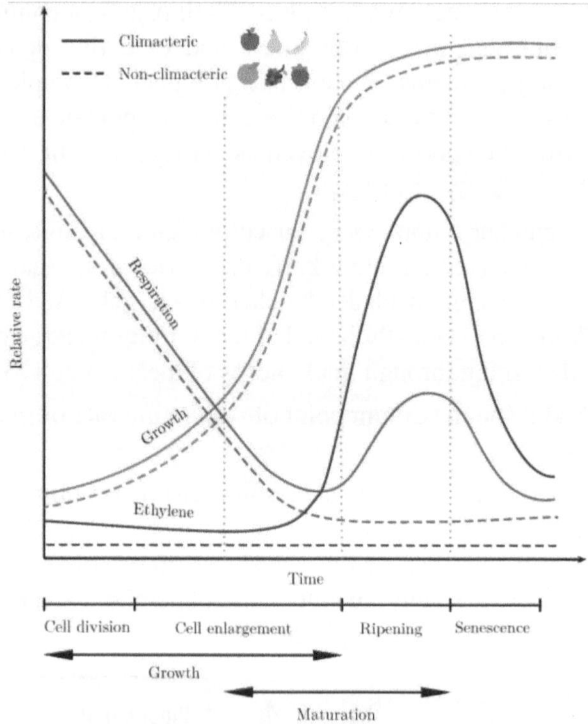

Plate 175: Climacteric vs Non-climacteric pattern

Climacteric period: It is the period in the development of some plant parts that involves a series of biochemical changes associated with the natural respiratory rise and autocatalytic production of ethylene. The climacteric period consists of (i) pre-climacteric (ii) pre-climacteric minimum (iii) climacteric rise (iv) climacteric peak and post climacteric phase (Plate 175). It is not true for all commodity i.e. climacteric fruit. Examples of climacteric fruit are apple, apricot, peach, mango banana, etc. and that of non climacteric fruits are citrus, cherry, strawberry etc.

Clinching of can: Covering of filled cans loosely by the lid through partially seaming by a single first roller action of a double seamer where the lid remains sufficiently loose to permit the escape of dissolved as well as free air from the contents and also the vapour formed during the exhaust process (Plate 176). Syn. Lidding of can. Clinching is the process of closing the filled can with an end loosely to permit inside air to be drawn out of the can and outside gas change to flow into the can. It results in two way transfer of gases. Clinching machine is absolutely same as a seaming machine, only the seaming rollers are loosely set.

Oxygen constitutes about 1/5 of air (atm.) i.e. 22% O_2. In a vacuumising chamber pressure sensitive switch for vacuum is set for 5-10 min i.e. about 99% vacuum that means 1% residual air i.e. 0.2 % residual O_2 against permissible level of 2%. Permissible limit of O_2 in can is 2%. Second roller from the double seaming machine is removed by L-key for clinching purpose.

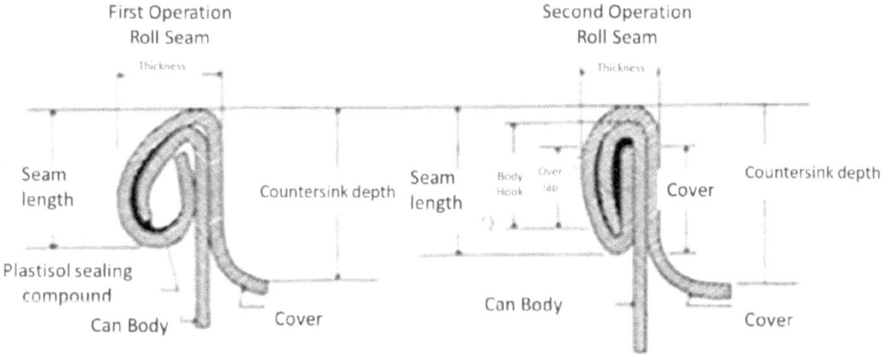

Plate 176: Illustration of first roller during cliching and second roller during double seaming

Closed and electronic auction/bidding: Under closed system, tenders or quotations are invited for bidding of their produce while by electronic auction, the bidding is done online or in one chamber or hall with the help of electronic device where the producer give the specification of composite sample of produce and produce is sold to highest bidder. This type of auction / bidding is done for marketing of fresh produce. This bidding system is regulated by Agricultural Produce Market Committee (APMC) at the hub of terminal markets.

Clostridium botulinum: It is most dangerous microorganisms in low acid food (food having pH > 4.6) and produces botulism toxin in food.

Clostridium spp: These bacteria are anaerobic and produce heat resistant spores. *C. botulinum* is important because it produce a lethal neurotoxin that paralyses the respiratory muscles. Historically botulism has been associated with under processed canned food. *C. perfringens* is usually associated with undercooked meats and sauces.

Clotted cream: A thick cream made chiefly in England by slowly heating whole milk on which the cream has been allowed to rise and then skimming the cooled cream from the top is called Clotted cream (Plate 177). It is also called Devonshire cream.

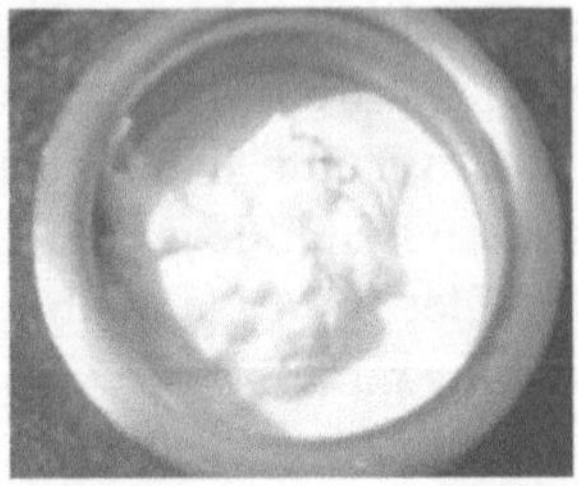

Plate 177: Clotted cream raised on the surface of milk by slow heating

Clotting: A sequence of complex chemical and physical reactions resulting in conversion of fluid blood into a coagulated mass is called clotting.

Cloudiness: If the structure of the vegetable used in pickling (onion pickle) is such that the vinegar is not able to penetrate deep enough into its tissue. Due to this, the activities of the bacteria and other microorganisms present in the tissues cannot be inhibited and so the fermentation starts from inside of the tissues, rendering formation of cloudy vinegar. The process is called cloudiness.

Cloudy jelly: A defective form of jelly having a cloudy appearance instead to clear and transparent body which may happen due to use of improper clarified juice, a immature fruits, over cooking, over cooling, slowly pouring into container and formation of air bubbles, non removal of scum before pouring and pre-mature gelation due to excess pectin in the juice. Syn. Foggy jelly.

CO_2 scrubbers: K_2CO_3, MgO, lime, ethanolamine solution 40% in water (it is corrosive), molecular sieves of aluminium calcium silicate and activated carbon are used to absorb CO_2 and H_2O. They are reactivated by heating.

Coagulase: An enzymes that cause coagulation (as of milk) is called coagulase.

Coagulate: It is mass of curdle, clot, solidified congeal. Proteins coagulate due to heat and or change of pH.

Coagulation: It is a process that utilizes a coagulant to neutralize electrostatic charge on particles and causes very small particles to attract forming large particles.

Coagulum: A mass of coagulated material is called coagulum.

Coated films: These films are coated with other polymers or aluminium to improve barrier properties properties or to impart heat sealability.

Cocoa: Powdered ground roasted cacao beans from which a portion of the fat has been removed or a beverage prepared by heating cocoa with water or milk

Cocoa bean: It is a tree having small yellowish flowers followed by fleshy pods from which cacao is obtained (Plate 178).

Plate 178: Cocoa beans

Cocoa Butter /Cacao Butter: It is a pale vegetable fat with a low melting point obtained from cacao beans. It is also called cocoa butter. It is solid at temperature below 30^{0}C.

Coconut: It is a drupaceous fruit of the coconut palm whose outer fibrous husk yields coir and nut contains thick edible meat and coconut milk.

Coconut oil: It is a fatty oil or semisolid fat extracted from fresh coconuts and used in making soaps and food products.

Co-current air flow: Movement of hot air in the drier in the same direction as the movement of the product which the hottest air contacts the wettest product here air at outlet end becomes cool and the final product may not be sufficiently dry (Plate 179).

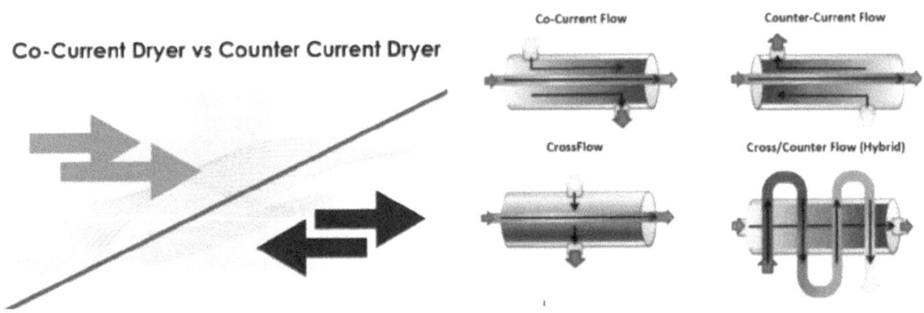

Plate 179: Co-current vs Counter currentair flow

Cod liver oil: Oil obtained primarily from the liver of the Atlantic cod and related fish. It is principally a mixture of the glycerides of many fatty acids, but its minor constituents, the fat-soluble vitamin A and vitamin D, give it its importance. It was once used to treat and prevent rickets, but the widespread fortification of milk with vitamin D help to eliminate rickets which is a significant public health problem. It is still used as a remedy for joint pain caused by arthritis and as a preventive measure of cardiovascular disease, although these benefits have not been proven scientifically. It is also used in feeds for poultry and other animals.

Codex Alimentarius Commission: It is a subsidiary body of the FAO of the United Nations and the WHO. The commission is entrusted with the elaboration of international standards of food to protect the health of consumers and to ensure fair practices in the food trade. It is abbreviated as CAC.

Coefficient of thermal expansion: It is the ratio of the change in volume to the °C change in the temperature. Coefficient of the expansion is very important in packing liquids like juice beverages. Therefore some head space is left in the packaging material to prevent bursting due to thermal expansion of the product during storage. Coefficient of thermal expansion of the glass is 0.001 cc/°C.

Coefficient of viscosity: The ratio of the tangential frictional force per unit area to the velocity gradient perpendicular to the direction of flow of a liquid is called coefficient of viscosity.

Coextruded film: It is multilayer film in which each distinct layer is formed by a simultaneous extrusion process. Five layer coextruded films have been used for packing fruit juices. Coextruded films are multilayered film formed by extrusion process (Plate 180).

Plate 180: Coextruded film

Coffee: A beverage made by percolation, infusion, or decoction from the roasted and ground seeds of a coffee plant **or** coffee seeds especially roasted and often ground **or** a dehydrated product made from brewed coffee.

Coffee Cake: Sweet rich bread often with added fruit, nuts, and spices that is sometimes glazed after baking.

Coffee Mill: A mill for grinding coffee beans is called coffee mill.

Coffee Ring: Coffee cake in the shape of a ring is called coffee ring.

Coffee Royal: A drink of black coffee and liquor is called coffee royal.

Coffeemaker: A utensil or appliance in which coffee is brewed is known as coffeemaker.

Coffeepot: A pot for brewing and serving coffee is known as coffee pot.

Cognac: It is a brandy distilled from wine made from varieties of grapes grown in the Cognac district of France.

Cohesiveness: Degree to which the sample deforms before rupturing when biting with molars. Cohesiveness is the strength of internal bond of product.

Colander: A perforated utensil for washing or draining food is called calander. It is also used to drain out whey during cheese manufacture (Plate 181). It is also called faisselle.

Colcannon: Potatoes and cabbage boiled and mashed together with butter and seasoning is called colcannon (Plate 182).

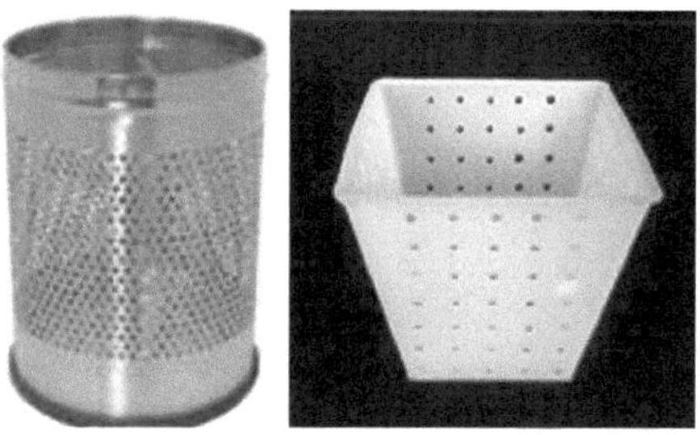

Plate 181: Colander

Plate 182: Colcannon

Cold chain: Low temperature environment through which a produce from field can reach to consumer in a fresh form. It starts from construction of cold storage in the producing centers, refrigerated trucks or rails for transporting, refrigerated ships for exporting them to foreign market and development of cold storage at retail shops.

Cold plasma: Cold plasma is a partially ionised gas comprising ions, electrons, ultraviolet photons and reactive neutrals such as radicals, excited and ground-state molecules. Cold plasma is a novel non-thermal food processing technology that uses energetic, reactive gases to inactivate contaminating microbes on meats, poultry, fruits, and vegetables. Cold plasma can inactivate the pathogenic microorganisms on the surface of low-moisture foods, by generating active species, ultraviolet radiation, and electric fields, which helps to extend the shelf life of foods while having minimal impact on food quality. Two sources of low temperature atmospheric pressure plasma: Dielectric Barrier Discharge in argon driven by repetitive short duration (ns - µs) high voltage pulses (a); A micro-jet using helium as operating gas, generating a cold plasma plume about 2.5 cm in length (b). The plasma can exist in two forms: diffusive and constricted. The majority of the electrical energy deposited in the NTP goes into heating the electrons rather than heating the background gas (Plate 183).

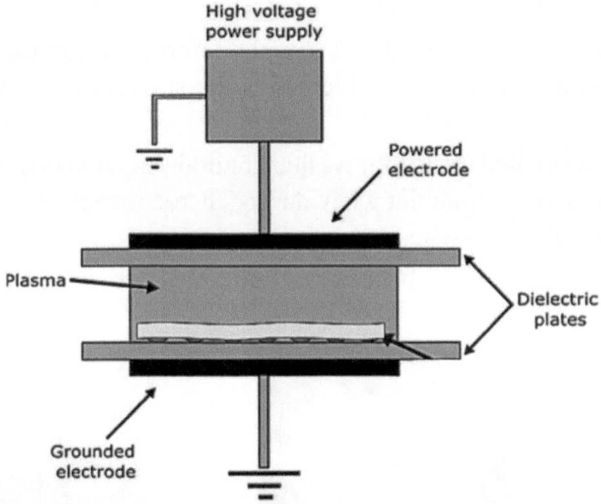

Plate 183: Schematic diagram of Dielectric Barrier in Cold Plasma

Cold plasma sterilization: Cold plasma produces different reactive oxygen species (ROS), superoxides required for disinfection without requiring any corrosive chemical like sodium hypocloride or additional UV rays. This process is safe to operator and environment friendly. Only Argon gas and Deionized H_2O is needed for disinfection. Plasma is energetically the fourth state of matter, apart from solid, liquid, and gas states.

Cold point in can: When a can is processed, all the points within the can are not at the same temperature and the zone of slowest heating is called the cold point of a container and this zone is most difficult to sterilize due to the lag in heating. The point in the can or mass of food which is last to reach the final or desired heating temperature is known as cold point within the can or mass of food.

Cold shock: If food products are cooled rapidly from an optimal temperature to 0^0C, the microorganisms present in foods may die which is known as cold shock.

Cold spot: The point in the can which have lowest temperature during heat processing is called cold spot or coldest point of the cans (Plate 184).

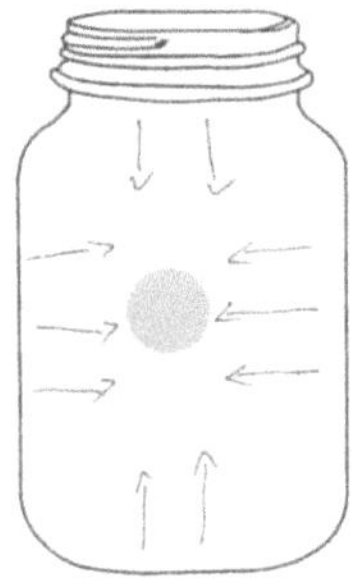

Plate 184: Cold spot / coldest point

Cold sterile bottling: It is done by micro- porous membrane.

Cold sterilization: Sterilization of heat sensitive substances by the application of ionizing radiation. Since, in this process the microorganisms are destroyed without raising the temperature, so it is called as cold sterilization. Such techniques are being development in the food and pharmaceutical industries.

Cold Storage: Storage of food in a cold place for preservation. It is an insulated storage with refrigeration to maintain a stable cold temperature for long term storage of perishable products. Refrigeration usually operates at 4 to 7°C.

Cold Store: A building made for cold storage of the food commodities is called cold store.

Coliform test: This test is used for the estimation of the coliform bacteria in the milk. The presence of such bacteria in the milk is an index of unhygienic method of the milk production.

Collagen: It is a group of fibrous proteins that occur in vertebrates as the chief constituent of connective tissue fibrils and in bones and yield gelatin and glue upon boiling with water.

Collaspe temperature: It is the temperature at which food material losses its structure and volume during storage and processing.

Colligative properties: Properties which depends on the number of molecules in the solution, is a function of concentration and molecular weight rather than just on the total percent concentration. The colligative properties include boiling point elevation, freezing point depression and osmotic concentration.

Color: It is a phenomenon of light (red, brown, pink, or gray) or visual perception that enables one to differentiate otherwise identical objects **or** the aspect of the appearance of objects in light sources which depends on spectral distribution of light and may be described in terms of hue, lightness, and saturation. Or It refers to the sensation arising from the activity of the ratina of the human eye and its nervous system. The colour of foods is usually due to the presence of natural colour like anthocyanins, carotenoids or chlorophyll and synthetic colour or due to discolouration like enzymatic browning and non-enzymatic browning.

Colorimetry: Colorimetry is concerned with the determination of the concentration of a substance by measurement of relative absorption of light with respect to a known concentration of the substance.

Colostrums: The first most milk after calving is called colostrums. It is lower in water, sugar, fat and higher in casein, albumin, globulin and ash than normal milk.

Colour codes for water bottles: Blue is for still water, green for sparkling and yellow and orange are for the flavoured varieties.

Colouring agents: Chemical added in the food items to make them attractive and appetizing are called colouring agents. Some foods have a tendency to lose their natural colour when processed and must be dyed back to make them more appealing to the consumer. They may be natural caramel, carotene and saffron or synthetic food colours like carbon black to impart blackness and titanium dioxide to intensify whiteness. The food colours used are mostly the coal-tar dyes designed as synthetic colours by WHO. The colours added to foods are a) natural, b) nature identical (synthetic) colours, c) artificial, d) inorganic colours. Ninety percent are artificial and do not contain any nutritional value. An example of this is banana ice cream which is dyed yellow and maraschino cherries which are dyed red and green. Natural colours are carotenoids from paprika, saffron, curcumin from turmeric, anthocyanin from grape skin and beet root.

Coloury: It is a descriptive term used for dried hops having a distinctive green ting.

Colza: The oil obtained from the rape seed is called colza.

Combi Top System of Packaging: Combi Top System of packaging is a new system of opening, pouring and reclosing of carton. It is developed in Germany. It consists of a reclosable opening and pouring device for aseptic carton and permit easy opening, pouring into a glass without spillage and protection of the product (usually fruit juice). This device comprises clip, closure, sealing tongue, vent hole, unsealing flap, channels and pouring lip.

Commercial maturity: It is the stage of the plant organ required by a market. Commercial maturity commonly bears little relation to physiological maturity and may occur at any stage during development/senescence. The terms immaturity, optimum maturity and over maturity related to these requirements.

Commercial sterile: It is also called practical sterile food or bacterially inactive food. It is a food in which all the pathogenic, spoilage causing and toxin forming organisms have been destroyed but may leave some microbe that are unable to grow under normal condition of storage.

Commercial sterilization: Commercial sterilization mean food is sterilized in such a way that it will kill most of the pathogenic microorganism but not all. Here, the food is not free from vegetative cells or bacterial spores but prevent their growth.

Communution: Fine division or shredding of the fruit or vegetables is called Communution.

Company or Voluntary standards for quality: These are established by various segments of the food industry. These standards generally represent consumer image and become symbol of product quality. These are used by private firms or supermarkets.

Composite container: A container made from more than one constituent material like paper, metal and plastic is called a composite container.

Composite scoring Test: In this test, specific characteristics of a product are rated separately. In this type test, rating scale is weighted in such a manner that the most important attribute of a particular commodity will account for a large part of the total score. This method is helpful in grading products and comparison of the quality attributes by giving indication of the faulty characteristics. The score quality attributes may be like 20 for colour, 20 for consistency, 40 for flavor, 20 for absence of defects and so comprising total score as 100.

Compressimeter: It is an instrument that measures the force required to compress a food sample.

Compression test: This test is carried out generally on containers to measure the ability of the container to resist external compressive loads applied to faces and when applied to diagonally opposite edges or corners. The compression stresses may be due to static condition during stacking or by dynamic stress during handling and transit. Therefore, during storage/ transport, packages are stacked to the maximum height possible. Consequently, the bottom layer is subjected to compression and hence has to bear the load of the packages stacked over it if the container is not strong enough with force of compression, then it damaged during stacking and transit. Thus the container used should have adequate compression strength for the given purpose. It is determined by standard compression tester (Plate 184a).

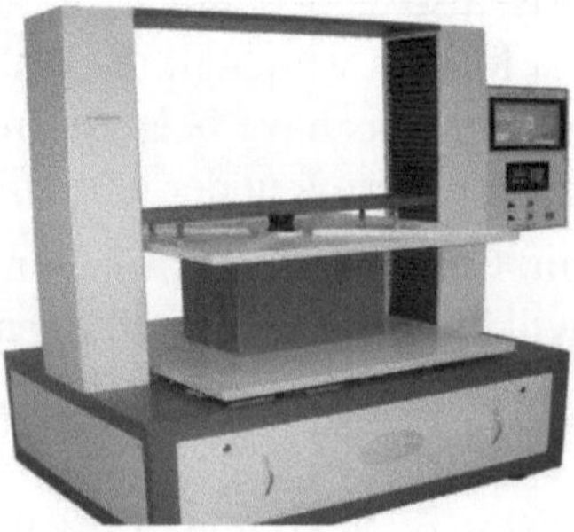

Plate 184a: Compression tester

Concentrate: A fruit juice or any liquid food from which water has been removed by heating or freezing is known as concentrate. It has TSS of equal to or more than 32°Brix.

Concentrated fuel food: Fats are known as concentrated fuel food and reserve fuel because 1 gm fat yield 9 cal energy while 1gm CHO yield 4 Cal energy. Fat are stored in adipose tissue.

Concentrated juice: Juices obtained by removal of major part of their water content by vacuum evaporation.

Concentrated oils: These are the oil from which unstable compounds of oil like terpene and sesquiterpenes are removed wholly or partially by fractional distillation, solvent extraction etc.

Concentrated tomato paste: Concentrated tomato paste has TSS of 32⁰B.

Concentration: The process of concentrating a juice is called concentration.

Concentration methods: Solar concentration, open kettles, flash evaporators, thin film evaporators, vaccum evaporators, freeze concentration, ultra filtration, reverse osmosis (RO), and diffusion membrane are the methods of concentration. Ultrafiltration resist macromolecule, RO resist micromolecules and allows passage of H_2O only.

Concentration of cloudy juices: Cloudy fruit juices contain variable amount so suspended particles comprising mainly of pectic substances cellulose and hemicelluloses, which increases the viscosity. For cloudy juices, centrifugal evaporators or agitated thin film evaporators are developed.

Concentration of pulpy juices: Pulpy fruit juices almost behave as Non-Newtonian Pseudo Plastic Fluids and the apparent viscosity of these fluids decreases with increasing shear rate. Therefore, evaporators having proper agitation are well suited for concentration of pulpy juices.

Concept of 12 D: It is enough heat or process lethality to achieve probability of survival of 10^{-12}. The concept is usually arbitrary but used universally. Or 12D concept is used mainly in low acid canned foods (pH >4.6) where *C. botulinum* is a serious concern. 12D concept refers to thermal processing requirements designed to reduce the probability of survival of the most heat resistant *C. botulinum* spores to 10^{-12} (Plate 185).

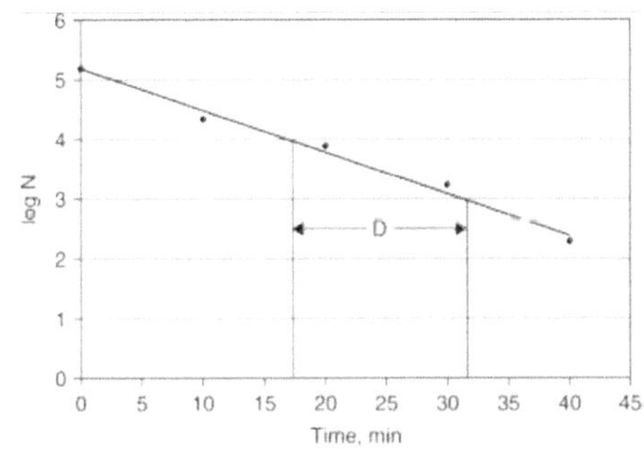

Plate 185: Decimal reduction time overview

Conche: It is a machine in which the chocolate is kept under agitation, so that the flavour is developed and chocolate becomes liquid.

Concrete: It is a waxy essence of flowers prepared by extraction and evaporation and used in perfumery.

Concurrent / Co current flow drier: Hot dry air and wet material enter at the same end and travel through the drier in the same direction (Plate 186).

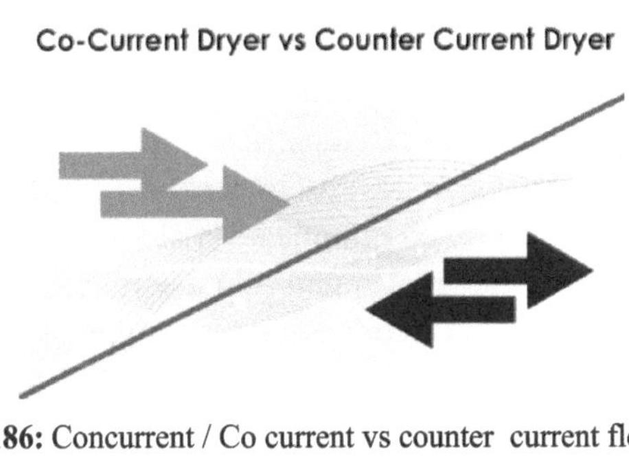

Plate 186: Concurrent / Co current vs counter current flow drier

Condensation: It is the process of condensing or a chemical reaction involving union between molecules often with elimination of a simple molecule (as water) to form a new more complex compound of often greater molecular weight **or** the conversion of a substance (as water) from the vapor state to a denser liquid or solid state usually initiated by a reduction in temperature of the vapour.

Condensation: It is the process of condensing or a chemical reaction involving union between molecules often with elimination of a simple molecule (as water) to form a new more complex compound of often greater molecular weight **or** the conversion of a substance (as water) from the vapor state to a denser liquid or solid state usually initiated by a reduction in temperature of the vapor (Plate 187).

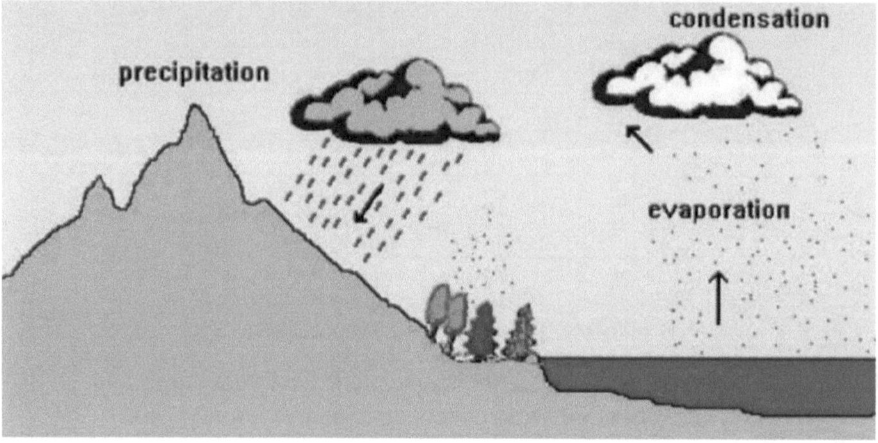

Plate 187: Condensation

Condensed milk: An evaporated milk with sugar added is called condensed milk. **or** It is a thick viscous product obtained by evaporating milk with or without addition of sugar. As per PFA it should contain atleast 31% milk solids including atleast 9 % fat and no other preservative than sugar.

Conduction: The process of transmission of heat through substances without any detectable motion of heated particles of the substances (Plate 188). The act of transmission of heat by means of a conductor or the transfer of heat through matter by communication of kinetic energy from particle to particle with no net displacement of the particles. In thick puree, thick sauce and soups, fruit juice concentrate, sweet potato etc with little or no free liquid, heating takes place by conduction. Conduction is transfer of heat by electron diffusion or photon vibrations.

Conductivity: The power of conducting or transmitting heat is called conductivity or it is the reciprocal of electrical resistivity.

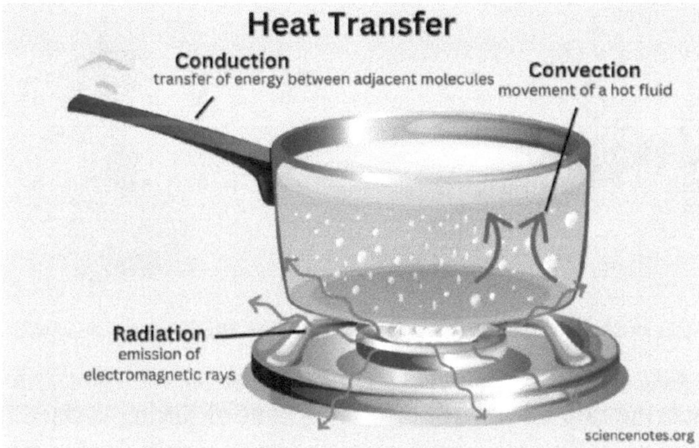

Plate 188: Illustation of heat transfer by conduction, convection and radiation

Confection: The act or process of confecting is called confection. It is a fancy dish or sweetmeat or a sweet food or it is a medicinal preparation usually made with sugar, syrup, or honey.

Confectioner: A manufacturer that deals in preparation of confectioneries.

Conidia: An asexual spore produced on a conidiophore of certain fungi is called conidia (Plate 189).

Conjugated protein: A compound of a protein with a nonprotein portion e.g. hemoglobin is a conjugated protein (Plate 190).

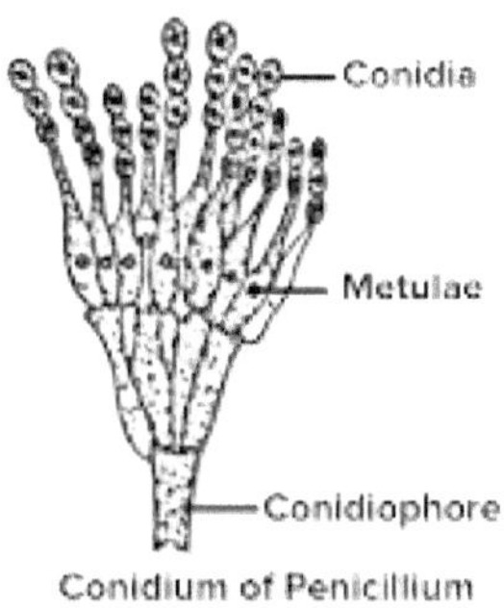

Plate 189: Conidia

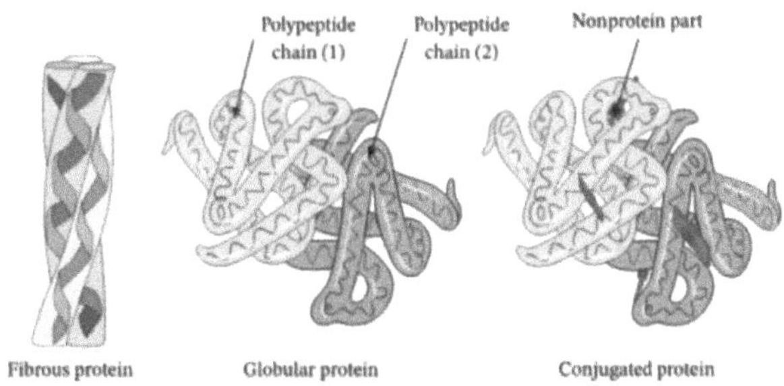

Plate 190: Conjugated protein

Consistency: It is the thickness / density or firmness of any thick liquid such as jam, jelly, ketchup etc. The term consistency and apparent viscosity is generally used for non-Newtonian fluids.

Consistometer: It is a device used for measuring spread or flow of the semi solid food in a specified time period (Plate 191 and 192).

Plate 191: Bostwick consistometer

Plate 192: Bench type consistometer

Constant rate period: Constant rate period is that part of drying process during which the rate of water removal per unit of drying surface is constant (Plate 193). The constant rate period during dehydration take place in the initial stage of dehydration when the food commodities have large amount of free water available for evaporation. Here, rate of removal of the moisture from the food to the surrounding is constant per unit time.

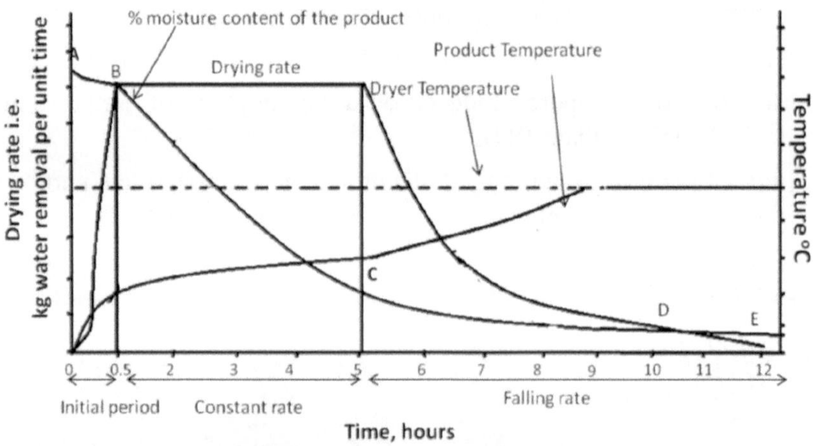

Plate 193: Drying Rate Curve showing constant rate period

Consumer or Grade standards for quality: These standards represent consumer's requirements of the product and generally based on the experience of the industry for consumers.

Consumer panel: Sensory evaluation panel selected from people who happen to be available at a test site and are willing to participate.

Consumer Price Index: An index measuring the change in the cost of typical wage earner purchases of goods and services expressed as a percentage of the cost of same goods and services during same period. It is also called as cost of living index.

Consumer Protection Act, 1986: The Act is enforced by the Department of Consumer Affairs, provides the constitution of district Forum/State/ National Commission for settlement of dispute between seller and buyer. This Act applies to all goods and services provide the rights of the consumers.

Container: One that contains (box or jar) or used for holding goods or a portable compartment in which freight is placed for convenience of movement.

Contaminant: Any biological or chemical agent, foreign matter or other substances not intentionally added to food, which may compromise food safety or food suitability.

Contaminated food: The food is said to be contaminated if food is injurious to health and contain filthy, putrid rotten odour of insect pests etc. and can affect a large number of the populations at a time and hazards may occur.

Continuous inspection: Here inspection and grading is done by one or more inspector(s) when plant is in operation. They make in-process checks on the preparation, processing, packing and warehousing of all products to assure sanitary requirements. The inspectors are present all the time in the plant (under contract) during different operations mentioned.

Continuous microwave dryer: This is an apparatus using microwaves for heating purpose and drying of the food commodities (Plate 194). This apparatus uses electromagnetic waves between infrared and radio waves. Microwave is comparatively short non ionizing electromagnetic wave having one millimeter to one meter in wavelength and has frequency between 300 MHz and 300GHz. These waves are used to cook or heat food in a microwave oven. Magnetron results in production of the microwaves.

Plate 194: Continuous microwave dryer

Control measure: Any action and activity that can be used to prevent or eliminate a food safety hazard or reduce it to an acceptable level.

Control point: Any step at which biological, chemical or physical factors can be controlled.

Convection: The process of transmission of heat from one place to other by the actual mass movement of the heated particles of a liquid or gas. Or The process of circulatory motion that occurs in a fluid at a non-uniform temperature owing to the variation of its density and the action of gravity. In most fruit/ vegetable juice, thin soup, fruit canned in water, syrup with large pieces, vegetables canned in brine or water etc. heating takes place by convection. It is transfer of heat by conduction in a moving medium such as fluid.

Convection oven: An oven having a fan that circulates hot air uniformly and continuously around food.

Convenient foods: The food products which reduce preparation time at home and are purchased by busy and high income consumer.

Conventional sterilants: Conventional sterilants used by UHT packaging machinery include H_2O_2, heat or UV light and ioning radiation.

Converter mash: It is a mash which is used to convert insoluble starch of the unmalted cereal into gelatinized and partly soluble starch (dextrin and maltose) is called converter mash.

Conveyors: These are used for conveying fruits, vegetables and can cases in the processing industry (Plate 195 and 196).

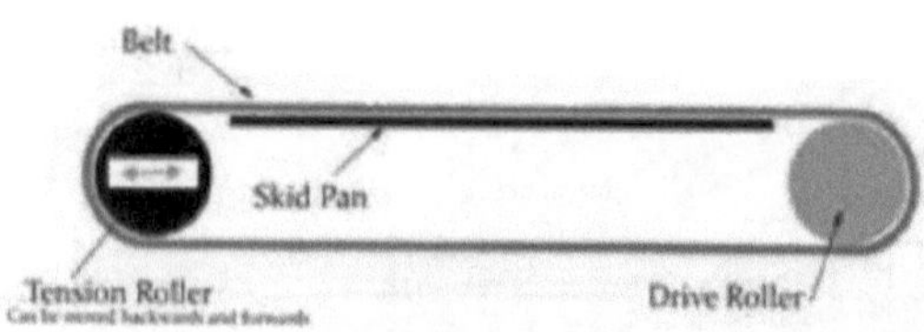

Plate 195: Pictorial diagram of conveyor

Plate 196: Conveyor for conveying fruit box

Cook: It is the process to prepare food for eating by means of heat. A person who prepares food to be eaten by other persons is also called cook.

Cook cheese: An unripened cheese made from curd that has been cooked to a soft consistency is called cook cheese. It is also called cooked cheese.

Cookbook: It is a book that has step-by-step procedures for preparation of food whose rationale is usually not explained.

Cookie: It is a small flat or slightly raised cake (Plate 197).

Plate 197: Cookie

Cooking: It is the art, science and craft of the preparation of large varieties of food items.

Cook–off: A cooking competition is known as cook off. Overheating by the use of fire is also called cook off.

Cool chain: It is a practice in which the produce is maintained in low temperature environment from field to consumer in a fresh form by using refrigerated trucks, rail, van or ships.

Cool storage: Storage at a temperature ranges from -2^{0}C to 15^{0}C depending in food.

COP: COP means cleaning out of place.

Cordial: A sparking clear fruit juice free from all pulp, peel, seeds and sweetened to medium sugar concentration. It contains at least 25% juice, 30% TSS, 1.5% acid and 350 ppm sulphur dioxide. The insoluble solids have been completely removed.

Core wash: The juice obtained by washing core. It contains rags and seed which are high in limonin and cause bittering. Core wash juice is very opaque and is used as a clouding agent in drink bases.

Corn Bread: Bread made with cornmeal is known as corn bread.

Corn chip: A piece of a dry crisp snack food prepared from a seasoned cornmeal batter.

Corn cob: The core on which the kernels of Indian corn are arranged is called corn cob.

Corn crib: A crib for storing ears of Indian corn.

Corn dog: A frankfurter dipped in cornmeal batter, fried, and served on a stick.

Corn meal: A meal ground from corn is called corn meal.

Corn oil: Yellow fatty oil obtained from the germ of Indian corn kernels and used chiefly as salad oil, in soft soap, and in margarine.

Corn pone: Corn bread often made without milk or eggs and baked or fried is called corn pone.

Corn salad: The herbs of the valerian family that is widely cultivated for its leaves used in salads and as a potherb. It is also called lamb's lettuce mache.

Corn snow: Granular snow formed by alternate thawing and freezing is called corn snow.

Corn starch: Starch made from corn and used in foods as a thickening agent, in making corn syrup and sugars, and in the manufacture of adhesives and sizes for paper and textiles. Corn flour contain higher amount of corn starch and so called corn starch.

Corn Sugar / Dextrose: It is dextrorotatory glucose

Corn Syrup: Syrup containing dextrins, maltose, and dextrose that is obtained by partial hydrolysis of cornstarch is called corn syrup.

Corn Whiskey: Whiskey distilled from a mash made up of not less than 80 percent corn.

Cornflakes: Toasted flakes made from the coarse meal of hulled corn for use as a breakfast cereal.

CornPak: It is biodegradable plastic and is marketed by Arcola Grain Product Incorporated and is made form ground puffed corn with soybean lecithin and traces of newspaper.

Cornstarch: Starch made from corn and used in foods as a thickening agent, in making corn syrup and sugars, and in the manufacture of adhesives and sizes for paper and textiles.

Corrosion: It is the process of gradual deterioration of a metal from its surface due to unwanted chemical or electrochemical interaction of metals.

Corrugated fiber board box (CFBB): CFBB consist of 3 layers of craft. Two outer layers are called facing or liner boards and an inner layer is called fluting structure or corrugated medium. This is called single walled board. Double walled board has 3 facing and 2 fluting (Plate 198).

Plate 198: CFBB

Cottage Cheese: A bland soft white cheese made from the curds of skim milk. It is also called as Dutch cheese/ pot cheese smearcase**.** It is a soft cheese, generally coagulated with lactic acid rather than rennin. The curd is not pressed, aged or ripened. It contains not more than 80% moisture and not less than 4% of the fat. It is prepared from skim milk.

Cottage Industry: An industry whose labor force consists of family units or individuals working at home with their own equipment or it is often a small and informally organized industry.

Cottage Pudding: A plain cake covered with a hot sweet sauce is called cottage pudding.

Coumarin: It is a toxic white crystalline lactone $C_9H_6O_2$ with an odor of new-mown hay found in plants or made synthetically and used in perfumery and as a parent compound in anticoagulant agents. It is prohibited food additive.

Counter Current Air Flow: Movement of hot air in the drier in the opposite direction of the material flow where hot dry air contacts the driest product first and a very dry product can be obtained in this drying process (Plate 199).

Counter flow heat exchanger: Two fluid streams flowing in opposite direction through the heat exchanger (Plate 200).

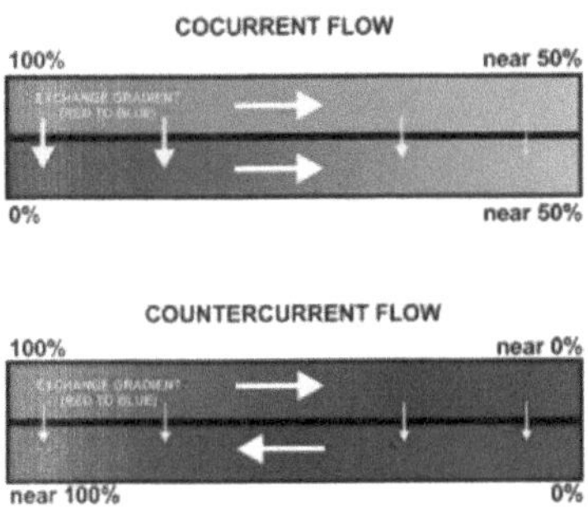

Plate 199: Air flow illustration

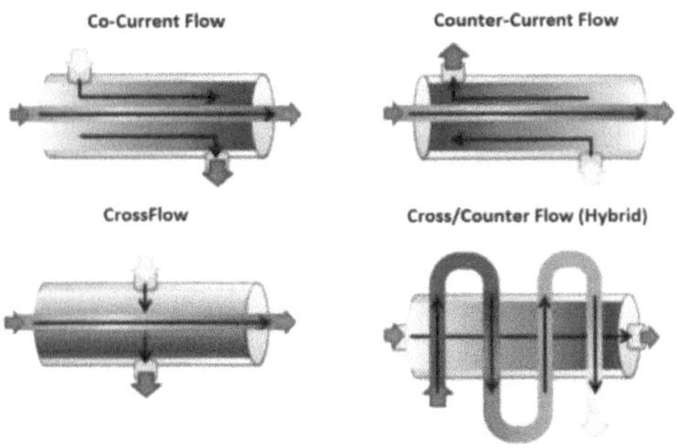

Plate 200: Illustration of heat exchangers

Cover up technique: During the adulteration of juice, it becomes necessary to add something else in order to cover up the adulteration. This technique involve the addition of amino acids, colours etc.

Cracker: A dry thin crispy baked bread product that may be leavened or unleavened is called cracker

Cream: The yellowish part of milk containing from 18 to about 40 percent butterfat is called cream. It is concentrated milk fat and a layer of fat globules. Or It is oily substance that contains fats and rises on the warm milk when left to stand. It is also manufactured on small or large scale, from milk by centrifugal action that separate milk into cream and skim milk. It is used for manufacture of butter and ghee.

Cream cheese: A mild soft unripened cheese made from whole sweet milk enriched with cream. It contains not more than 55% moisture and not less than 33% of the fat.

Cream of tartar: A white crystalline salt $C_4H_5KO_6$ used in baking powder.

Cream puff: A round shell of light pastry filled with whipped cream or a cream filling (Plate 201).

Cream separation: It is the process of separation of the cream from milk either by gravity or centrifugal method. The basic principle of cream separation is based on difference in specific gravity of milk fat and milk serum.

Cream separator: It is a small machine, hand operated or power driven, that uses centrifugal action to separate whole milk into lighter cream and skim (separated) milk that contains the heavier particles and fed to cattle or used for human consumption. It is common for small scale consumption (Plate 202).

Plate 201: Cream Puff

Plate 202: Cream Separator

Cream soda: A carbonated soft drink flavored with vanilla.

Creamery: It is an establishment or a place where cream is prepared from the milk supplied by the producer, part of the cream prepared is then used for making butter and ghee etc.

Credence foods: The food products whose quality is not known both before and after its consumption are called credence foods.

Crib: A building for storage is called crib. It is also called bin.

Critical control point (CCP): A step at which control can be applied and is essential to prevent or eliminate a food safety hazard or reduce it to an acceptable level.

Critical limit: Critical limit is a value which separates acceptability from unacceptability.

Critical mass: A size, number, or amount large enough to produce a particular result is called critical mass.

Critical moisture content: Critical moisture content is the minimum moisture content that will sustain a rate of flow of free water to the surface of material equal to maximum rate of removal of water vapour from the material under the drying conditions. The critical moisture content occurred between constant and falling rate periods. It is the moisture at which rate of drying changes.

Critical point (temperature): It is the temperature of any gas above which in can not be liquefied by compression.

Critical point in ERH: In ERH, it is stage at which the product just becomes lumpy.

Critical ray: It is the ray that travels parallel to the surface of the prism and represents the minimum angle at which the scattered light coming through the sample can strike the prism. Scattered light from the surface of the prism can enter the dark zone and the dark zone is calibrated to a Brix scale because °B and the angle of the refraction are determined by the sample density (Plate 203).

Croissant: A flaky rich crescent shaped (shape of moon at any stage between new moon and first quarter and between last quarter) roll is called croissant.

Chroma: Chroma is differentiation / identification of vivid colours. It is used differentiates (same colour) light red from dark red in term of degree (how much these differ from each other) and so on for other similar colours (Plate 204).

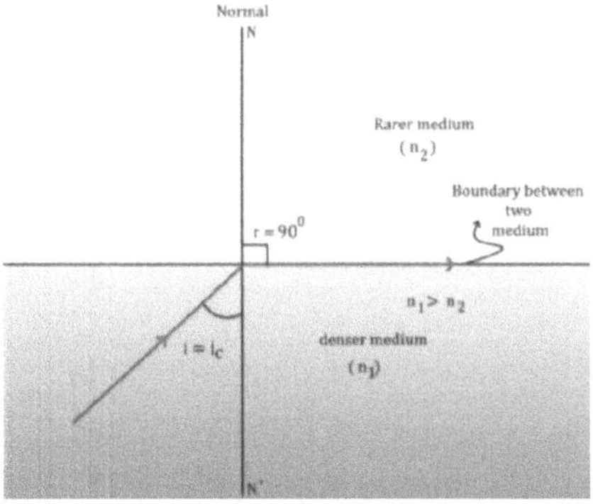

Plate 203: Critical angle of refraction

Plate 204: Hue, Croma and value

Cross flow heat exchanger: In cross flow heat exchanger, two fluid streams flow in perpendicular direction to that of heat exchanger (Plate 205).

Plate 205: Cross flow heat exchanger

Cross-contamination: Cross contamination is the transfer of harmful microorganisms from one item of food to another via a non-food surface such as human hands, equipment, or utensils. It may also be a direct transfer from a raw to a cooked food item.

Crown corking: It is a unit operation used to seal bottles with crown corks after filling with hot juice, RTS etc following by processing. The closing of the bottles with the help of crown corks is known as crown corking (Plate 206).

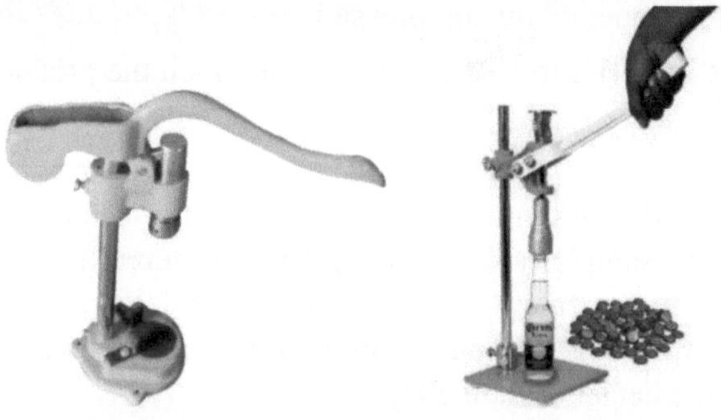

Plate 206: Crown corking machine and corking of bottle

Crown pad: These are the pads which are used to prevent the crown rot of the banana.

Crumb: It is a small fragment of baked bread having porous structure.

Crush: This type of fruit beverage contains at least 25% juice or pulp and 55 % TSS. It is almost similar to squash and contain about 1% acid and is dilute before serving. In other word it is a drink or pulp prepared by crushing fruits e.g. orange crush.

Crust formation: During the dehydration of the food products a hard impermeable layer is formed at the surface resulting in a reduced drying rate. This phenomenon is known as case hardening. It is also called case hardening. This is formed if dehydration is carried out at high temperature.

Cryogenic freezing: Freezing of the food at very fast freezing rate is known as cryogenic freezing. It is freezing of the food products at very low temperature (<-60°C). It is also known as ultrafast / ultra super freezing. The refrigerant used in cryogenic freezing is liquid nitrogen and liquid carbon dioxide. Ultra super fast freezing of foods is the best technique of freezing preservation. It may be achieved by direct dipping, cycling dipping, and contact freezing or by liquid N_2, vapour freezing

Cryopreservation: Preservation of seeds or pollen by immersing them in liquid nitrogen at a very low temperature (-196^{0}C) and relatively low moisture content.

Cryoscopic constant: The freezing point depression is elevated by $DT_f = Km$, where m is molality of the solution and K is cryoscopic constant. It is also called as molal freezing point depression constant.

Crystalline jelly: When crystal appears through out the surface of the jelly due to recrystalization of sugar or when sugar does not get sufficient time to dissolve. Crystal formation in jelly is due to high concentration of sugar or little inversion of sucrose.

Crystallization: It refers to a process of solid – liquid separation, in which solid particles are formed within a homogenous phase. The process of formation of the crystals is called crystallization.

Crystallized fruit: When candied fruit is coated with crystals of sugar either by rolling it in finely powdered sugar or by allowing the sugar crystals from dense syrup to deposit on it is called crystallized fruit.

CTC tea: It stands for crush, tear and cut tea.

Cull fruits: The unwanted, small, sorted out, ungraded, unripened, undersized or oversized and irregular fruits are called cull fruits.

Curd tension: The force required for cutting or breaking the curd at a normal atmospheric pressure and temperature is called curd tension.

Curd: It is the product obtained from pasteurized or boiled milk by souring, natural or by harmless lactic acid or other bacterial culture. It is also called dahi.

Cure: The preservation of the meat by salting, smoking or by pickling for storage is called cure.

Curie: It is the older unit used to indicate the intensity of radioactivity of radionuclide; replaced by Becquerel {1Curie (Ci) = 3.7×10^{10} Becquerel (Bq)}.

Curing (Preservation): A pre-treatment of the raw materials before being processed (mainly for pickle preparation), done by treating the material with dry salt or brine (8-12%) for a sufficient time to facilitate processing and to control the microorganisms and fermentation of the material. During curing, the vegetable lose their raw flavor and become good textured, firm and crisp. The curing is also done in case of fish with salt for keeping.

Curing: A post harvest treatment of tuber and bulb crops by exposing them to relatively high temperature and high humidity to facilitate the drying of upper skin and suberization of the outer tissue, periderm formation and healing of injured surface reduces the moisture loss through transpiration, prevent microbial attack, and reduce rotting and results more shelf life. Or **Curing** is a wound healing operation to replace the damaged periderm (a corky layer over the surface of root crops). Sprout inhibitors /suppressants inhibit periderm formation. Curing is an effective operation to reduce the water loss during storage from hardy vegetable like onion, garlic, sweet potato etc.

CUT-OUT: Examination of cans containing fruit /vegetable to determine their quality, whose measurements are made of their weight of drained solids, and the weight, density and TSS of the syrup or covering liquid, is called Cut Out. The syrup which is added to the fruit at a density of 45°Brix will be diluted by the water content present in the fruit and will generally cut out in the region of 25-30°Brix. Similarly an original syrup of 40 °Brix usually cut out at 21-26°Brix.

C Value: Thermal conductance, or C-factor, is the rate of heat flow through a specified thickness of insulation and is the reciprocal of the R-value. It then follows that the lower the C, the better the insulator and the C-factor for flat insulation is the K-value divided by the thickness of the insulation.

Cyanogenic glycosides: Cyanogenic glycosides give hydrocyanic acid / hydrogen cyanide upon hydrolysis by enzyme β glucosidase in limabean. Safe level of cyanide content is 10-20 mg/100 g of pulse.

Cyclamate: An artificially prepared salt of sodium or calcium used formerly as a sweetener (Plate 207).

Cycle (thermodynamic): When a system undergoes a series of processes and returns to its initial state is called a cycle.

Cyclodextrin: A class of complex cyclic sugars that are products of the enzymatic decomposition of starch and that can catalyze reactions between simpler molecules (Plate 208).

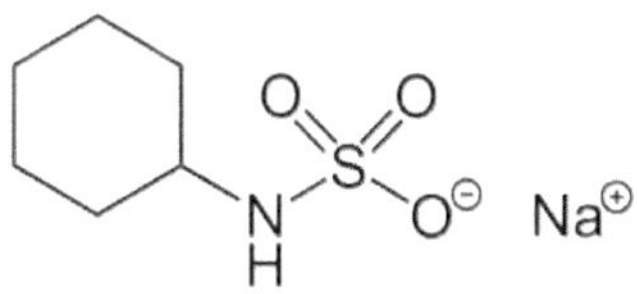

Plate 207: Cyclamate

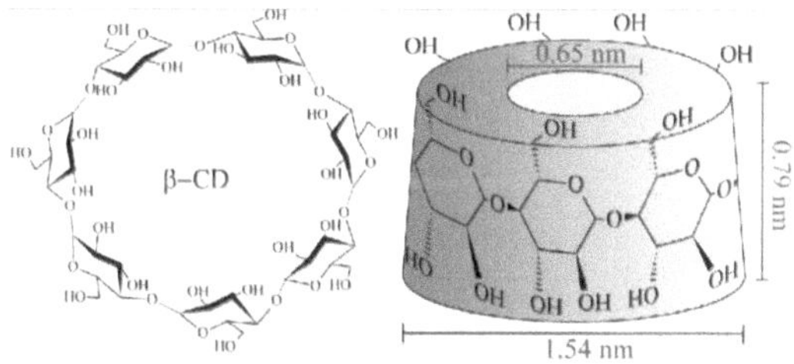

Plate 208: Cyclodextrin

Cysteine: It is crystalline sulfur containing amino acid $C_3H_7NO_2S$ that is readily oxidizable to cystine.

Cytokinins: A category of plant growth regulator, chemically N^6- substituted adenine derivatives which promote cell division, regulate nucleic acids, prevent abscission, senescence of fruits and inhibit root initiation e.g. Zeatin, Zeatin riboside, isopentenyladenine and dihydrozeatin found in plants and synthetic compounds kinetin, 6-benzylamino purine (BAP) or Benzyladenine (BA), 6 (benzylamino) -9-(2-tetrahydropyranyl)-9H-purine (PBA).

D

Dahi: It is the product obtained from pasteurized or boiled milk by souring, natural or by harmless lactic acid or other bacterial culture. It is also called Curd (Plate 209).

Plate 209: Dahi / curd

Dairy: It is a room, building, or establishment where milk is kept and butter / cheese is made or it is the department of farming / farm that is concerned with the production of milk, butter, and cheese or an establishment for the sale / and distribution of milk and milk products.

Dairy husbandry: It deals with care, breeding, feeding, milking of the diary cattle and the production and sale of the milk. It is the economical management of a dairy, including looking after the dairy cattle breeding and proper disposal or sale of milk and milk products.

Dairy products: These are the milk preparation, including butter, cream, cheese, ghee, yoghurt, pasteurized milk, tonned milk, butter milk etc processed and prepared in dairy. Ice cream and kulfi or sweets are not included in this.

Damage: Damage is a physical spoilage, often a partial deterioration, are subjectively judge and very difficult to measure and usually reported as per cent damage.

Daminozide: It is a growth regulator which is used for red colour development in apple.

Dang Stage: It is the stage of date palm fruit when it starts softening from tips. This stage is also called Rutab stage.

Danger point in ERH: In ERH, it is a point which has 5% lower RH than the critical point.

DCPA: It is also called 2, 4 –D and is abbreviated as 2,4 dichlorophenoxy acetic acid. It is a growth regulator and is used for increasing the berry size of the grapes.

Dead period: It is a period (about 24 hours) for which the wort is held at a temperature of about 41°C.

Dead weight: It is the weight of the dressed carcase.

Deaeration: The process of removal of O_2 along with other gases dissolved in juice to prevent oxidative deterioration done by high vaccum, by boiling or adding some antioxidant or inert gases like N_2 which helps to improve the flavor, retain vitamin C in better way, increase the efficiency of heat exchange, improve the uniformity of the container liquid by removing frothing.

Deaerator: The equipment used for the removal of the oxygen from the fruit juice.

Deamination: The process of removal of the amino group from a compound is called deamination.

Death rate curve: It is logarithmic order of death. It is plot of log N/No vs time. The slope of this curve gives D-value. Where, No = the initial number of the microorganisms while N = the final number of the microorganisms remained survived.

Deboner: It is an instrument which is used for removing bone from the meat product.

Deborah number: It is a dimensionless measure of the time which can be used to relate a time constant (t) which is a characteristic of the fluid to the time (t) which is the characteristics of the process to which the material is subject.

Decaffeinated: The coffee / tea from which the caffeine has been removed.

Decantation: It is a method of separating solid from the liquid, by allowing the solid particles to settle down and pouring off the liquid from the top.

Decanter: The equipment used for the removal of suspended material from the juice.

Decarboxylation: The process of the removal or elimination of carboxyl from a molecule is called decarboxylation.

Decortication: It is the method of removal of the husk from seed; oilseeds are generally decorticated before extracting the oil. The mechanical decorticator is used for extraction of oil. In the equipment two roller moves in inward and opposite direction and thus results in breaking of the hard seed coat. Mechanical decorticator can be used for decortications of apricot / peach stones.

Deep bed drying: The drying which takes place in a drying zone and the layer of the grains for drying is >20cm.

Deep fat frying: It is the process in which food is totally immersed in hot oil and cooked vigorously by convection currents and cooking in uniform on all sides of the foods is called deep fat frying (Plate 210).

Plate 210: Deep fat frying of chips

Dcfatting: It is the process of removal or draining of the fats from the chips or French fries. Here the chips or French fries are passed over a vibrating screen.

Degree Baume: Degree Baume (°Be) is the empirical concentration scale.

Degree lactometer °L: Lactometer is used to check the specific gravity of the milk and is expressed in term of degree lactometer.

Degree of saturation: It is also known as percentage humidity or percentage absolute humidity. It is the ratio of absolute air humidity (H) to the specific humidity of the saturated air (Hs) at the same dry bulb temperature.

Degree proof: The unit which expresses the strength of alcohol as C_2H_5OH in terms of percentage of alcohol present in wine. Degree proof = 2 x % of alcohol by weight. So a wine having 40^0C proof strength contains 20% alcohol by volume (P= 1.75 x % alcohol). Alcohol content in beverage industry is expressed as degree proof. In North America proof mean ½% alcohol by volume whiskey.

Degree salometer (⁰S): The concentration of the salt brine is measure in degree salometer (Plate 211). 20°S = 5% w/w salt solution or 4°S = 1% w/w salt solution. Degree brix /degree salometer is industry term for measurement of concentration of sugar/salt, respectively.

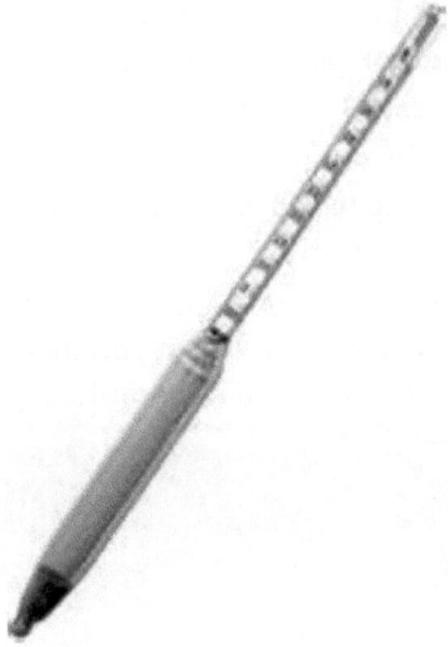

Plate 211: Salometer

Degreening of citrus: The degreening of citrus is due to decomposition of chlorophyll in the rind and is induced by low night temperature. Degreening of citrus can also be done be by treatment with ethylene (1-10 ppm) at 15-25^{0}C, treatment with ethephon [(2-chloroethyl) phosphoric acid, 2-CEPA)].

Degreening: The treatment of oranges with ethylene under controlled conditions to hastens the loss of chlorophyll and this process is known as degreening. or The process of decomposing the green pigment in fruit by applying ethylene (1000-2000ppm) or similar metabolic inducers to give a fruit its characteristic colour as preferred by consumer, generally followed in citrus fruit but also practiced in mango, banana, tomato etc.

Dehulling: The removal of the outer covering from the intact grains of seeds.

Dehumidification: Removal of the moisture from the air without change in its dry bulb temperature is called dehumidification.

Dehumidifier: It is equipment used to control humidity in a specified areas or space.

Dehydration: A drying process where moisture of food is removed by artificially produced heat under controlled conditions of temperature, relative humidity (RH) and air flow. Synonymous of dehydration is desiccation. Mechanical dehydrator is used for dehydration purpose (Plate 212).

Plate 212: Dehydration of fruits

Dehydration Ratio: It is the ratio of weight of raw material to the weight of dehydrated matter.

Dehydro brining: A storage procedure of perishable where the foods are at first dried partially and then salted which helps to the upto the free water present in food. These foods have good shelf life without any refrigeration.

Dehydro canning: An efficient storage procedure of the perishables where the products are dried to half their original weight prior to their canning.

Dehydro freezing: An efficient storage procedure of the perishables where half the weight (50% moisture is removed) of the produce is reduced by warm air drying prior to freezing

Deliquescent: These are the substances which absorb moisture from the air to form the solution.

Denaturation: The alteration of proteins structure due to heat, light or change in pH and which is responsible for decrease in solubility of proteins is called denaturation. Protein denaturation is responsible for toughness of the frozen foods.

Densimeter: It is an instrument used for measuring the relative density of the substance.

Density: It is expressed as mass of a substance per unit volume. or It is the quantity per unit volume, unit area, or unit length.

Dents: Mechanical injury causing a significant reduction in the internal volume of the can or deformity of the seam side which may lead to flipper type of spoilage.

Deodorization: Removal of odour form fats/oil by heat and vaccum or by adsorption into activated characoal is called deodorization. The elimination or prevention of the offensive odor from the food commodity is also called deodorization

Descriptive Flavour Analysis Panel (DFAP): Thoroughly trained panel that works as a team to precisely describe in words the "flavor of sample".

Descriptive tests: Sensory testing designed to provide information on selected characteristics of food sample. These tests are used to identify / describe the sensory characteristics (colour / appearance, texture, body, consistency, taste and flavor) of the food and express them on the standard agreed scale. The panelist's record the impact of the quality attributes in relative term. These tests are used to describe the aroma and flavor intensity characteristics (flavour profile tests) of foods. Descriptive tests are also used to describe the texture profile tests like texture, body and consistency of foods.

Desiccant: It is a drying agent e.g. Calcium chloride.

Desiccation: It is a method of preservation of a food by drying process where moisture of food is removed by artificially produced heat / without heat and may /may not be under controlled conditions of temperature, relative humidity and air flow.

Designer drug: It is a synthetic version of a controlled substance (heroin) that is produced with a slightly altered molecular structure to avoid having it classified as an illicit drug.

Designer food: The food product formulated to contain / enhance levels of the phyto chemicals is called designer foods. These foods are supplemented with food ingredients that are naturally rich in disease preventing substances.

Detartarization: The elimination of potassium bi-tartarate from the grape beverage industry.

Detectors: A detector is a device that measures the change of composition of the effluent. There are different detectors which can be used in GC. A non selective detector responds to all compounds except carrier gas, a selective detector responds to a range of compounds with a common physical and chemical properties whereas a specific detector responds to a single chemical compound. Detectors may be concentration dependent and mass flow dependent. Mass flow dependent detectors usually destroy the sample and the signal is related to the rate at which solute molecules enter the detectors.

Detoxify: To remove a harmful substance (poison or toxin) from the food commodity.

Developed flavour: These are the flavour compounds that are formed during food processing. It may by either solid or volatiles.

Development: The series of processes from the initiation of growth to death of a plant part is known as development.

Devonshire cream: A thick cream made chiefly in England by slowly heating whole milk on which the cream has been allowed to rise and then skimming the cooled cream from the top is called Devonshire cream. It is also called Clotted cream.

Dew point: It is the temperature at which a vapor (as water) begins or would begin to condense e.g. if outside temperature is 78°C, the *dew point* is 63°C. It is the temperature to which air and its water vapour must be cooled till saturation at constant humidity to start condensation of water. At dew point, the air is saturated with water. The dew point is also that temperature where the vapour pressure of water equals to the partial pressure of the water in the air. *When the temperature is below the freezing point of water*, the dew point is called the frost point, as frost is formed via deposition rather than condensation. The dew point is the temperature the air needs to be cooled to (at constant pressure) in order to achieve a relative humidity (RH) of 100%.

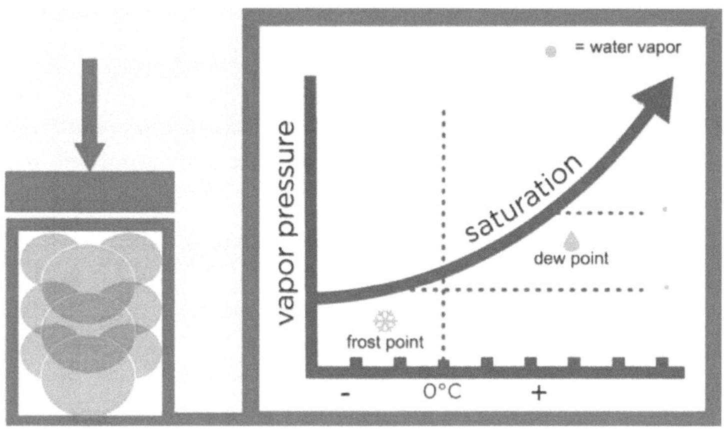

Plate 213: Dew point measurement

Dew point hygrometer: It is an instrument for finding out the dew point by recording the temperature at which vapour being cooled in a silver vessel begins to condense.

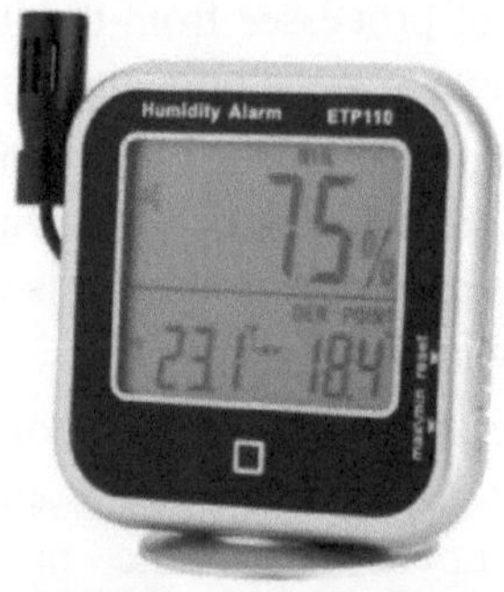

Plate 214: Dew point hygrometer

Dew point temperature (DPT): The temperature at which the water vapour begins to condense is called the dew point temperature. At 100% RH, all three temperature i.e. dry bulb temperature, wet bulb temperature and dew point temperature are equal. It is also called dew point depression.

Dextrin: These are water-soluble gummy polysaccharides $(C_6H_{10}O_5)_n$ obtained from starch by the action of heat, acids, or enzymes and used as adhesives, as sizes for paper and textiles, as thickening agents (as in syrups), and in beer. Dextrin is intermediary product by breakdown of starches.

Dextrose: It is dextrorotatory glucose.

Diastase: It is an enzyme that helps in conversion of starch into dextrin and maltase.

Dibis: Juice extracted from the date palm pulp is called dibis.

Dielectric material: A dielectric material (e.g. air, glass) has high resistance to flow an electric current but does allow an electrostatic or magnetic field to pass through it. A perfect vacuum forms perfect dielectric.

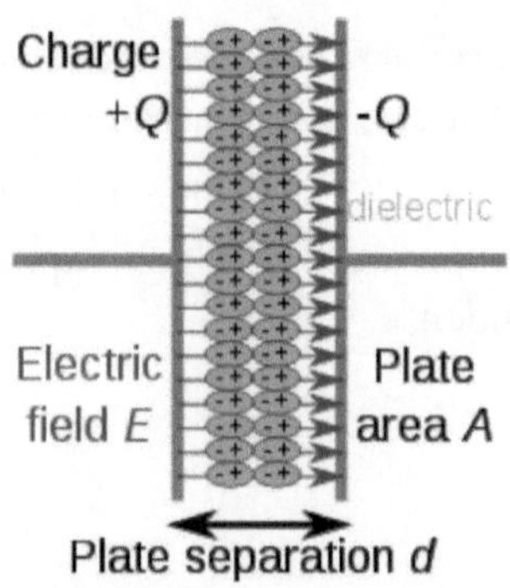

Plate 215: Dielectric material

Diet: Food and drink regularly provided / consumed. It is the kind and amount of food prescribed for a person or animal for a special reason.

Dietary disease: A disease which is caused by the deficiency of nutrients and vitamins is called dietary disease.

Dietary Fibre: It is mostly indigestible material in food that stimulates the intestine to peristalsis. It is also called bulk roughage. The dietary fibre has the ability to enhance the saliva production and activate the production of HCl and digestive hormones in the stomach (Plate 216).

Plate 216: Dietary fibre

Dietary law: Any of the laws observed by Orthodox Jews that permit or prohibit certain foods.

Dietary supplement: A product taken orally that contains one or more ingredients (as vitamins or amino acids) that are intended to supplement one's diet and are not considered food. If vitamins and minerals in a diet are 50-150% of the RDA, then it is called as Dietary Supplement.

Dietetics: The science or art of applying the principles of nutrition to the diet is called dietetics.

Dietetic drink: The drink which can be given to the patients suffering from diabetes e.g. sorbitol based drinks.

Difference tests: Sensory test designed to determine whether detectable differences exist between products. These tests are used to determine any difference between or among the samples. These tests are used to evaluate food on the basis of simple difference, directional and quantitative difference or quality preference. The difference tests commonly used in food evaluation are paired comparison tests, duo-trio and triangle tests.

Diffraction: Diffraction is a modification which light undergoes in passing by the edges of opaque bodies or through narrow openings and in which the rays appear to be deflected.

Diffusion: The process whereby particles of liquids, gases, or solids intermingled as the result of their spontaneous movement caused by thermal agitation and in dissolved substances move from a region of higher to lower concentration (Plate 217). The movement of the molecules is by virtue of its kinetic energy.

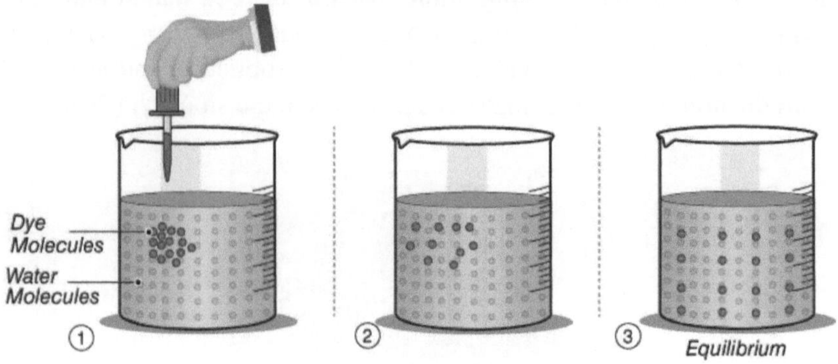

Plate 217: Diffusion

Diffusivity: It is the constant of proportionality in Fick's law and has the dimension L^2T^{-1}. It is the ratio of molar flux of components to its concentration gradients.

Digestion: The mechanical and chemical breakdown of food to simple substances which can be absorbed and used by the body cells.

Dilatant fluids: Dilatant fluids are those whose apparent viscosity increases with an increase in the shear rate. The viscosity of the substance increases with increasing velocity gradient. They are uncommon but suspensions of starch and sand behave in this way. Dilatant fluids are also called as shear thickening fluids.

Dill Pickle: Cucumber pickle, prepared by fermentation of cucumbers in dilute brine flavoured with dill herb spice (Anethum graveolens).

Dioscorine: It is a toxic alkaloid present in certain species of yam.

Dioxin: Any of several persistent toxic heterocyclic hydrocarbons that occur as by-products of various industrial processes and waste incineration. The compounds is created as by products in the process and used to manufacture pesticides, preservatives, disinfectants and paper processing. They can also be formed when materials such as plastic, papers, and wood are burned at low temperatures. There are several dioxins but small number is amongst the most toxic substances known. Dioxins are obiquitous environmental contaminants and are generally present in very low concentrations in all foods.

Dip – a - Sauca Packaging: Dip -a – Sauca packaging hold 200 g of flame grilled chip and has a dip lid at the top to hold tomato sauce. The pack is flat bottomed, making it stable and has an integral die-cut handle for easy handling. The pack is made of recyclable food grade board and has a protective varnish coating on the outside.

Direct heating: It can be achieved either by steam injection or steam infusion. Here, the air is in direct contact with a flame or combastic gases.

Disaccharide: It is a class of sugars (as sucrose) that yields on hydrolysis two monosaccharide molecules. The concentration of disaccharides in fruit is generally determined by the difference between reducing and non reducing sugars before and after inversion (Plate 218).

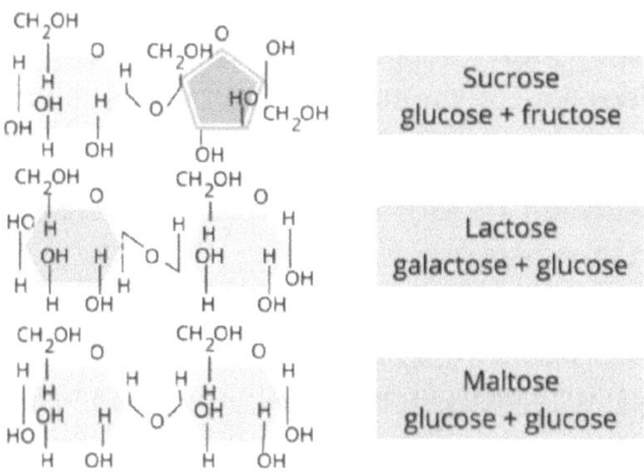

Plate 218: Examples of Disaccharides

Discolouration: The browning of food products due to biological causes like enzymatic reactions, non-enzymatic reactions or metallic reactions is known as discolouration.

Disgorging: Disgorging is a process for removing yeast form sparkled bottle by the pressure created by the fermentation process.

Disgorgement: It is the process of refrigeration of the inverted bottles and then freezing of the neck as a plug of the ice in ice calcium chloride mixture. Then the bottles are uncorked to remove the yeast sediments with gas pressure. This process is called disgorging. After riddling, disgorgement process is done. Volume of wine lost during disgorgement is then replaced by addition of dosage syrup, which is acidified with citric acid and lightly sulfited. Depending on residual sugar in champaigne; they are given different names like Brut if residual sugar is 2-10 g/l, Extra dry if residual sugar is 10-20 g/l, Dry if residual sugar is 20-40 g/l, Semi dry if residual sugar is 40-60 g/l, Sweet if residual sugar is 80-100 g/l (Plate 219 and 220).

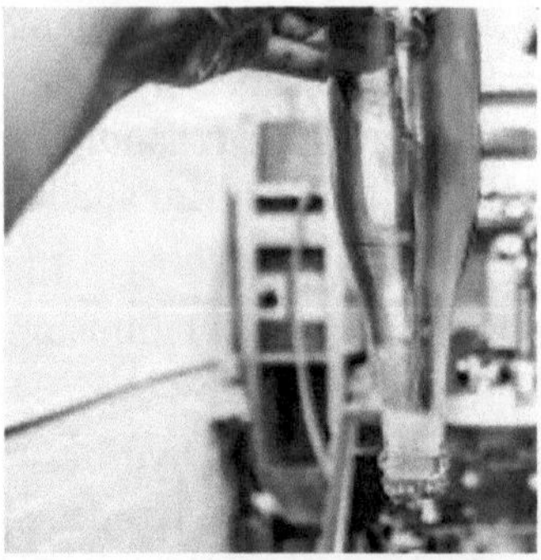

Plate 219: Process of Disgorgment

Plate 220: Process of remuage or riddling

Disinfection: The reduction of the number of micro-organisms in the environment by means of chemical agents and/ or physical methods to a level that does not compromise food safety or suitability.

Disintegration: It is a process of size reduction.

Dissolved air floatation (DAF): It is used to separate suspended pulp and waxes from the waste water.

Distillation: The process of purifying a liquid by successive evaporation and condensation is called distillation. This process involves vaporization and subsequent condensation of the liquid solution (Plate 221).

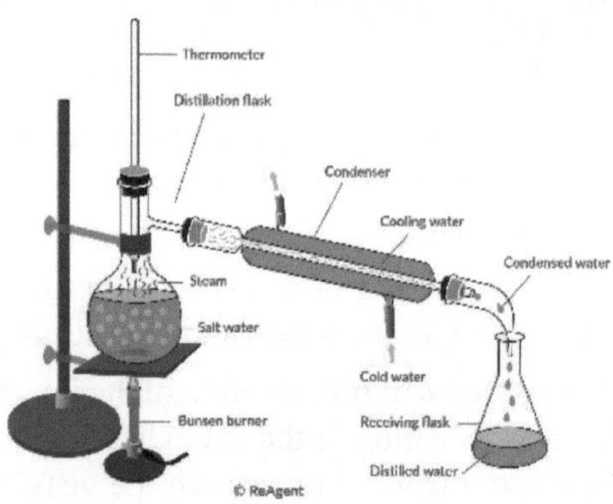

Plate 221. Distillation unit set up

Distiller: One that distills alcoholic liquors is known as distiller.

Distiller's grains: A by-product of whisky manufacturing similar to Brewers grains consisting of the remains of the malted barley. It is used as animal feed.

Distillery: The workstation where distillation of alcoholic liquors or other solvent is done is called distillery.

Distillery waste: Waste generated from distillery is known as distillery waste. It is a type of residue which is obtained from the fermentation and distillation of beer and distiller spirit.

Divided table top tester: This type of tester is useful with packages which are difficult to sling. The package is held in desired position of fall and the trap door opened thereby projecting the package on to the floor. The height of drop, the position of fall and the type of floor can be altered at will.

Dockage: This term is used in grading of grains and oilseeds. It is the process of removal of the unwanted material from grains and oilseeds during grading like weed seeds, weed stems, chaff, under-developed seeds etc. or It is a place (platform) used for the loading or unloading of materials (Plate 222).

Plate 222: Dockage sieve

Dockering: The process of reduction of blistering and pillowing in some food products like potato. It can be reduced by perforating the product with opening. This process is called dockering.

Doka stage: Fruit of date palm when full grown hard and show yellow or red colouration marks this stage. It is also called Khalal stage (Plate 223).

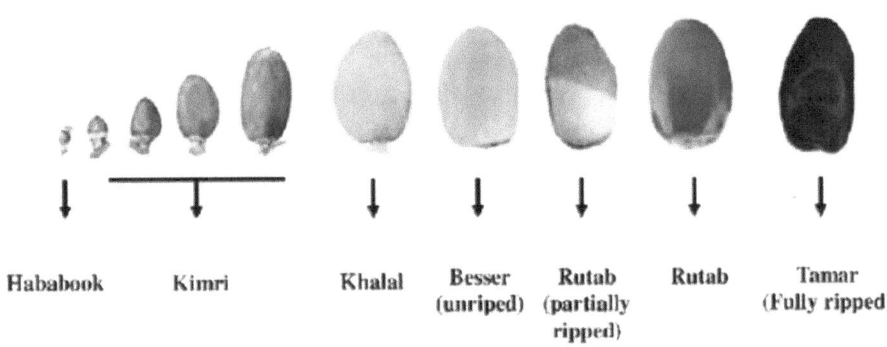

Plate 223: Different stages of date fruits

DOPA [dihydroxy phenylalanine]: A phenolic amino acid $C_9H_{11}NO_4$ occurring naturally in broad beans or prepared synthetically from tyrosine.

Double fortified salt: The fortification of the iodised salt with iron is called double fortified salt. Or the fortification of the salt with iodine and iron is called double iodised salt (DFS). Iron fortified iodised salt help to fight against anaemia and goitre. It is double fortified salt.

Double seaming machine: It is a machine used for sealing of the cans. This operation is done after can flanging. It is also called double can seamer (Plate 224).

Double toned milk: It is similar to toned milk except that in this milk the fat content is 3% and SNF is 9%.

Dough: Dough is a complex mixture of flour constituents, water, salt, and other ingredients depending upon the requirement of the end products (Plate 225). It is a mixture that consists essentially of flour or meal and a liquid (as milk or water) and is stiff enough to knead or roll. It is the principal intermediate stage in the transformation of wheat into different end used products. It is thickened, uncooked mass of combined ingredients for bread rolls and biscuit but usually applied for bread.

Plate 224: Double seamer

Plate 225: Dough making

Dough Conditioner: An agent that is used for the conditioning of the dough is called dough conditioner.

Downstream: The movement of the food in the direction from mouth of a stream toward the latter stages of a usually industrial process or the stage of movement of marketing after manufacture.

DR plates: Mean Double Reduced tinplates.

Draft beer: Unpasteurized beer which is held under refrigeration is known as draft beer. Draft beer has better flavor than pasteurized beer. It has a very short life and so kept in refrigerator.

Drag flow: The forward jumping action of the extruder and results from relative motion between food and barrel.

Drained weight in canned product: It is the net weight of the fruit and vegetables in canned cans free from the sugar syrup or brine. It is obtained by emptying the contents of the container over the mesh sieve and thus draining the syrup or brine for a specific time.

Dredge: It is an apparatus in the form of an oblong iron frame with an attached bag net used for gathering fish and shellfish (Plate 226).

Plate 226: Dredge

Dredging: It is a process of passing a food through a fine dry or powdery substance in order to coat food by sprinkling.

Dressed Carcase: It is butcher's term used for the animal body, after the removal of head, hide and offal, ready to sold, preserve and export as meat.

Drop test: This test helps to measure the ability of the container and packaging material to provide protection to its contents and to measure the ability of the container to with stand rough handling and also provide information useful in improving the design of the container. Common types of apparatus are divided table top tester and hoist type apparatus (Plate 226a).

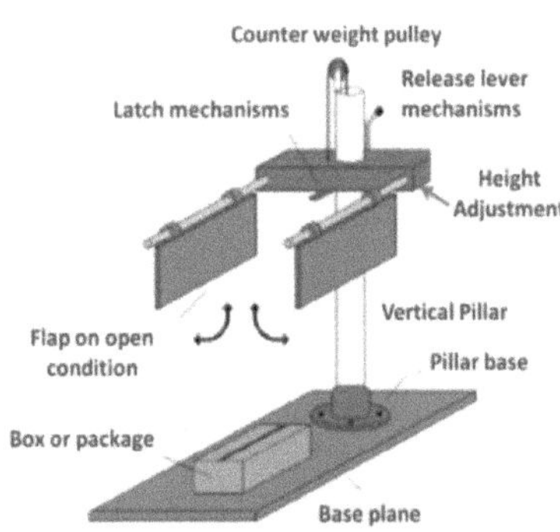

Plate 226a: Drop test illustration

Drosometer: It is an instrument used for measuring the dew.

Drug: A substance intended for use in the diagnosis, cure, mitigation, treatment, or prevention of disease or a substance other than food intended to affect the structure or function of the body. If vitamins and minerals in a diet are more than 150% of the RDA, then it is called as Drug and can not sold as food or supplement.

Drum test: This test helps to evaluated loaded shipping container against certain hazards of handling shipment. This is use to test boxes or crates made of metal, wood, fibre board or combination of these.

Dry air: The dry air is considered as a mixture of nitrogen and oxygen neglecting the small percentage of other gases.

Dry batch dry (DBD) process: This process is used for dehydration of apples, grapes, pears, apricots and peaches. In this process the fruit is first washed, halved, and pitted. It then sulphured lightly by dip in bisulphite solution and dried to 50% weight reduction in a conventional tunnel dehydrator. Then blanched in steam blancher and dried in tunnel dehydrator to finish dehydration.

Dry bulb temperature: It is the temperature of the air recorded by a thermometer when the moisture present in the air does not affect it (Plate 227).

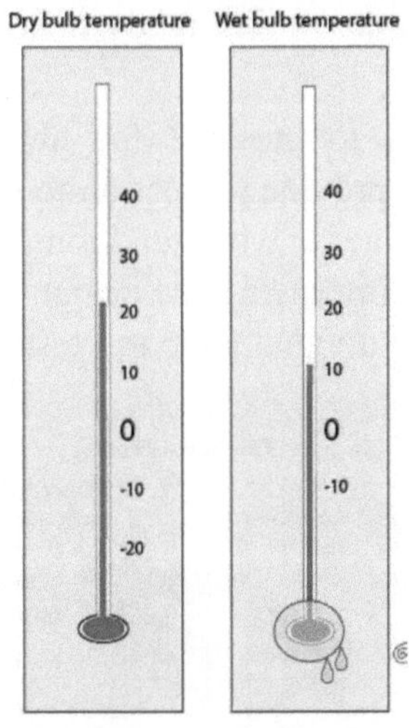

Plate 227: Dry / wet bulb temperature

Dry cow: A cow not producing milk.

Dry cured bacon: Bacon preserved by being rubbed with dry salt as opposed to being soaked in brine.

Dry heat: It is a kind of the heat used in roasting, grilling, toasting, broiling, baking and frying.

Dry matter content: Food solids (total) with out moisture content, accomplished by even drying of the samples usually at 70⁰C for 12 hrs.

Dry milk: Whole milk dehydrated to the extent of 97% by spray drying or vacuum drying is called dry milk.

Dry milking: It is also dry hand milking. It is complete milking to ensure that the udder of the cow has been completely empty or dry.

Dry storage: Storage of food commodity without extremes of temperature change, at about 20⁰C and below 50% relative humidity.

Dry weight basis nutrient content: It is the ratio of nutrient content of fresh sample x 100 to that of dry matter content of the fresh sample.

Dry wine: Dry wine contains little or no unfermented sugar and alcohol content about 11-14%.

Drying: It is a method of preservation of foods by reducing the moisture content to the point at which microorganisms fail to survive (on them) and the action of enzyme is also checked as well and in this situation, osmotic pressure due to increased dissolved solids acts as active agent for preservation. It is the process of removal of water or volatile liquids from the solid or semi solid food by means of thermal energy. Sun / solar drying is also called drying.

Drying methods: There are different drying methods like Drum drying, spray drying vaccum shelf drying, vaccum belt drying atmosphere drying, kiln drying, tunnel drying, sonic drying, ball drying, electromagnetic drying, osmatic drying, foam mat drying, puff drying, slush drying, pneumatic drying Azeotropic drying, microwave drying, hot oil immersion, thin process drying.

Drying rate: It is the amount of the moisture evaporated per unit time or per unit area.

Drying ratio: It is the ratio of fresh material taken for drying to the dried material.

Dryness: Degree to which the sample feels dry in the mouth.

DSC: It stands for differential Scanning Colourimetry. It is used in thermal analysis of foods.

Duhring's rule: According to this rule linear relationship exist between the boiling point of a solution and the boiling point temperature of water at the same pressure.

Duopoly market: It has two sellers of a commodity in the market. These are **imperfect markets.**

Duo-trio test: In this test, the panelists are provided with three samples. Out of these three samples, two samples are identical and the other one is different. Here at first the panelists are provided with the reference sample and then the other two samples are given in random order to find that out of two samples which one matches with the reference sample.

D–Value / Decimal Reduction Time: Decimal reduction time or D value is the time required to kill or destroy of 90% of the microorganisms in the preserved product at a specific temperature (Plate 228). When the D is determined at 250°F, it is often expressed as Dr. It is a measure of the death rate of the microorganisms. or It is the percentage of digestible organic matter in the dry matter of animal feed such as hay, silage or dried grass. D-value, or decimal reduction time, is the time it takes to reduce a microbial population by 1 logarithm, or 90% of its initial value, under specified conditions.

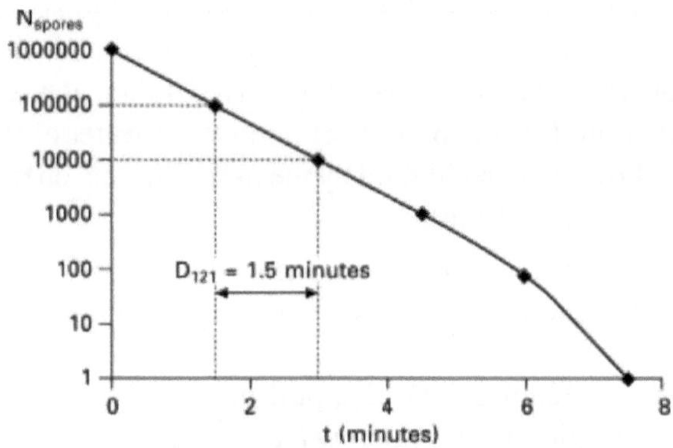

Plate 228: Decimal Reduction Time / D value

Dynamic flavour Profile (DFP): The graphical depiction of a food product combining both approaches (dimensions of flavour at a time; the spider web or descriptive attributes *vs.* intensity and the flavour profile curve addresses intensity *vs.* tasting time) is called the Dynamic flavour Profile (DFP). The Dynamic Flavor Profile Method uses a three-dimensional model to show the changes in flavor attributes over time. The DFP method requires trained panelists familiar with the nomenclature involved. A simple approach is used in flavour description based on similarities in chemical structure (similar flavour types are similar in structure). The graphical depiction of the DFP is somewhat complex.

E

E numbers: Each food additive used in the food industry has been given a specific number code for their identification. These numbers are called as E numbers. These E numbers are used within the European Union. The word prefix "E" stands for Europe. They are commonly used on the food label through out the European Union. These are not used in Australia and New Zealand. The different food additives are given different E numbers depending upon the type of food additive like E100-E199 (food colours), E200-E299 (preservative), E300-E399 (antioxidants), E400-E499 (thickners, stabilizers, emulsifier), E500-E599 (pH regulator, anticaking agent), E600- E699 (flavour enhancers), E700-E799 (antibiotics), E900-E999 (miscellaneous like waxes, synthetic glazes, packaging gases, sweetners, foaming agents), and E1000-E1099 (additional chemicals).

Earling: The young lamb of one to two years' age is called earling or yearling.

Ebullioscopic constant: The boiling point elevation is expressed by $Dt_b = Km$, where m is molality of the solution and K is ebullioscopic constant. It is also called as molal boiling point elevation constant. The unit of an ebullioscopic constant is **K Kg mol^{-1}**.

ECO Mark Act: The Ministry of Environment and Forest has instituted a labeling of environment friendly products, on a national basis. BIS also started Eco Mark Scheme.

Economic spoilage: The spoilage of food by microorganisms which poses no danger to the public health is called economic spoilage.

Edam cheese: A yellow pressed hard cheese of Dutch origin usually made in flattened balls and often coated with red wax is known as Edam cheese (Plate 229).

Eddy current separation: An electric current induced by an alternating magnetic field is called Eddy current separation.

Edible box: Box made up of popcorn, chocolate or any other edible material is known as edible box and has short shelf life (Plate 230).

Plate 229: Edam cheese

Plate 230: Edible chocolate box

Edible films: Recently, edible polysaccharide films are gaining importance in certain packaging application. Polysaccharide film look like plastic films are processed from polysaccharide beads by application of heat and moisture. In Japan, edible polysaccharide films are already widely used for packaging of processed meat such as ham and poultry products. The processors can wrap the meat in edible film before smoking and steaming. The surface structure and texture are improved and yields are better as less moisture is lost during processing. Edible films can extend the shelf life of the fresh fruits and vegetables. An edible packaging will gradually absorb the moisture given off from the product, preventing it from accumulating inside the package. This would lead to reduced risk of mould growth. Eventually, the moisture absorbed may cause the film to dissolve.

EDTA (ethylenediaminetetraacetic acid): It is a white crystalline acid $C_{10}H_{16}N_2O_8$ used as a chelating agent, a preservative, and in medicine as an anticoagulant and in the treatment of lead poisoning. It is tetraprotic acid.

Effectiveness of heat exchanger: It is the ratio of the actual heat transfer to maximum possible heat transfer.

Effluent: Effluent is a liquid/water of industrial waste, which may be the outflow from sewage during purification. The effluent is a rich source of organic carbon and contains plant nutrients in varying amounts. Its use in agriculture is very limited because of its high BOD (Biochemical Oxygen Demand). Many distilleries have now installed biogas plants to produce methane which in turn reduces BOD and can be used for pre-sowing irrigations with beneficial effects on in crops. The effluent can as well be used for irrigating crops with appropriate dilution.

Efflux Tube viscometer: The instrument is used to measure the viscosity of tomato puree (Plate 231).

Egg Benedict: Poached eggs and broiled ham placed on toasted halves of English muffin and covered with hollandaise (Plate 232).

Egg beater: A hand-operated kitchen utensil used for beating, stirring, or whipping egg is known as egg beater. A rotary device is use for these purposes (Plate 233).

Egg cream: A sweetened drink made with milk or cream and other ingredients; specially a drink consisting of milk, a flavoring syrup, and soda water is called egg cream.

Egg roll: A thin egg-dough casing filled with minced vegetables and often bits of meat (as shrimp or chicken) and usually deep-fried (Plate 234).

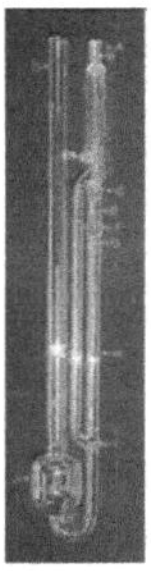

Plate 231: Efflux tube viscometer

Plate 232: Egg Benedict

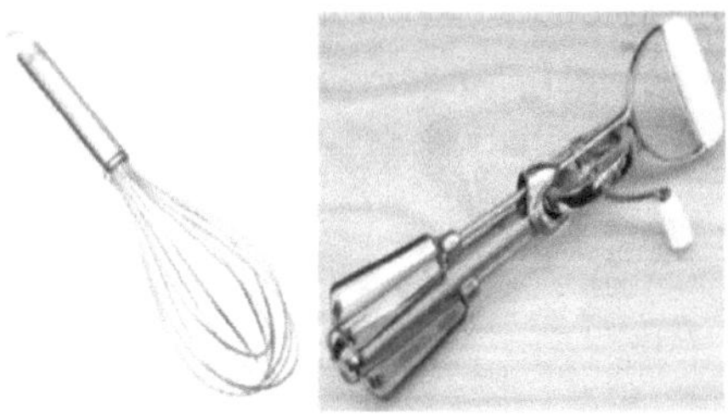

Plate 233: Egg beater

Plate 234: Egg roll

Egg timer: A small sandglass used for timing the boiling of eggs is called egg timer.

Elasticity: It is the speed at which the material returns to its original shape after deformation.

Electron volt (eV): It is the energy gained by an electron in moving through a potential different of 1volt. Or It is a unit of energy equal to the energy gained by an electron in passing from a point of low potential to a point one volt higher in potential (1 eV= 1.60×10^{-19} joule).

Electronic Nose: An electronic nose (E-nose) is used to evaluate the maturity and to monitor the shelf life of the food commodities. The electronic nose is a computer based technology and in fact, an electronic nose functions like a black box and it knows nothing until it is taught (Plate 235). It has simply a series of sensors that respond to volatile components of the headspace above a sample. The experimenter optimizes operating conditions and teaches the electronic nose what to recognize through teaching sets and data libraries. The software of the electronic nose uses a series of algorithms and chemometric methods to provide meaningful data from the sensor response as it is a correlation technique.

Plate 235: Electronic nose device "Z Nose"

Electrostatic sprays: Electrostatic sprays are based on the principles that all fine particles have same electrical charges and thus repel each other and so maintain uniformity of application.

Ellagic acid: It is a crystalline phenolic compound $C_{14}H_6O_8$ with two lactone groupings that is obtained from oak galls and some tannin and is used medicinally as a hemostatic. It is just like bioflavonoid and found in grapes.

Elution: The process of removal of adsorbed material from an adsorbent by means of a solvent is called elution.

Elutriation: A process to purify, separate, or remove by washing.

Emaciation: It is the wasted condition of the body mainly due to lack of food, wherein the fat reserves and muscles are used for sustenance of life to some extent.

Embossing: It is used in can end covers. It is process of giving impression of prints on covers. It is the process to raise the surface of into bosses. It is the coding of can's end with the coding machine.

Emissivity: Emissivity is the property of a material and it ranges from 0 to 1; which measures how much energy a surface can emit with respect to an ideal emitter (e=1) at the same temperature. It is the relative power of a surface to emit heat by radiation. Or It is the ratio of the radiant energy emitted by a surface to that emitted by a black body at the same temperature.

Emmental cheese: A hard cheese characterized by elastic texture, mild nut like flavor, and large holes that form during ripening. It is cooked cheese. It is also known as Swiss cheese (Plate 236).

Plate 236: Emmental cheese

Empty calories: These are pure carbohydrates which provide only calories i.e. energy. These do not contain any vitamins, proteins and minerals. Sugars are known as empty calories food. An empty calorie is met in connection with slimming diet.

Emulsification: The process of conversion of two or more immiscible liquids into an emulsion is called emulsification.

Emulsifiers / Surface-active agents: Surface active agents are the additives that promote the formation of an emulsion. These additives are used to stabilize oil-in-water, water-in -oil, gas-in-liquid and gas-in-solid emulsions. The emulsifying agents used in food production are amphophilic substances; they are able to stabilize a system which is naturally unstable in oil or water. Natural origin emulsifier is Lecithin and synthetic is mono and triglycerides, bile acid, fatty acids & their derivatives can be used as emulsifying agents. Synthetic surface active agents include defoaming compounds and detergents e.g., propylene glycol monostearate and monosodium phosphate. These compounds contain both polar and non-polar groups and thus are drawn to the interface between the two phases of an emulsion to coat the surface of the droplets.

Emulsion: A physically stable system containing two mutually insoluble liquid materials is called an Emulsion. It is colloidal dispersion of a liquid in another liquid with which it is immiscible. or A system (as fat in milk) consisting of a liquid dispersed with or without an emulsifier in an immiscible liquid usually in fine droplets of larger than colloidal size.

End point: It is a point in a titration at which a definite effect (color change) is observed.

Endoenzyme: It is intracellular enzyme that acts within the cells in which these are produced.

Endo-osmosis: The reverse flow of water caused by cell being put in a solution of lower solute concentration is called endo-osmosis. Or the movement of the water in the cell / a body through a semipermeable membrane is called endo-osmosis.

Endothermic reaction: Reaction in which heat is absorbed is called endothermic reaction.

Endotoxin: Endotoxin is a toxic heat stable lipopolysaccharide substance present in the outer membrane of gram negative bacteria that is released from the cell upon lysis.

Energy: The Capacity to do work is known as energy. Energy exists in various forms like kinetic, potential, thermal, chemical, electrical, and nuclear—and can be converted from one form to another. For example, fuel-burning heat engines convert chemical energy to thermal energy; batteries convert chemical energy to electrical energy. Though energy may be converted from one form to another, it may not be created or destroyed; that is, total energy in a closed system remains constant. All forms of energy are associated with motion. A rolling ball has kinetic energy, for instance, whereas a ball lifted above the ground has potential energy, as it has the potential to move if released. Heat and work involve the transfer of energy; heat transferred may become thermal energy.

Engineering: The application of science and mathematics by which the properties of matter and the sources of energy in nature are made useful to people.

English gin: It is similar to Holland gin except that it is carefully rectified until only a slight grain flavour remains in it.

Ennoblement: It is synonymous term of fortification or enrichment.

Enology: It is the study of a science that deals with wine and wine making. Enology growing making and consumption of wine.

Enrichment: Juice, squashes, syrup etc, are some times fortified with vitamins to enhance their nutritive value, to improve taste, texture or colour and to replace nutrient lost in processing. Ascorbic acid and beta-carotene imparts an attractive orange colour. Citric acid is often used for all type of beverage and phosphoric acid for cola type of drinks. It is also known as Fortification. It is also known as ennoblement.

Enrobing: It is also known as battering. It is blending of batters i.e. flour, leavening (rising agents) and seasoning. The enrobing process comprises the application of chocolate or compound coatings to cover centres, such as confectionery units, confectionery bars, plain biscuits, biscuit sandwiches, granola bars, or other baked items such as cakes or croissants.

Plate 237: Enrobing

Enterotoxin: A toxin produced by certain stains of *Staphylococcus aureus* during their growth which causes gastroenteritis / gastrointestinal symptoms or inflammations of the lining of the intestinal tract. It causes commonly occurring food poisoning or cholera.

Enthalpy: The total heat content of the substance is known as enthalpy. It is the sum of the internal energy of a body or system and the product of its volume multiplied by the pressure. Enthalpy (H) is the measurement of energy in a thermodynamic system. The quantity of enthalpy equals to the total content of heat of a system, equivalent to the system's internal energy (E) plus the product of volume (V) and pressure (V). H–E+PV.

Entrapment: The entrapment technique of immobilization is based upon occlusion of biocatalyst within the lattice of a polymer matrix tight enough to prevent the leakage of biocatalyst, while allowing diffusion of the substrate into a product out of matrix. This technique is of three type (i) fibre entrapment, (ii) gel entrapment (iii) microencapsulation.

Entropy: Entropy is an index of unavailability or degradation of energy. Or It is a measure of the unavailable energy in a closed thermodynamic system that is usually considered to be a measure of the system's disorder. It is property of a substance or system's state like pressure, volume, or internal energy. It is the degradation of the matter and energy in the universe to an ultimate state of inert uniformity. Entropy is the measure of a system's thermal energy per unit temperature that is unavailable for doing useful work.

ENZA: New Zealand Apple and Pear Marketing Board – Brand name – 'Enza' (1948).

Enzymatic browning: Many fruit and vegetables have a tendency to turn brown when damaged or when cut, surfaces are exposed to air e.g. Apples, banana, potatoes etc and this is due to enzymatic reactions. The formation of brown colour is due to the action of the enzyme phenol (also known as polyphenol oxidase, tyrosinase or catechoase) on phenolic substances. Normally, the phenolic substrates are separate from phenolase in the intact tissues and browning does not occur. When foods containing such substances are cut and exposed to air rapid browning of the cut surface takes place. Polyphenols (in cells) + oxygen in cells cause browning after cutting or bruising only (Plate 238).

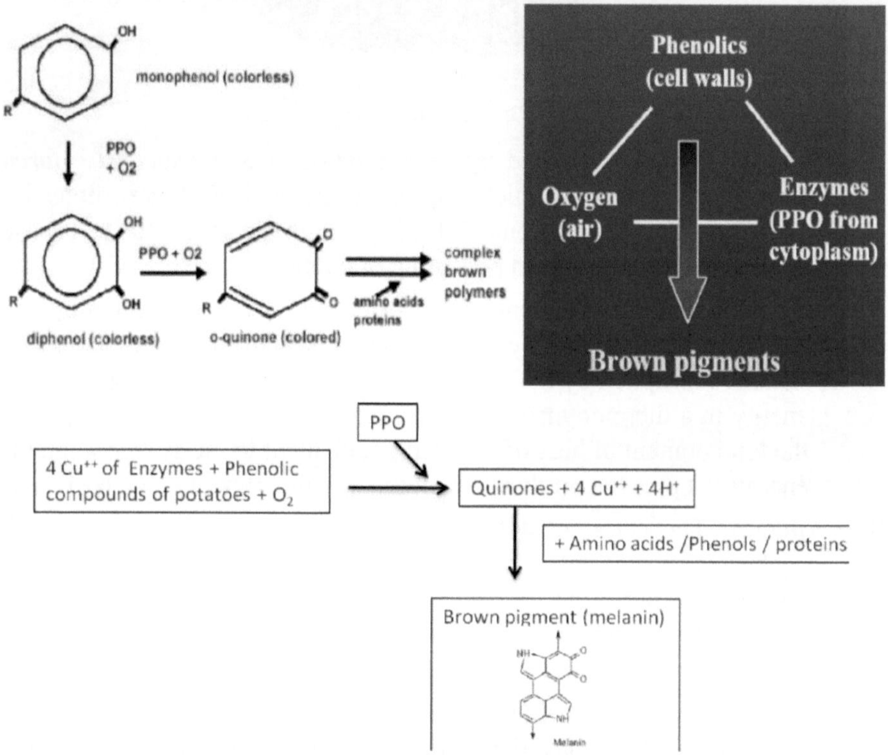

Plate 238: Mechanism of enzymatic browning

Enzyme: Enzyme(s) is /are organic biocatalysts which initiate / accelerate and control biological reactions (Plate). All enzymes are proteins but all proteins are not enzymes (exception is ribozyme) (Plate 239).

Enzyme activity: The enzyme activity is expressed by the unit of enzymes per mg of the protein.

Enzyme inhibitor: A molecule that prevents an enzyme from catalyzing a reaction is called enzyme inhibitor.

Enzyme technology: The use of the isolated and purified enzymes as catalysts in industrial processes is called enzyme technology.

Enzyme unit: It is the amount of the enzyme which will catalyze the transformation of 1 micro-mole of the substance per minute.

Enzymology: The scientific study of the enzyme is called enzymology.

Equal: It is an artificial sweetener containing aspartame, dextrose and malt dextrin (Plate 240).

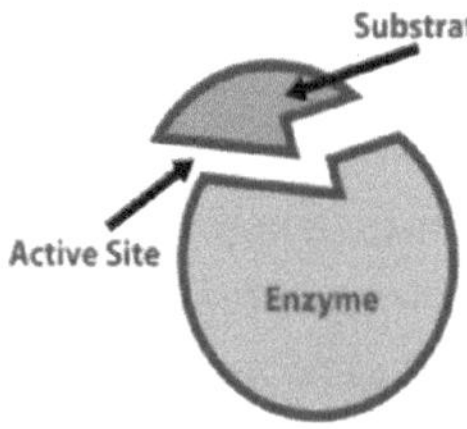

Plate 239: Enzyme working mechanism

Plate 240: Equal sweetner logo

Equilibrium constant: It is a number that expresses the relationship between the amounts of products and reactants present at equilibrium in a reversible chemical reaction at a given temperature.

Equilibrium moisture content: If the moisture of the sample is in equilibrium with surrounding then it is called equilibrium moisture content (Plate 241).

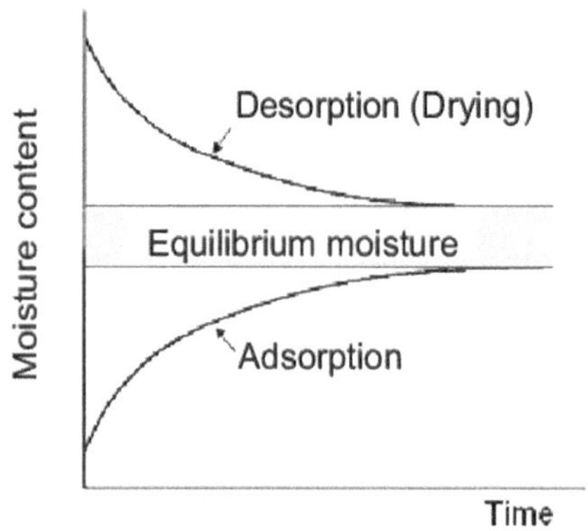

Plate 241: Equilibrium moisture content

Equilibrium relative humidity (ERH): It is defined as percentage of relative humidity at which the moisture content of the sample is in equilibrium and results neither loss nor gain of the moisture content by the sample. In ERH, the stage at which the product just becomes lumpy is called Critical point. A point which has 5% lower RH than the critical point is called Danger point. A portion of curve between danger point (D) and critical point (C) is called Safety range (Plate 242).

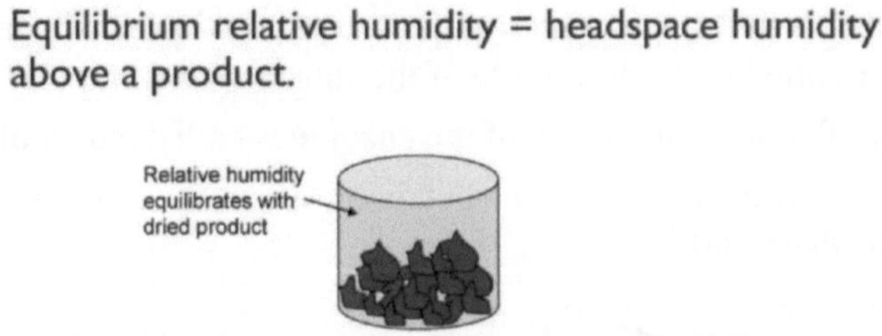

Plate 242: Equilibrium relative humidity illustration

Ergonomics: An applied science concerned with designing and arranging things for people use so that the people and things interact most efficiently and safely is called ergonomics. It is also called biotechnology human factors / engineering human factors. In ergonomics, human factor is a rapidly expanding discipline that formally addresses the relationship between peoples and technology. In Europe this discipline is called ergonomics (Plate 243).

Plate 243: Ergonomics

Ergotism: The poisoning of wheat and rye with the fungal growth is called ergotism.

***Escherichia coli*:** Most strain of *E. coli* a universal intestinal inhabitant is harmless to their human and animal hosts. However, several strains are capable of causing food borne infections (Plate 244). It is an enterobacterium (Escherichia coli) that is used in public health as an indicator of fecal pollution (water or food) and in medicine and genetics as a research organism and that occurs in various strains that may live as harmless inhabitants of the human lower intestine or may produce a toxin causing intestinal illness.

Essential amino acids: These are the amino acids (AA) which can not be produce / synthesis by the human bodies itself and required to be taken from the outside sources. These are 10 in numbers (Plate 245).

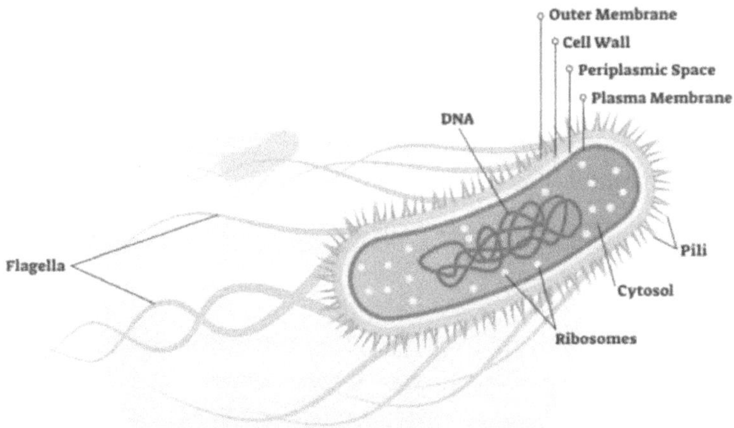

Plate 244: E. coli

Essential AA	Nonessential AA	Conditionally essential AA
Arginine	Alanine	Cysteine
Histidine	Asparagine	Glutamine
Isoleucine	Aspartate	Hydroxyproline
Leucine	Glutamate	Proline
Lysine	Glycine	Taurine
Methionine	Serine	
Phenylalanine	Tyrosine	
Threonine		
Tryptophan		
Valine		

Plate 245: Essential amino acids

Essential Commodities Act, 1955: This Act is administered by the Ministry of Consumer Affairs, Food and Public Distribution through the State/ Union Territory. The aim of this Act is to regulate the manufacturer towards quality aspect, commerce and distribution of essential commodities like food. Number of orders has been listed in the provision of the act. Twenty two essential commodities are Rice, Wheat, Atta, Gram Dal, Tur (Arhar) Dal, Urad Dal , Moong Dal, Masur Dal, Sugar, Gur, Groundnut Oil, Mustard Oil, Vanaspati, Sunflower Oil, Soya Oil, Palm Oil, Tea, Milk, Potato, Onion, Tomato and Salt. The ECA Amendment also brings about the following benefits: Creation of a competitive agricultural market and reduction of agri-waste due to increased investment in cold-storage facilities. Ensuring price stability for farmers.

Essential fatty acids: Linoleic, linolenic and arachidonic acid are essential fatty acids. Human body can not synthesis these and so called essential fatty acids. However, human is able to synthesis arachidonic acid from former two fatty acids. Vegetable oils (except coconut and olive) contain first two fatty acids. Omega 3 (alpha linolenic acid) and omega 6 (linoleic acid) are essential for health. These belong to polyunsaturated fatty acids and are known as essential fatty acids or EFA. These are also called vitamin F (Plate 246).

COMMON NAME (Abbreviation)	NUMERIC FORMULA (chain length:double bonds)	LOCATION OF FIRST DOUBLE BOND (relative to omega end of chain)
Linoleic acid (LA)	18:2	Omega-6
Alpha-linolenic acid (ALA)	18:3	Omega-3
Arachidonic acid (AA)	20:4	Omega-6
Eicosapentaenoic acid (EPA)	20:5	Omega-3
Docosahexaenoic acid (DHA)	22:6	Omega-3

Plate 246: Essential fatty acids with Numeric formula

Essential oil distillation apparatus: It is an apparatus used for the extraction of the essense form the oil cake left after oil extraction. It extracts the essence by the generation of the steam. The essential oil after condensation is collected by the collector provided.

Essential oil: A class of volatile oils that give plants their characteristic odors and are used in perfumes and flavorings for aromatherapy.

Ester: A compound formed due to reaction between acid and alcohol is called ester. Fats are ester of an alcohol (glycerol) and fatty acid such as stearic acid, oleic acid etc.

Esterification: It is the reaction between acetic acid and ethyl alcohol in the presence of sulphuric acid to form ethyl acetate.

Ethanol: It is a colorless volatile flammable liquid C_2H_5OH that is the intoxicating agent in liquors and is also used as a solvent and fuel. It is also called ethyl alcohol or grain alcohol

Ethyl cellulose: The thermoplastic substance used in preparation of plastics and lacquers is ethyl cellulose.

Ethylene: A ubiquitous, volatile hormone, synthesized in plant parts under stress and also induced in plant (e.g. for floral initiation) by the application of auxins or certain growth inhibitors. It increases responsible rate, hastens fruit ripening, stimulate root initiation in cuttings, promotes abscission, stimulates floral initiation, inhibits elongation, breaks rest in buds and seeds and inhibit lateral bud development. 2-chloroethylphosphonic acid (Ethephon) is a synthetic ethylene releasing compound.

Ethylene absorbants: The chemical which absorb the emitted ethylene from around the fresh perishable in the package and thus delay the senescence e.g. celite with KMuO4, purifil etc. The examples of Ethylene absorbent are $KMnO_4$, ethylsorb, Purafil, squalene, phenylmethyl silicon absorbed in celate, vermiculite, silica gel or alumina pellets.

Ethylene antagonist: CO_2, Ag, Norbornadiene are ethylene antagonist.

Ethylene diamine tetraacetate: It is a salt of EDTA

Ethylene dibromide (EDB): It is a colorless toxic liquid compound $C_2H_4Br_2$ that is used as a fuel additive in leaded gasolines, that has been found to be strongly carcinogenic in laboratory animals and that was used formerly in the U.S. as an agricultural pesticide.

Ethylene glycol: A thick liquid diol $C_2H_6O_2$ used as an antifreezing agent and in making polyester fibers.

Ethylene inhibitor: AVG, AOA are ethylene inhibitors.

Ethylene oxide: A colorless flammable toxic compound C_2H_4O used in synthesis of ethylene glycol and in sterilization /fumigation.

Eutectic point: Point in a solution below which no unfrozen solution exists in a solution is called Eutectic point (Plate 247). It is the point of the lowest melting / freezing point in the solution. Point in a solution below which no unfrozen solution exists in a solution is called Eutectic point. Such unfrozen liquids are called metcryotic liquids. The eutectic point is when a solid solute, a solid solvent, and a liquid mixture all exist in the same phase. The eutectic point is the lowest temperature at which the liquid phase is stable at a given pressure (Plate 145). A eutectic system is a homogenous, solid mixture of two or more substances that form a super-lattice that melts or solidifies at a temperature lower than any of the individual ingredients' melting point. The precise ratio between the components forms a eutectic system. The word "eutectic" is derived from the Greek words "eu," means "good" and "tectic," means "melting."

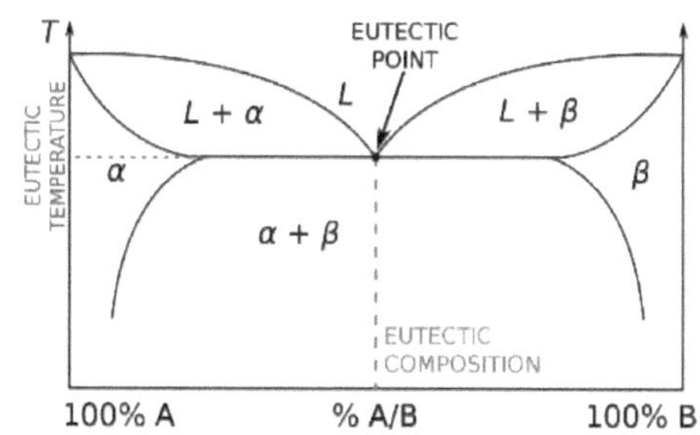

Plate 247: Graphical representation of Eutectic point

Eutectic temperature: It is the temperature where solution can exist in equilibrium with both crystalline solvent and crystalline solute.

Evaporated milk: Unsweetened milk concentrated by partial evaporation is called evaporated milk.

Evaporation: It is an operation that converts a liquid into a vapour, leaving behind a concentrated solution or a high boiling liquid.

Evaporative concentration: Evaporation is probably the oldest method of concentration and even today the most favourable and widely used method for concentration of liquid foods. Hence it is called evaporative concentration. Evaporation is the partial removal of water from liquid food by boiling. Separation is achieved by exploiting the differences in volatility between water and solutes.

Evaporative cooling: Each mole of the water which is vaporized required heat transfer equivalent to the heat of the vaporization. If the rate of the mass transferred (evaporation) is too rapid; then heat transfer to equal the heat of vaporization, the material will drop in temperature. This process is called evaporative cooling. Evaporative cooling is the process of removing heat from a surface due to the evaporation of water. Water can absorb a lot of heat without changing phases, allowing it to remove heat from the surface it is on. When water turns from liquid to gas, it's called evaporation (Plate 248).

Evaporator: It is an apparatus used for drying by heating and evaporating the liquid part of the substance (Plate 249).

Evaporimeter: It is an apparatus for measuring the evaporation rate at a given time interval (Plate 250).

Evapotranspiration: Loss of water from the commodity both by evaporation and by transpiration is called evapotranspiration.

Exhaust box: Exhaust box is used for removal of air in the canning line commercially (Plate 251).

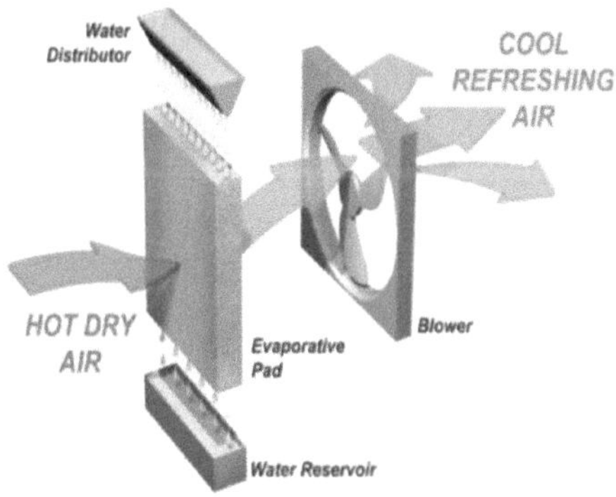

Plate 248: Evaporative cooling

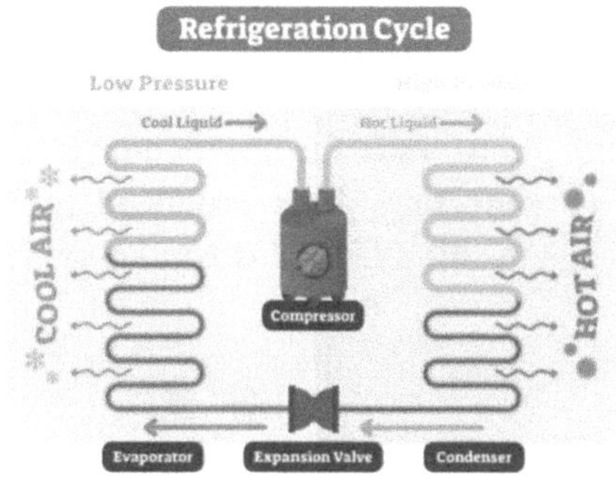

Plate 249: Evaporator

Plate 250: Evaporimeter

Plate 251: Exhaust box

Exhausting: Removal of air from the head space of filled container either by heat treatment or mechanical method is called exhausting. It is removal of air from within and around food, jars and canners. Blanching exhausts air from the food tissue.

Exoenzyme: An extracellular enzyme is called exoenzyme. It is extracellular enzyme that acts outside the cells in which these are produced.

Exo-osmosis: It is the movement of water out of the cell or body through a semipermeable membrane. Exosmosis is the movement of water and other molecules from a higher concentration to a lower concentration through a semipermeable membrane. Endosmosis is the opposite of exosmosis; it is the movement of water and other molecules from a lower concentration to a higher concentration through a semipermeable membrane.

Exothermic reaction: Reaction in which heat is released is called exothermic reaction (Plate 253). In thermochemistry, an exothermic reaction is a "reaction for which the overall standard enthalpy change ΔH□ is negative." Exothermic reactions usually release heat.

Exotoxin: Exotoxin is a soluble poisonous substance produced during growth of a microorganism and released into the surrounding medium. Toxin consisting of glycoproteins and is excreted by the bacteria outside its cells into the surrounding item (foods) causing acute gastroinstinal infection. It is produced by anaerobic spore forming bacteria gram positive like *Clostridium perfringens* (Botulinum) (Plate 254).

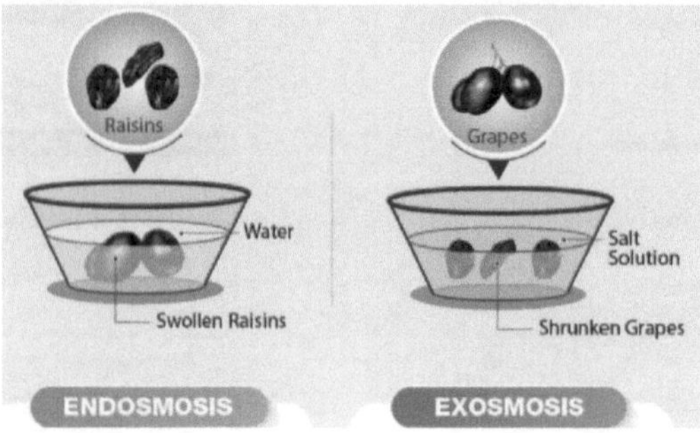

Plate 252. Exo-osmosis vs Endo-osmosis

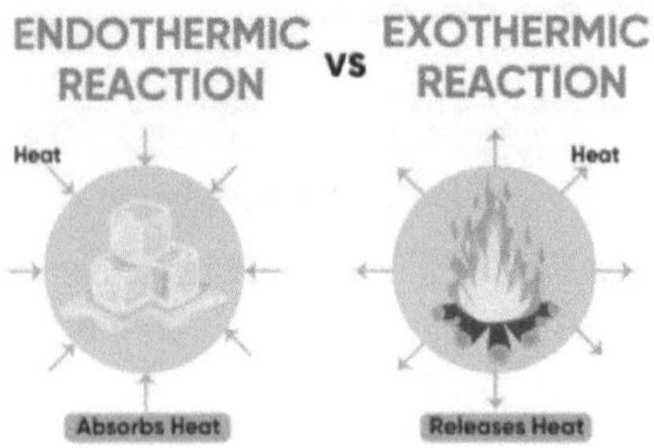

Plate 253: Endothermic vs exothermic reactions

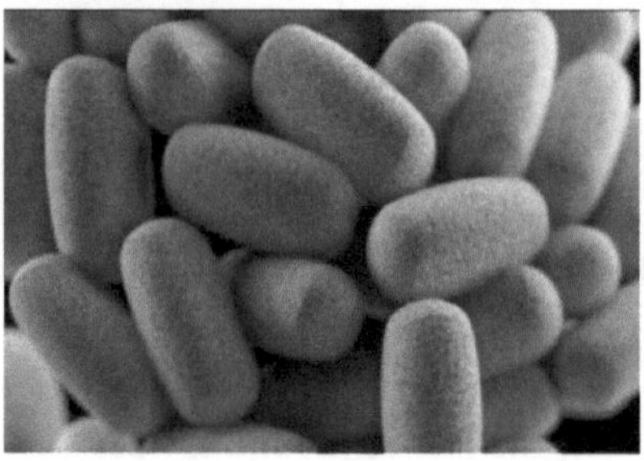

Plate 254: Clostridium perfringens

Experience foods: These are the foods whose quality is ascertained by experience after consumption of the substance.

Explosion puffing: It is the first low cost dehydration process for fruits that will produce relatively large pieces of dehydrated products which reconstitute rapidly.

Explossive puff drying: It is used for preparation of the puffed rice and wheat breakfast cereals. For a product to be explosive puff dried, it must be partially dried to a moderate low moisture level. The product to be puffed is pressured in a chamber by heating. The puffing of the product when the pressure in the chamber is release during heating process. The sudden change in the pressure results the expansion of the food and results porous structure of the food material (Plate 255).

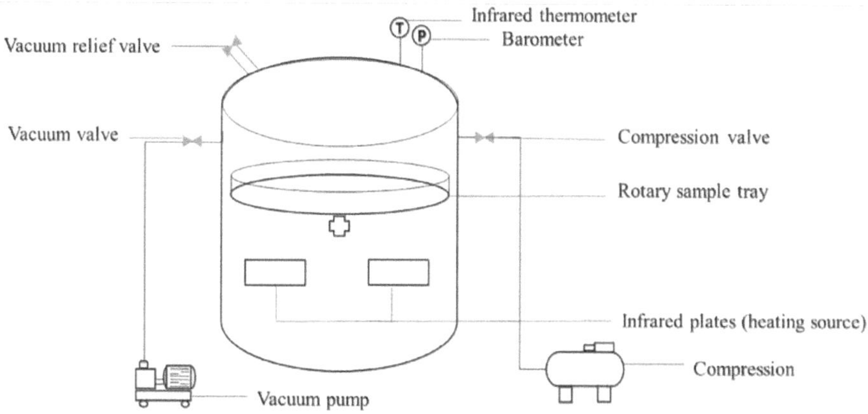

Plate 255: Schematic diagram of explosion puff drying system

Export: The process by which commodity is conveyed / send from one country or region to another for a purpose of trade is called export.

Export (Quality Control & Inspection) Act, 1963: The Act was framed by the Department of Commerce, Government of India to promote and regulate the export trade through Export Inspection Council (EIC), Export Inspection Agency (EIA), and Agricultural & Processed Food Products Export Development Authority (APEDA). Under this Act, exportable commodities have to be notified for compulsory pre-shipment inspection, but in some circumstances it is exempted from pre-shipment Inspection.

Exporter: A wholesaler who sells to merchants or industrial consumers in foreign countries.

Expression: The process of separation of liquids from solids by applied pressure is called expression.

Extra Lean food: Food having less than 5 gm fat, 2 gm or less saturated fat and less than 95 mg of the cholesterol per serving and per 100 gm.

Extraction: The act or process of extracting juice or pulp.

Extra–virgin: A virgin olive oil that is lowest in acidity and highest in quality is called extra virgin.

Extrinsic factor: The vitamin B complex is known as extrinsic factor. Or These are factors of production systems which include factors like amount of pesticides used, type of packaging material used, specific processing technique or the use of biotechnology to modify the product properties.

Extrusion cooking: Extrusion is a process that combines several unit operations including mixing, kneading, shearing, heating cutting, shaping and forming. It involves compressing and working of a material to form a semisolid mass under a variety of controlled conditions and then forcing it to pass through a restricted opening such as a shaped hole or slot at predetermined rate. It is a process of cooking moistened starchy and / or proteinaceous food material by combination of heat, pressure and mechanical shear. Single screw or twin screw extruder is used for extrusion cooking (Plate 256 and 257). It is a high temperature short time process.

Extrusion: The act or process of extruding a product is called extrusion.

Plate 256: Twin screw extruder

Plate 257: Extruded product

F

F_0 value: It is used to express F-value when z =10°C / 18°F and the temperature is 121.1°C. It is the number of minutes at a specific temperature (250°C) required to destroy or kill a stated number of microorganisms and has specific z-value. It is also called reference value.

Facing or lining board: The layer of the paper on which the flute is embedded is called facing or lining board (Plate 258).

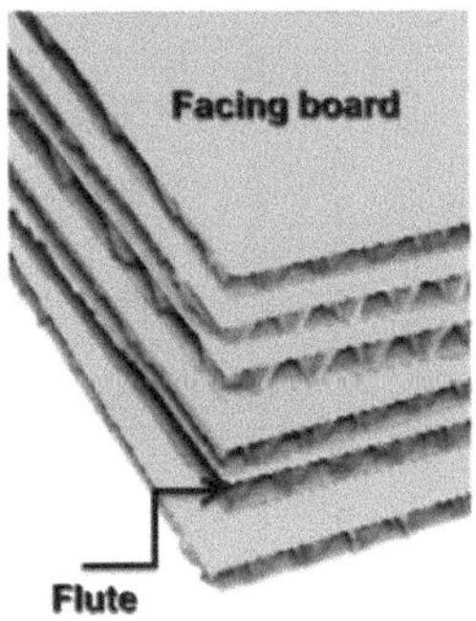

Plate 258: Layers on paper board

Facultative anaerobic: Organisms that grow either with or without oxygen.

Facultative parasites: Facultative parasites can adapt to live on dead organic matter and often have a wide host range and can easily be cultured on artificial media. These organisms may be aerobic or anaerobic but under changed conditions, they changes to anaerobic or aerobic.

Falling rate period: It is that part of the drying process during which the drying rate varies with time and the instantaneous drying rate per unit surface or weight of dry material continuously decreases. Falling rate takes place when most of the moisture has been evaporated during constant rate period. The falling rate period during dehydration take place in the final stage of dehydration when the food commodities have large amount of free water available for evaporation. Here, rate of removal of the moisture from the food to the surrounding is continuously decreasing / falling per unit time.

False color: False color in an image (photograph) of an object that does not actually appear in the object but is used to enhance contrast, or distinguish details. Contrast is known as false colour.

FAO: Food and Agricultural Organization of the United Nations. It is established in 1945 to improve the production, distribution and consumption in member countries, to increase the nutritional level of the peoples.

Farina: A fine meal (powder) of vegetable/cereal grains which is used chiefly for puddings.

Farinograph: It is an objective testing instrument that is used to measure the resistance of shearing rods moving through a batter or dough and records results. It is used to measure the protein strength of the wheat flour (Plate 259 and 260).

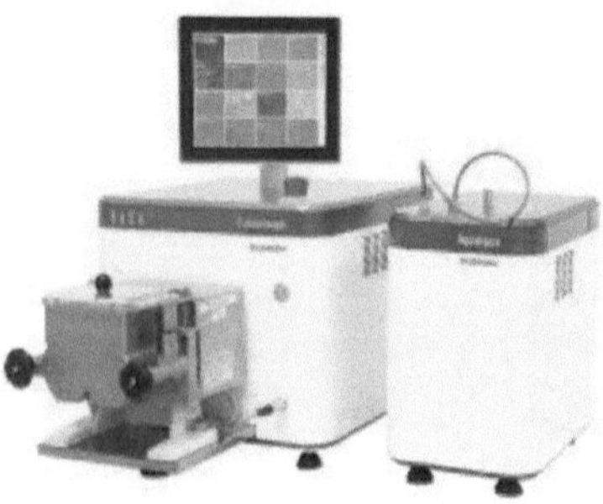

Plate 259: Farinograph tester / dough analyser

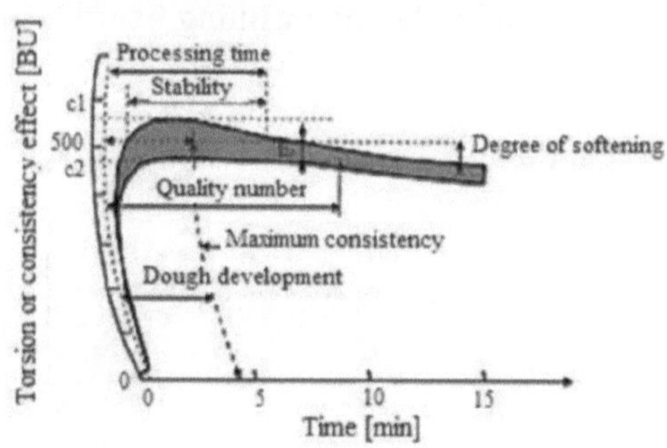

Plate 260: Farinograph curve

Fat: The compounds of carbon, hydrogen, and oxygen that are glycerides of fatty acids, are the chief constituents of plant and animal fat and are a major class of energy rich food, and are soluble in organic solvents but not in water. These are long chain fatty acids esterified to glycerol, which is solid at room temperature.

Fat free food: Less than 0.5 gm of fat per serving or per 100 gm

Fat soluble vitamins: These are usually associated with the lipids of the natural foods e.g. vitamin A, D, E and K.

Fatty acid: The saturated aliphatic monocarboxylic acids $C_nH_{2n+1}COOH$ (acetic acid) including many that occur naturally in the form of esters in fats, waxes, and essential oils. These are saturated or unsaturated monocarboxylic acids (palmitic acid) usually with an even number of carbon atoms that occur naturally in the form of glycerides in fats and fatty oils. Or Fatty acids are organic compounds containing long hydrocarbon chain and terminal carboxyl (- COOH) groups, may be saturated (i.e. all single bonds) or unsaturated (one or more double bonds). Fatty acids combined with glycerol to form fats and oil. Fatty acids are organic acid composed of carbon, hydrogen and oxygen (Plate 261).

Favism: It is haemolytic anaemia and is due to deficiency of glucose-6-phosphate dehydrogenase by consumption of faba bean or broad beans (Plate 262). Individuals who are susceptible to favism are low in glutathione in the erythrocytes. Glucoside in faba bean is Vicine and covicine and one amino acid derivative like dihydroxy phenyl alanine (DOPA).

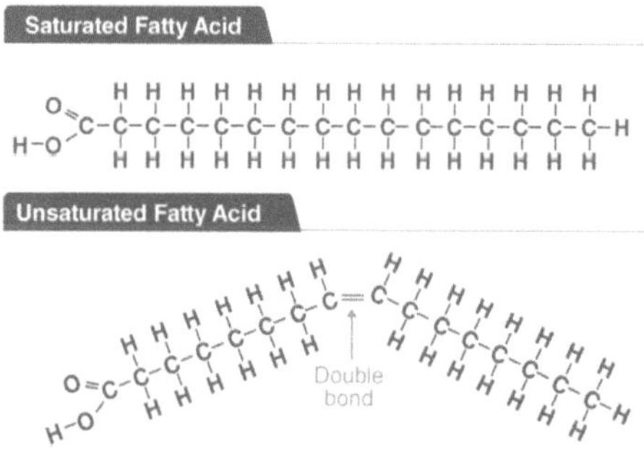

Plate 261: Saturated vs unsaturated fatty acids

Plate 262: Faba bean causing favism

FCC: It stands for Food chemical codex, 1931. The *Food Chemicals Codex* is a compendium of internationally recognized standards for determining the purity and quality of food ingredients.

FCR: It stands for fruitlet core rot in pineapple. Fruitlet core rot (FCR) is a worldwide fungal disease of pineapple, which occasionally occurs on susceptible cultivars in Taiwan and their occurrence in Taiwan reported for the first time for FCR fungus (Plate 263).

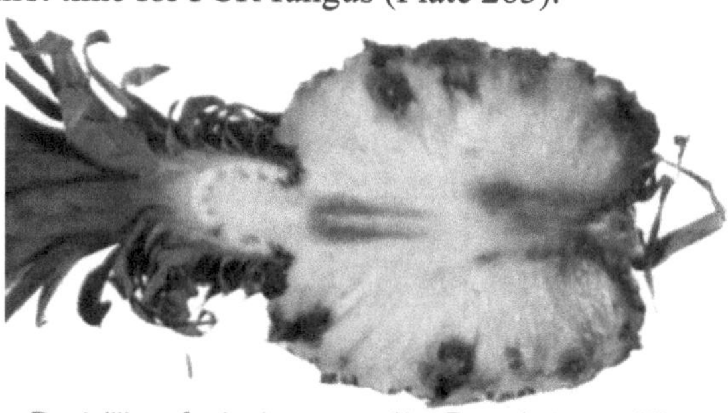

Penicillium funiculosum and/or Fusarium moniliforme

Plate 263: FCR in pineapple

FDA: It stands for food and drug administration (1906). The *FDA* is responsible for protecting the public health by ensuring the safety, efficacy, and security of human and veterinary drugs, biological products, and medical devices; and by ensuring the safety of our nation's food supply, cosmetics, and products that emit radiation. FDA also has responsibility for regulating the manufacturing, marketing, and distribution of tobacco products to protect the public health and to reduce tobacco use by minors.

Fehling's test: It is a deep blue alkaline solution used to detect the presence of aldehyde or any groups containing the aldehyde functional group -CHO. Glucose consists of an aldehyde group while fructose consists of a ketone functional group (Plate 264).

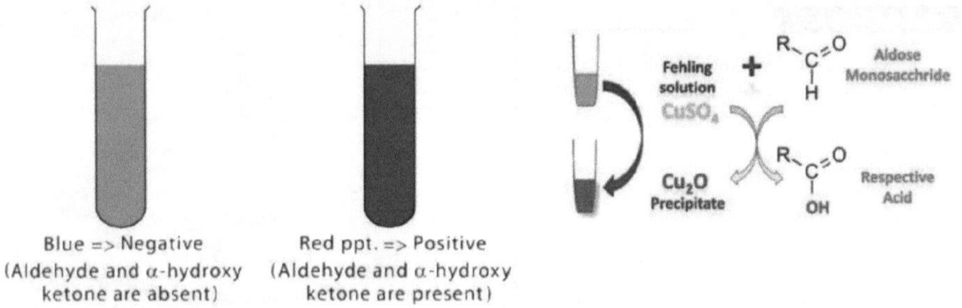

Plate 264: Fehling's Test

Fenni: A wine or a fermented product prepared from cashew apple is called feni.

Fermentation: It is the anaerobic process by which sugar in juice are converted to alcohol by yeasts to make wine, the oxidation process by which alcohol is converted to vinegar by certain bacteria etc. Or It is a chemical reaction carried out by many types of the microorganisms to obtain energy. In fermentation microorganisms break down complex organic compounds into simpler substances. It is an enzymatically controlled anaerobic breakdown of an energy-rich compound (carbohydrate to carbon dioxide and alcohol / organic acid). Or It is a process that allows respiration to occur in the absence of oxygen. Biologically, it allows cells to obtain energy from molecules (e.g., glucose) anaerobically. Alcoholic fermentation occurs when yeast cells convert carbohydrate sources to ethanol and carbon dioxide. or

Fermentation: The fermentation can be broadly defined as the process in which the carbohydrate or like compounds are broken down, under anaerobic or aerobic conditions. The basic purpose of micro-organisms is to breakdown the organic compounds to get energy for their metabolic activities. Fruit wines referred to as Soma. Wines are made principally from fruits like grape or extract of other fruits of various types (table wine, sparkling, dessert, aperitif or vermouth and pop wines) while the beers are prepared from grains like barley, rice and corn. The principal carbohydrate in grain is starch which can not be fermented directly. This process is brought about by starch splitting enzymes by germinating the grains followed by drying and storage until needed. Distillation of beer produces a variety of liquors (whiskey, rum, gin, vodka), while that of wine produces the brandy.

Fermented fruit beverage: A fruit juice which has undergone alcoholic fermentation by yeast containing varying amount of alcohol e.g. grape wine, cider, berry wines.

Fermented milk: It is also called cultured buttermilk. It is a refreshing beverage. It is cultured with LAB to bring about souring and cause change in texture, aroma and taste.

Fermenter: An organism causing fermentation is called fermenter.

Fermentor: A container / an apparatus where fermentation process is carrying out (Plate 265). It is the fundamental unit of industrial microbiology where controlled environment is provided for the growth of microorganisms to obtain the desired product. Lab scale fermentor is presented in plate.

Fermi: It is a unit of length equal to 10^{-13} centimeter.

Feta cheese: Feta is Greece's most famous cheese, a pure white, aged curd cheese that crumbles easily (Plate 266). The feta is a sheep's milk cheese, or a mixture of sheep's and goat's milks. Outside of the EU, it can also be made of cow's milk. It is formed into bricks and salted and cured in a brine solution. The cheese is semi-hard, with a flavor that can range from mild and milky to salty with a very tangy acidity. The excess salt if present in the cheese can be removed by dipping cheese in water.

FFS Machine: It is a machine which form, fill and seal plastic pouches / sachets from form film roll/web in a single continuous operation used for packaging milk, RTS beverages, powders, biscuits etc (Plate 267).

Plate 265: Lab scale fermentor

Plate 266: Feta cheese

Plate 267: FFS machine

Fiber: It is the mostly indigestible material in food that stimulates the intestine to peristalsis. It is also called bulk roughage.

Fibrometer: Instrument used to estimate tenderness of green beans.

Ficin: A protease obtained from the latex of fig trees. It has antihelmintic and protein digestive properties. Ficain also known as ficin, debricin, or higueroxyl delabarre is a proteolytic enzyme extracted from the latex sap from the stems, leaves, and unripe fruit of the American wild fig tree *Ficus insipida* (Plate 268). The chemical formula of ficin is CH_2FI_2N (Plate 269).

Plate 268: Ficin enzyme in fig

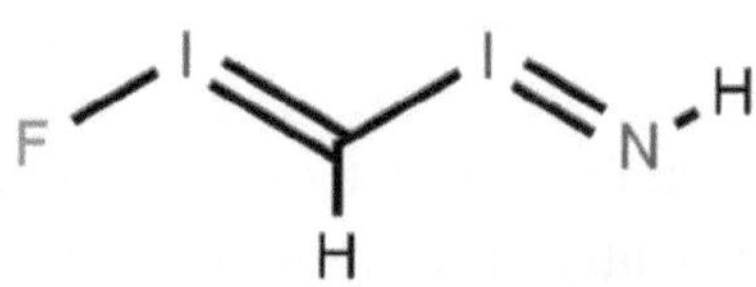

Plate 269: Ficin chemical structure

Fick's Law: Fick's law of diffusion states that mass flux per unit area of a component is proportional to its concentration gradient. It is similar to Fourier's law of heat conduction. Gas exchange equation in MAP is a modification of the Fick's Law.

Field heat: It is the initial heat of the material (biological objects) and in case of fresh fruit and vegetables, it is the heat of these produce after harvesting from fields which can be determined by the thermal energy removed form the produce in cooling.

Fielding process: Fielding process is for vinegar production and is similar to "let alone" process. The oldest method of vinegar production is the 'let alone' method in which wine left in open vats became converted to vinegar by acetic acid bacteria entering it from the atmosphere. Later the wine was put in casks and left in the open field in the 'fielding process'.

Filler: A substance added to a product to increase bulk, weight, viscosity, opacity, or strength is called filler.

Filter press: It is an apparatus used for filtration of oil or juice (Plate 270). It is a special equipment used to remove finely suspended particles from the oil / juice.

Filtration: It is the only successful way for the complete removal of microorganisms where the liquid is filtered by positive or negative pressure through a previously sterilized bacteria proof filter made of asbestos pads, sintered glass, diatomaceous earth, unglazed porcelain, membrane pads or similar material. This method has been used successfully for filtration of fruit juice, beers, soft drinks, wine, and water. It also removes the dirt particles or unwanted materials.

Plate 270: Oil filter press

Finger foods: The food products mostly used as snack foods are called finger foods.

Firming agents: These are the agents added to keep the tissue of fruits and vegetables crisp. Firming agents are added in order to precipitate residual pectin of food, thus strengthening the structure of the food and preventing its collapse during processing.

First law of thermodynamics: It concerned with principle of energy, according to it energy can neither be created nor destroyed (Plate 271).

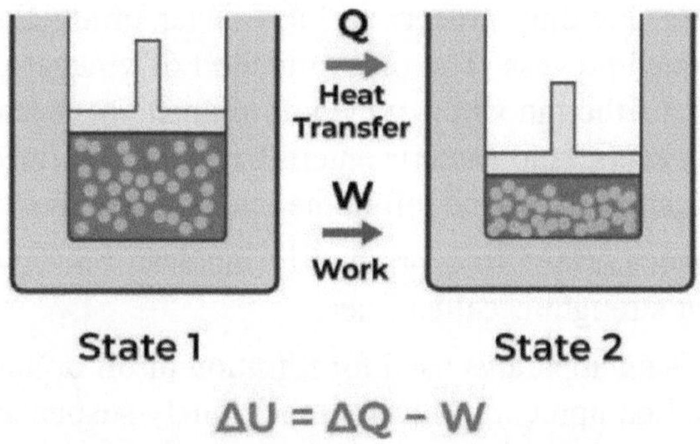

Plate 271: Illustration of First law of thermodynamics

First order transition: The physical state of the material that changes isothermally from one state to another (e.g. solid to liquid, liquid to gas) by release or absorption of latent heat of melting and crystallization.

Fish farming: The breeding and rearing of fish in ponds, lakes, tanks etc. ultimately harvesting for table purpose.

Fish finger /stick: A small elongated breaded fillet of fish is known as fish finger / stick.

Fish Garth: It is an enclosure on the river for preserving fish. It is also called fish weir.

Fish guano: It is a phosphatic fertilizer made from the unmarketable fish and fish waste products after oil extraction. It is also called fish meal. It usually contains 7-8 % nitrogen and 4-8 % phosphoric acid.

Fish meal: Ground dried fish and fish waste used as fertilizer and animal food.

Fish oil: Fatty oil from the bodies of various fishes (menhaden or sardines) that contains large amounts of unsaturated fatty acids and used in making various products.

Fish pick: It is a hand held device used to remove the fish from the net's web.

Fish protein concentrate: A protein rich food additive made from ground whole fish.

Fisherman: One who adopt fishing as an occupation or a ship used in commercial fishing.

Fishery: It is particular place in water established for growing and catching fish or taking other sea animals.

Fishhook: The barbed hook used for catching fish is called fishhook (Plate 272 and 273).

Plate 272: Fish-hook

Plate 273: Fish-hook used for catching fish

Fishing: The sport / business of catching fish called fishing.

Fishmeal: It is one of the best poultry feed, prepared by crushing clean, dried, undecomposed whole fish or fish cutting (with or without oil extraction). Most of the Indian fishmeal contains 45 – 55% proteins.

Flake: A small / thin flattened piece or layer of processed food is called flake. Or Dried pulp of fruits and vegetables in single layer is also called flake.

Flank: It is a fleshy part of the side between the ribs and the hips (Plate 274).

Flank steak: It is a pear shaped steak cut muscle of the beef flank (Plate 275).

Flame photometer: It is a type of spectrophotometer in which a spray of metallic salts in solution is vaporised in a very hot flame and subjected to quantitative analysis by measuring the intensities of the spectral lines of the metal present (Plate 276).

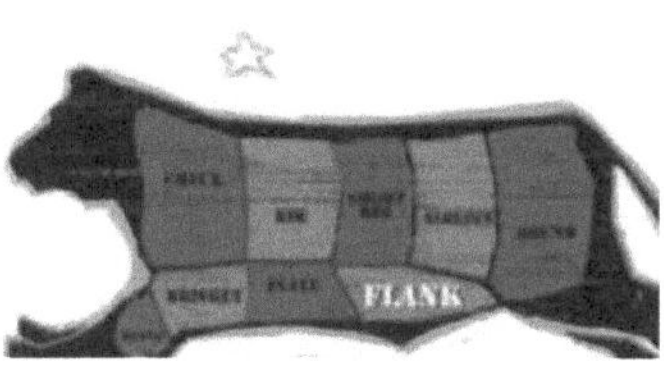

Plate 274: Flank

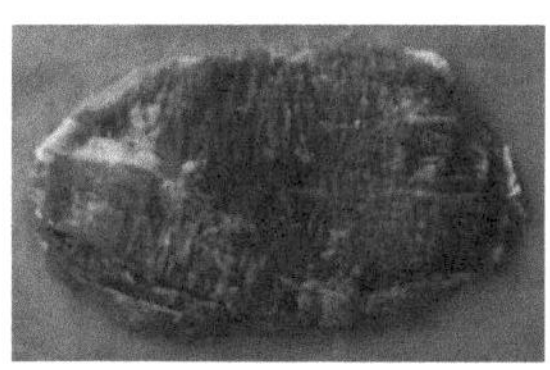

Plate 275: Flank steak

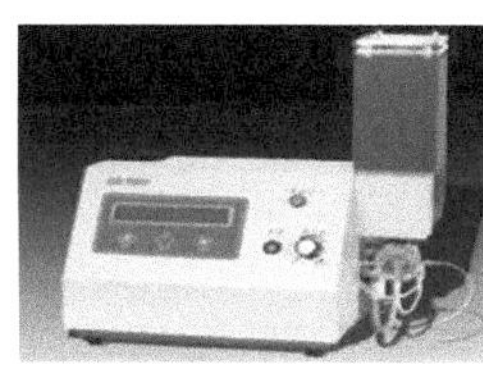

Plate 276: Flame photometer

Flash 18 processes: It is also known as Smith Ball Process for low acid foods. Low acid foods require heating well above 100°C for their sterilization. If filling of such sterilized food is attempted in the containers at such temperature under atmospheric pressure, then violent boiling of the product during can fill and sealing occurs. This flash 18 process avoid this problem of the violent boiling by placing the entire canning line within a room like a pressure chamber between 18-20 psi pressures. Under this pressure, water does not boil below a temperature of 124-127°C. Therefore low acid foods can be presterilized by HTST technique and pumped to the filling line within the pressurized room without boiling over the filler.

Flash evaporation: It is a process where hot liquid is introduced into a chamber which is at an absolute pressure where the boiling point of the liquid is below the liquid temperature. The liquid will not boil immediately upon exposure to the low pressure, vapour is released and the liquid temperature will drop to the boiling point of the liquid at the given absolute pressure. Flashing done for essence recovery is the best method. Flash evaporation (or partial evaporation) is the partial vapor that occurs when a saturated liquid stream undergoes a reduction in pressure by passing through a throttling valve or other throttling device (Plate 277).

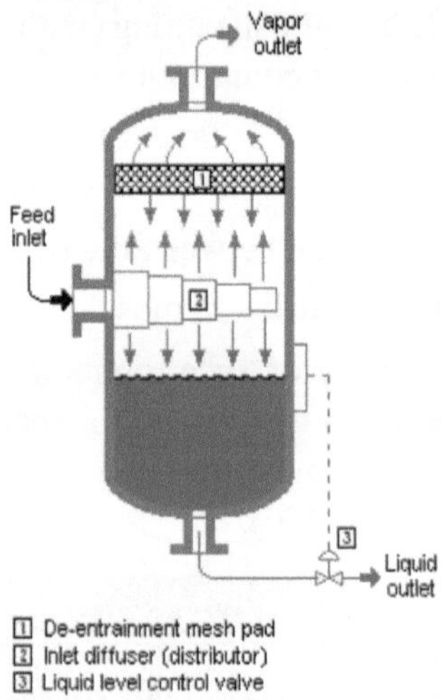

Plate 277: Flash evaporation device

Flash pasteurization: A type of pasteurization in which a thin film of juice is allowed to flow over a hot plate (heated to 82 to 85°C) and cooled immediately thereafter. Here, the juice is heated rapidly to a temperature of about 5.5°C higher than the pasteurization temperature and kept at this temperature for about a minute. This method is specially used for canning of natural orange juice but can also be used for grape and apple juices. It is similar to HTST method. In this type of pasteurization a thin film of the juice is allowed to flow over a hot plate and then cooled immediately.

Flat can: A can of normal appearance where both ends are well collapsed, flat or concave. A can with flat ends is the best priority for purchasing of canned foods.

Flat sour spoilage: Spoilage of low acid foods (corn, pea, etc.) by thermophillic microorganisms (*Bacillus stearothermophilus* and *B thermoacidurans*) due to under processing of the can, here acid is produced but no gas is evolved, and some disagreeable odour may be present. Flat sour is characterized by a lower pH due to formation of Lactic acid, lack of gas formation and occasionally a black colour due to iron content of food. Since the can do not swell, the process was labeled "Flat sour". Flat sour spoilage occurs in acidic foods like tomato by *B coagulans*.

Flavedo: It is the yellow portion of the citrus peel in which oil glands are embedded (Plate 278).

Flavone: Flavone is an organic compound with the formula C 6H 4OC 3H(Ph) O. A white solid, flavone is a derivative of chromone with a phenyl (Ph) substituent adjacent to the ether group (Plate 279). Natural flavones possess antioxidant, antiinflammatory, antitumor, antiallergic, neuroprotective, cardioprotective, and antimicrobial activities. Flavones are a class of flavonoids that are a subject of increasing interest because of their biological activities.

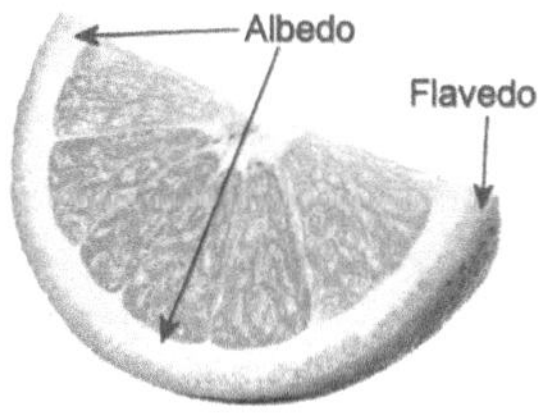

Plate 278: Flavedo

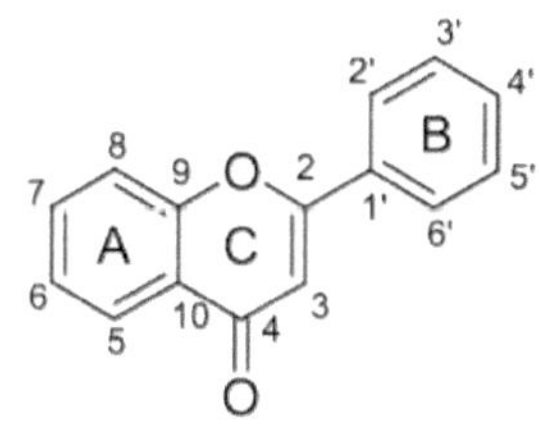

Plate 279: Flavone

Flavonoids: These are 15-C compounds (C6-C3-C6) consisting, of the two aromatic rings linked by an aliphalic three carbon chain and often synthesized in the chloroplasts. The hydroxylation pattern and degree of polymerisation of the C6-C3-C6 unit result in different polyphenolic compounds. The major flavonoids are anthocyanins, flavonols and flavones. These are white and cream coloured plant pigments e.g. pigments of potatoes and cauliflower etc.

Flavour: Flavour is a combination of taste, smell/ aroma and feeling (hot, cold, astringency, bite etc. especially in spices, wine and coffee). In short, it is combination of taste and aroma. Flavour embraces the senses of taste, smell and a composite sensation known as mouth feel. Taste is due to sensation felt by tongue. Taste is limited to sweet, sour, salty and bitter. The dimension of these can be measured chemically and can be related to the consumer's preferences. In short, it is combination of taste and aroma and is a composite term for any food.

Flavour absorbing agents: Substances used to suppress the odour of the products. Activated carbon, sodium bicarbonate, vitamin E, molecular sieves are the flavour absorbing agents.

Flavour enhancer: These are naturally occurring or synthetic flavouring agents. The flavour enhancers do not possess any flavour themselves but they intensify the flavour of other substances through synergistic effect.

Flavour potentiometer: The compound that enhances the flavour of the other compounds without adding its own unique flavour are called flavour potentiometer.

Flavoured milk: A type of milk which is standardized to a certain fat percentage with skim milk to which added flavouring materials, chocolate, fruit syrups and vitamin concentrates etc.

Flavouring agents: Flavour is one of the most important attribute of food. It is combination of taste and smell (aroma). Flavouring agents have been classified as natural, synthetic and originated from products as a result of reaction from heat, enzymatic and microbiological formation. There are approximately 1100 to 1400 natural and synthetic flavourings available to food processors. Flavourings give foods a more acceptable taste, restore lost flavours due to processing and in some cases will improve natural flavours. Spices, herbs, plant extracts and essential oils are widely used flavouring agents of natural origin. Esters, aldehydes, ketones, alcohols and ethers having characteristic fruity odours can be easily synthesized. Natural flavours are difficult to produce and based on 100% natural fruit extracts, distillates, essential oils and natural aroma chemicals. Flavour modifiers or enhancer do not possess any flavour themselves but they intensify flavours of other substances through a synergistic effect. The best known and most widely used flavour enhancer is mono-sodium glutamate (MSG). Flavour modifiers should not be permitted in baby foods.

Fleam: It is veterinary equipment used for bleeding.

Flexible cans: The retort pouches are known as flexible cans.

Flexible films: Film bags are used for packaging of liquid, semi-liquid, soft or powdered food material. These films are based on form-fill-seal method / concept. The efficiency of sealing is improved by cooling the pack before product displacement and after sealing operation. Polymeric plastic materials are extremely useful for food packaging. Flexible packaging material consists of multilayer laminated sheets in which a combination of polyester film, BOPP film, polyethylene and metal foils, ensure printability, moisture resistance and aroma retention.

These materials act as better barrier to the passage of gases and odours and protect the flavour and aroma of the food. These are generally heat sealable, suitable for high speed filling, printing and add little weight to the product. These films have thickness of less than 0.25mm.

Flip: It is a mixed drink consisting of sweetened spiced liquor with beaten eggs.

Flip top can: It is a 'can' with one end having a very small, simple device which when operated flips open the can (Plate 280).

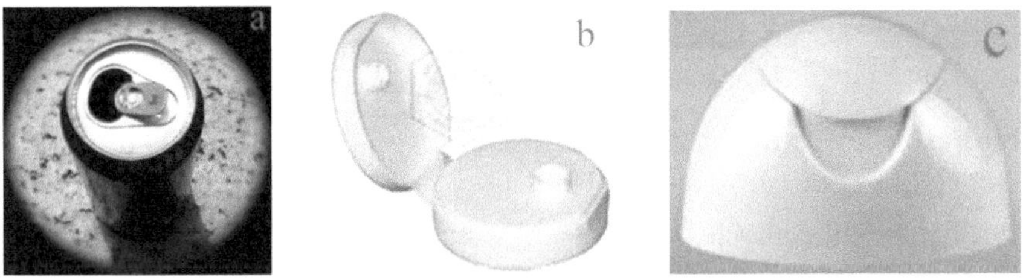

Plate 280: Flip top can (a), lid (b) and cap (c)

Flipper: A mild positive pressure due to hydrogen swelling inside the can result swelling of the lid which can be brought down to its original position with finger pressure and lid remains in that place even after removal of finger pressure . It is the initial stage of the hydrogen swell.

Flitch: A side of bacon which is salted and cured.

Flocculation tank: It is a part of effluent treatment plant. Flocculates of microorganisms are allowed to be formed for the degradation of organic waste present.

Flour: A product consisting of finely milled wheat and a similar product made from another grain or food product (dried potatoes or fish) to break up into particles.

Fluid flow: Moving a fluid product from one point to another with varying degrees of turbulence.

Fluid mechanics: A branch of mechanics dealing with the properties of liquids and gases at rest and in motion. Fluid mechanics has a wide range of applications in mechanical and chemical engineering, in biological systems, and in astrophysics (Plate 281).

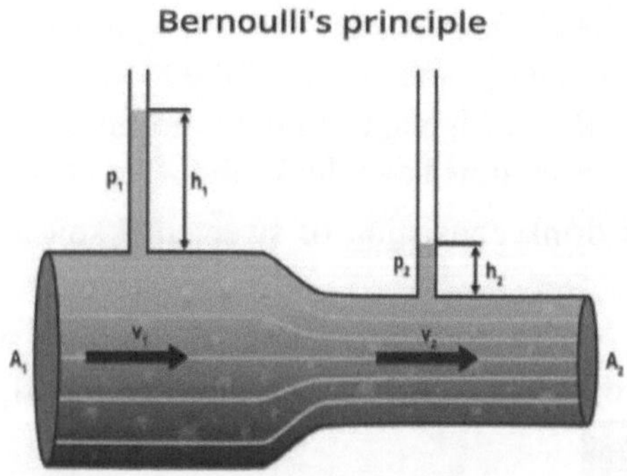

Plate 281: Fluid mechanics

Fluidic: A device that depends for operation on the pressures and flows of a fluid in precisely shaped channels.

Fluidity: It is the physical property of a substance that enables it to flow.

Fluidization: It is an operation in which fine granular solids are transformed into a fluid like state through contact with a fluid. Here the air velocity becomes higher than the critical velocity (Plate 282).

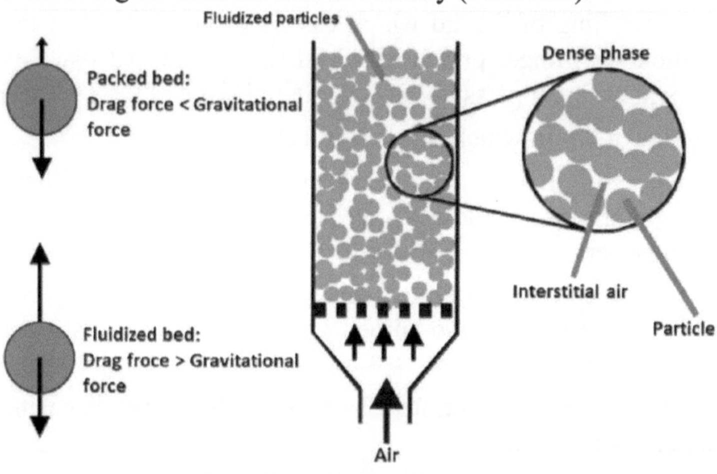

Plate 282: Fluidization

Fluidize: The process that causes flow of the substances like a fluid. A substance is said to be fluidized when it start behaving like a liquid.

Fluidized bed: A bed of small solid particles suspended and kept in motion by an upward flow of a fluid (as a gas). It is also called fluid bed.

Fluidized bed drying: The pneumatic conveyor dryer are used for drying. In fluidized bed drying (FBD), heated air is blown up through the food particles (semi dry) with just enough force to suspend the particles in a gentle boiling motion (Plate 283). It is used for drying grains, peas, potato granules (semi-dry). It is similar to principle and construction to the fluidized bed freezing.

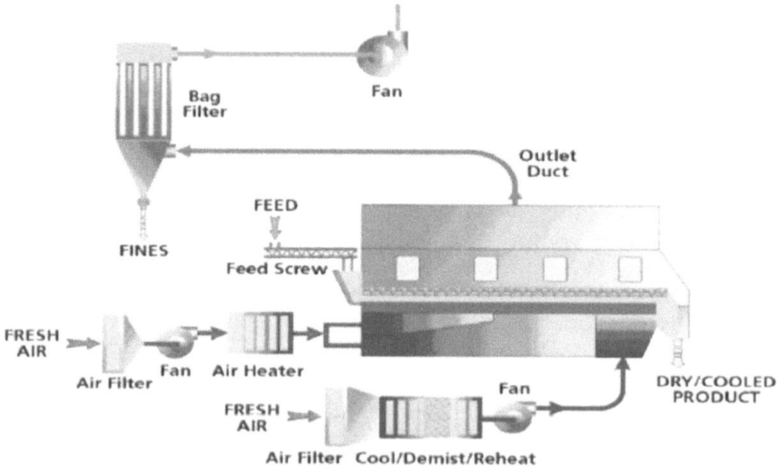

Plate 283: Fluidized bed dryer sketch diagram

Fluidized bed freezing: It is the modification of the air blast freezing. The food stuff is fluidized to form a bed of particles and then frozen.

Fluting: Papers obtained from semi-chemical pulp process are used which provide good rigidity to the board but in India only mostly kraft paper in used. Its main function is to provide the necessary cushion, rigidity to the board, contribute to resistance to stress, especially after conversion to box. The grammage of fluting medium may be in the range of 80-150 gms. or The spiral layer / corrugated medium in the corrugated fiber board is called fluting (Plate 284). Strength of board is affected by fluting contour. Highest quality fluting materials made form semi chemical.

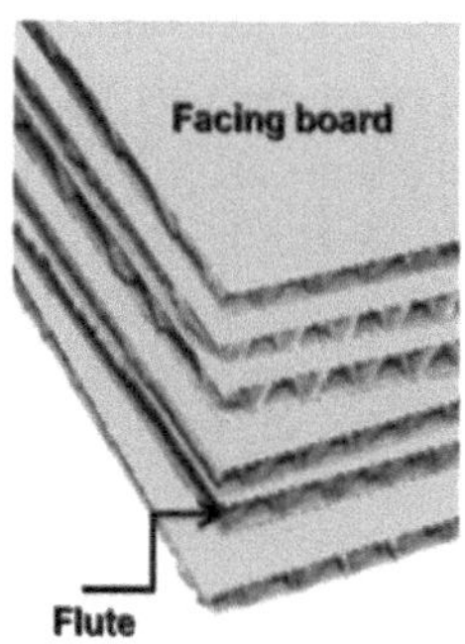

Plate 284: Fluting on the board

Fluting contour: Fluting contour is height of each flute and number of flutes.

Foam: A light frothy mass of fine bubbles formed in or on the surface of a liquid or a stabilized froth produced chemically or mechanically. It is a colloidal dispersion in which gas is in dispersed phase and liquid in the continuous phase (liquid may be converted into solid).

Foam mat drying: Foam mat drying is used for drying liquid or pureed material as thin layers of stabilized foam by heated air at atmospheric pressure. The foam mat dried juice products are quite low in density, with many gas bubbles remaining in the dry product. The foam mat dried products have limited storage life due to porous structure.

Foaming agents: These agents are used to enhance the formation of foam in the pulp or other food material and help in enhance the drying rate. CMC, alginates, Egg albumin, sanspas (*Asparagus ascenders*) are the example of the foaming agents.

Folded oil: The concentrated citrus oil without d-limonene is called folded oil. These are added back to citrus juice to increase the flavour with out decreasing oil level.

Folic acid: A crystalline vitamin $C_{19}H_{19}N_7O_6$ of the B complex that is used in the treatment of nutritional anemia's. It is also called pteroylglutamic acid. It belongs to a group of compounds known as pterins. Deficiency causes acute pernicious anemia, synthesized by bacteria in the intestine of animals. The synonyms of the vitamin are folacin, pteroyglutamic acid, anti-anemia factor, fermentation factor.

Fomite: It is a dish that may be contaminated with infectious organisms and serve in their transmission.

Fondant: It is a mixture of the invert sugars prepared by heating sugars to a temperature of 115-120°C and adding glucose or weak acid to accelerate the inversion.

Food: Any commodity produce or harvest in organic or inorganic form to be fit to eat / drink (other than drug and water) be a particular society; and contributes materially towards growth and repair of tissue is known as food. Food consist of three essential component of life together O_2 and H_2O, which provide the human body with organic and inorganic substances to sustain its biological viability. Food nourishes body, remove hunger and give satisfaction. Or Food material consisting essentially of protein, carbohydrate, and fat used in the body of an organism to sustain growth, repair, and vital processes and to furnish energy. Or Food may also be defined as any thing eaten/drink, absorbed by body to be used as energy source for body building, regulating and protective material.

Food additive: Food additive is defined as a substance or a mixture of substances other than the food stuff incorporated in food either directly or indirectly as a result of any phase of production, processing, storage or packaging to improve or maintain the nutritional value, enhance quality, consumer acceptability and facilitate the preparation of materials e.g., acid, alkali, bleaching agents, emulsifying & sterilizing agents, flavours, colours, nutrient supplements, clarifying agents, humectants, chemical preservatives etc.

Food adulteration: Mixing of a food item with material of inferior quality or of an undesirable nature is known as adulteration. It comes under 402 section of the adulteration act (Plate 285).

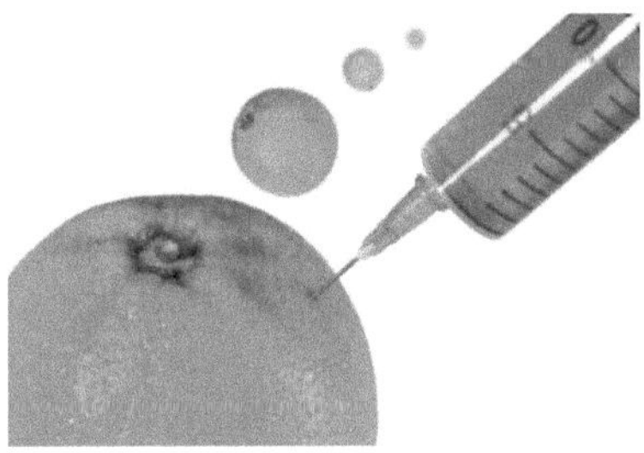

Plate 285: Food adulteration illustration

Food adulteration Act: It is the Prevention of Food Adulteration Act, 1954. The main objective of the prevention of food adulteration act is *to prevent contamination of all food items and beverages except drugs and water.*

Food allergen: A substance that induces allergy after food consumption is called food allergen.

Food and Nutrition: It is the science of foods, the nutrients and other substance present in the food, their action and relationship to health and diseases.

Food Chain: Sequence of the stages and operation involved in the production, processing, distribution, storage and handling of a food and its ingredients from primary production to consumption.

Food conversion ratio: It is a measure of the efficiency of the animal in converting food into flesh. It is the number of kgs of food consumed by an animal required to produce a live weight gain of 1 kg. A small ratio indicates high efficiency. It is also called feed conversion ratio.

Food court: An area within a building (as a shopping mall) set apart for food concessions.

Food guide: It is guide which help individuals to select foods to meet their nutritional requirements.

Food hygiene: All conditions and measures necessary to ensure the safety and suitability of food at all stages of the food chain.

Food infection: Food infection is the infection of the food by a bacteria or organisms and the disease is produced by the toxin which is produced after food consumption within the body.

Food Intoxication: Some bacteria produce a toxin in a food prior to being eaten and causes the poisoning on consumption of the food by the toxin (already produced) is called intoxication.

Food poisoning: Illness or diseases caused due to intake of foods contaminated by the microorganisms and their poisonous metabolites and those resulting from infection of the host through the intestinal tract. It is an acute gastrointestinal disorder caused by bacteria or their toxic products or by chemical residues in food.

Food preservation: Food preservation can be defined as the science which deals with the process of prevention of decay or spoilage of food, thus allowing being stored in a fit condition for future use. Preservation is also defined as the method / process of extending the shelf life of fresh as well as processed products by means of low temperature, drying, anaerobic condition, chemicals, irradiation and fermentation etc. without losing its colour, texture, flavour and nutritive value. It includes a set of methods and techniques to transform raw ingredients into food for consumption by humans or animals.

Food process engineering: Food process engineering is the operation of process in which food is manufactured, modified and packaged. It is quantitative science in which accuracy and precision, measurement, mathematical reasoning, modeling and prediction are all important.

Food processing: Food processing is application of techniques in a systematic manner for preventing losses through preservation, processing, packaging, storage and distribution and ultimately to ensure greater availability of a wide variety of foods. It is a set of methods and techniques used to transform raw ingredients into food for consumption by humans or animals.

Food pyramid: An ecological hierarchy of food relationships in which a chief predator is at the top, each level preys on the next lower level, and green plants are at the bottom. Food pyramid, is a visual representation (in shape of pyramid) of optimal number of servings of food for a person to eat daily from each basic food group (Plate 286). The food pyramid first evolved in Sweden in the 1970s and was adopted by the U.S. Department of Agriculture (USDA) in 1992.

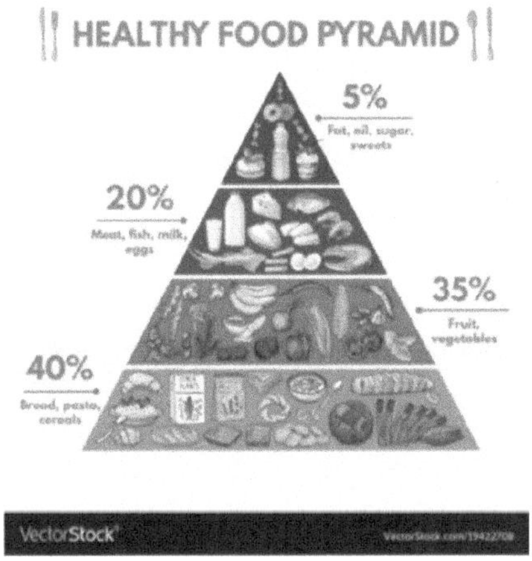

Plate 286: Food pyramid

Food quality standards: These are the body of rules directly concerning foodstuffs, whether they take the form of official, semi-official or factory form, and whatever the aspect treated, from food ingredients to retail marketing.

Food Safety and Standards Act: An act to consolidate the laws relating to food and to establish the Food Safety and Standards Authority (FSSA) of India for laying down science based standards for articles of food and to regulate their manufacture, storage, distribution, import, to ensure availability of safe and wholesome food for human consumption and for matters connected therewith or incidental thereto. FSSA, 2006 bring the different pieces of legislation pertaining to food safety under one umbrella which will override the PFA, 1955 and various Quality Control Orders under Essential Commodities Act, 1955.The aim is to better coordinate and integrate food safety controls across India to give highest level of health protection.

Food safety assurance systems: The food safety is one of the most important components of assurance system required for global expansion of trade between developed and developing countries. Assurance that food will not cause harm to the consumer when it is prepared and/or eaten according to its intended use is known as food safety concept. It is set of conditions and measures taken during the food chain like production, processing, storage and distribution in order to ensure that, the product consumption does not represent a risk to the human health. There are mainly five assurance systems that are mostly adopted to ensure food safety and quality like Good Agricultural Practices (GAP), EurepGAP, Good Manufacturing Practices (GMPs), Good Hygiene Practices (GHP) or Sanitation, Hazard Analysis Critical Control Points (HACCP) and Total Quality Management (TQM).

Food safety management system: It is adoption of Good Manufacturing Practices, Good Hygiene Practices, Hazard Analysis Critical Control Point and such other practices as may be specified by regulation for the food business.

Food safety: Food safety is defined as the assurance that the food will not cause any harm to the consumer when it is prepared and/or eaten according to its intended use. Implementation of Good Agricultural Practices and Good Manufacturing Practices are primary steps in reducing the risks associated with fresh fruits and vegetables.

Food Science and technology: It is the production of food with the help of applied science and advanced technologies.

Food science: This is a profession in which biological and physical sciences are used to learn the nature of foods, causes of their deterioration, spoilage and principles underlying processing and improvement of foods for the consuming public.

Food security: It means that all peoples at all times must have an economic access to safe and nutritious food.

Food spoilage: Food spoilage is the deterioration of food resulting in undesirable sensory quality loss. Spoilage can occur by microbial or physic-chemical means.

Food stamp: A government issued coupon that is sold or given to low income persons and is redeemable for food.

Food standards: Food standards give precise criteria to ensure that products are fit for their stated purposes. They provide common frames of reference for defining the product.

Food technology: Application of food science for the selection, preservation, processing, packaging, distribution and use of food commodities. Or Food technology is the science and technology of conservation of the desirable qualities of food in more or less stabilized form to permit their widespread distribution. It is application of principles and facts of science, engineering and mathematics to the processing, preservation, storage and utilization of food.

Food web: The totality of interacting food chains in an ecological community is called food web.

Food with more nutrition: Food containing at least 10% more nutrition or calories than the regular or reference food. These foods are altered by fortification or enrichment.

Food with reduced (less) nutrition: Food containing at least 25% less nutrition or calories than the regular or reference food. Low nutrition food should not be used as reference food.

Foodie: A person having an avid interest in the latest food dish is called foodie.

Foodstuff: A substance having some food value and is the raw material of food before or after processing.

Foodways: It is the eating habits and culinary practices of a people, region, or historical period.

Foot candle: It is thc intcnsity of light of onc standard candlc at a distancc of one foot from the light source (Plate 287).

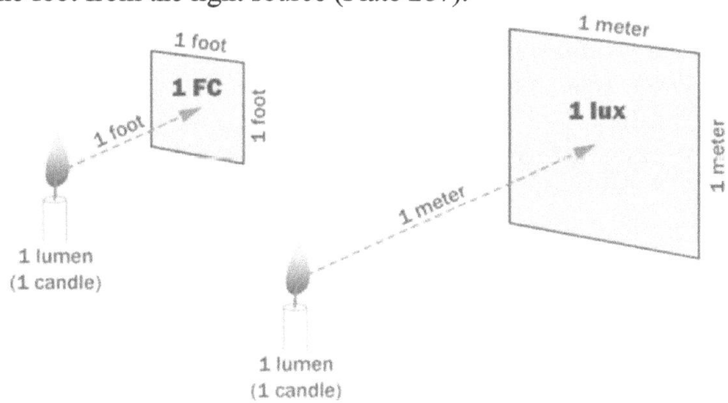

Plate 287: Foot candle

Form Fill Seal Packaging: The form / fill / seal machines are an extension of pre-formed bags and are popular because of its major application in food industry. These are high speed automatic machines, continuous in operation, and form a complete packaging line by themselves. The bags are continuously formed from roll stock, filled and heat sealed at a single station.

Form fill seal technology: Liquid foods particularly are packed by such technology. The rolls of films are directed towards machine which forms the pouches, simultaneously filling and sealing these filled pouches. Now solid foods are also being packed by this technology.

Formaldehyde test: It is the simplest technique used to detect the addition of the pulp wash in the juice. It measures the amount of the primary amines. It is also called formol test.

Formalin: Forty percent solution of formaldehyde is called formalin.

Fortification: Juice, squashes, syrup etc, are some times fortified with vitamins to enhance their nutritive value, to improve taste, texture or colour and to replace nutrient lost in processing. Ascorbic acid and beta-carotene imparts an attractive orange colour. Citric acid is often used for all type of beverage and phosphoric acid for cola type of drinks. It is also known as enrichment. It is also known as ennoblement. Iron fortified iodised salt help to fight against anaemia and goitre. It is double fortified salt.

Fortified wine: A wine (sherry) to which alcohol in the form of grape brandy has been added during or after fermentation. Wine are fortified with brandy and produced by distillation of regular wine. or Fortified wine to which distillate of wine called as brandy is added and contain 19-21% alcohol.

Forward feed evaporator: In such evaporator, the liquid is fed to first effect and flows through remaining effect in same direction as vapour (Plate 288 and 289).

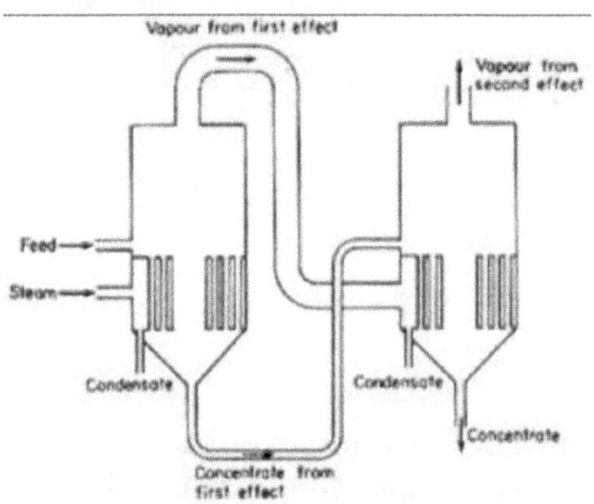

Plate 288: Forward feed evaporator

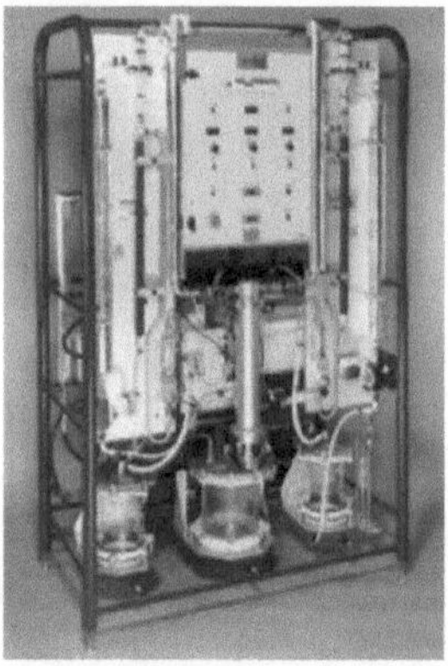

Plate 289: Modular evaporator

Forward or future markets: Here a transaction takes place for a standardized commodity with a promise to pay and deliver a commodity at some future date.

Foss milkotester: It is semi-automatic instrument used for determination of the fat content of the milk and calibrated by using Gerber test.

Fouling: It refers to the dirt and scale accumulated in the faces of the heat transfer surfaces. It reduces the heat transfer rate. Fouling of membrane is caused by plugging of the small pores, selective plugging of the larger pores and cake formation of the adsorbed macromolecules and particles. It results in reduction of permeation rate. Fouling is the deposition of burn layer of organic matter in the hot surface of the evaporators is a severe problem during concentration of fruit juices particularly pulpy and cloudy juices.

Fourier equation: It is related to heat transfer by conduction.

Fourier number: It represents the ratio of the rate of conduction of heat across a characteristic length to the rate of storage of heat in a volume of L^3. It is defined by $F_o = \alpha t/L^2$.

Frankfurter: It is a cured cooked sausage (of beef or beef and pork) that may be skinless or stuffed in a casing (Plate 290).

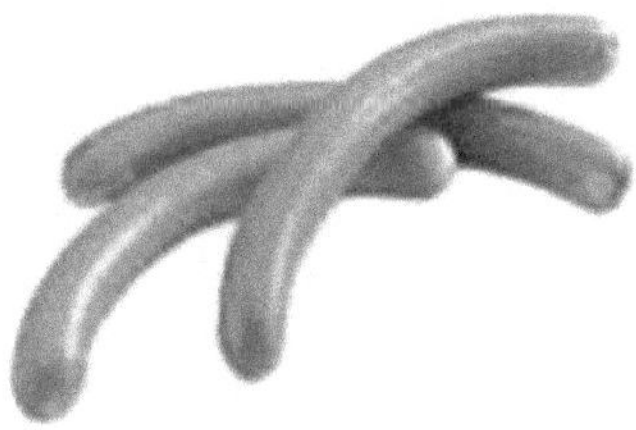

Plate 290: Frankfurter

FPLC: It stands for fast Protein Liquid Chromatography. It is used for separation of proteins using column chromatography.

Frappe: It is a partly frozen drink of fruit juice or a thick milk shake. It is a whipped mixture of a foaming agent in sugar syrup. Frappe can be manufactured by using hydrolysed casein.

Free / No /Zero Nutrition food: Foods contain no /zero or trivial (of little worth) level of nutrients like fat, saturated fat, cholesterol, sodium, sugar or calorie.

Free fatty acids: The fatty acids obtained by the breakdown of fats under adverse condition are called free fatty acids (FFA). Thus, the measurement of FFA is an index of the fat quality.

Free moisture content: Total average moisture content minus EMC for prevailing condition of drying. It may include both bound and unbound moisture.

Freezants: The agents that causes freezing are called freezants. These are liquid N_2 (boiling point is -196^{0}C), liquid CO_2 (B.Pt = -120^{0}C). Other freezants are Ammonia (Quite toxic and non-recoverable) and Freon (non toxic and recoverable).

Freeze burn: Freeze burn is a defect when ice crystals evaporate from the surface area of the perishable food products. It is characterized by a dry grainy, brownish spot on the surface and the tissue become dry and tough.

Freeze concentration: Freeze concentration results in higher retention of volatile flavor constituents in the concentrate than any other concentration process, including freeze drying. During freeze concentration, latent heat is removed from the product and water in the crystalline form is separated from the dissolved solids. By other concentration processes except reverse osmosis, latent heat is added to the product and water vapor is removed (Plate 291).

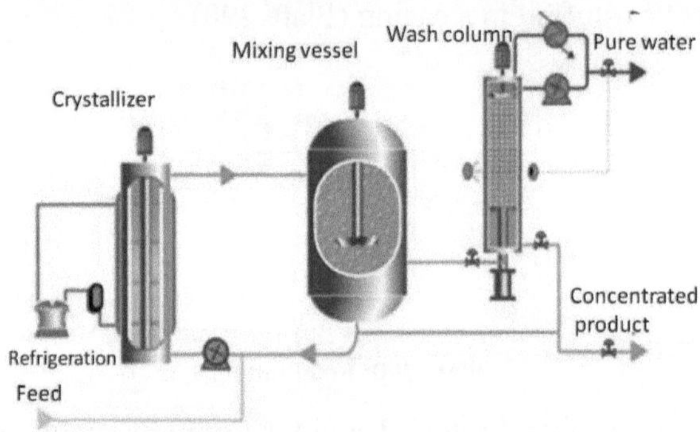

Plate 291: Freeze concentration

Freeze dried: A food product produced by freeze-drying.

Freeze dry: To dry a food in a frozen state under high vacuum for long preservation.

Freeze dryer: It is an apparatus used for drying high value food. Here food is dried while food is in frozen condition. In this, food is first frozen at -18°C on trays in the lower chamber of the freeze drier and then the frozen material is dried under vacuum (0.1 mm Hg) in upper chamber at 20-30°C (Plate).

Freeze drying: A process of drying in which the products are frozen quickly and the moisture there in is removed by sublimation and high vacuum. So, it is the removal of the water from a substance by direct sublimation from frozen state to the vapour state. It is also called lypholization. In this process the food is first frozen at -18°C on trays in the lower chamber of the freeze drier and then the frozen material is dried under vacuum (0.1 mm Hg) in upper chamber at 20-30°C. The loss of volatile flavour may occur by this drying (Plate 292 and 293).

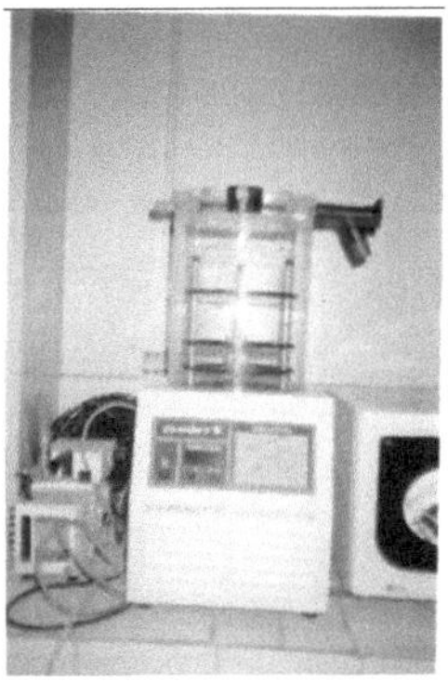

Plate 292: Freeze dryer

Plate 293: Freeze dried mushrooms

Freeze injury: A post harvest (PH) disorder where tissue degeneration occurs due to low temperature storage causing water soakedness, tissue browning and collapsing and glossy appearance.

Freeze thaw stability: The ability of a starch thickened product to be frozen and thawed without developing a gritty and crystalline texture is called freeze thaw stability.

Freezer: A compartment, room, or device for freezing food or keeping it frozen.

Freezer burn: Freezer burn is a term for the moisture lost from frozen food. A drawback of air blast freezing is the dehydration of food in an unwrapped condition, which is known as freezer burn (Plate 294). This occurs due to ice changing directly in water vapors molecules without going through the liquid state. Freezer burn results in discoloration change in texture and off flavor. It is a defect of stored frozen food which appears usually as dry, air freezing and in all condition of frozen storage. Light-colored spots developed in frozen foods as a result of surface evaporation and drying when inadequately wrapped or packaged. It irreversibly alters the colour texture, flavor and nutritive value of frozen foods.

Freezer storage: Storage at temperature below 0^0C and preferably below -18^0C.

Freezing: It is the low temperature storage method for long term preservation of foods and other biomaterials. It involves crystallization of most of the water and some of the solute by reducing the product temperature to -18^0C or lower.

Freezing point depression: If salt, sugar, glycol, alcohol are added to water the solution will freeze at lower temperature than pure water. Freezing point depression is caused by dissolved substances. Freezing point depression (DT) is directly proportional to the concentration of solute in water (M), DT= -1.86 x M (^{0}C). For ideal solution containing 1 mole of any solute, the freezing point will be -1.86^{0}C. The decrease in freezing point of solution over that of the water is known as freezing point depression (Plate 295).

Plate 294: Freezer burn

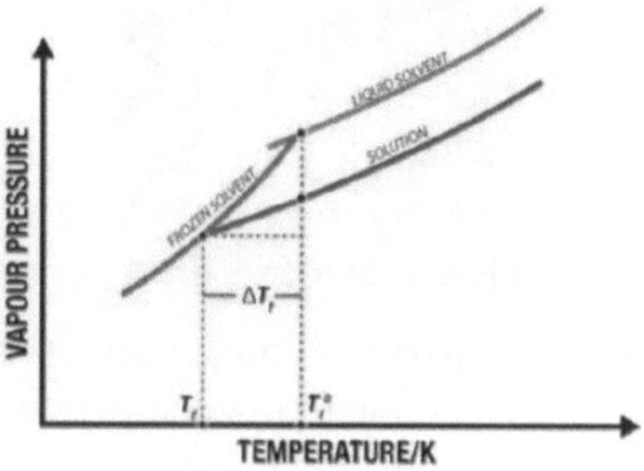

Plate 295: Freezing point depression

Freezing point: The freezing point of product is highest temperature at which ice crystals have stable existence in product. It is the temperature at which the liquid is in equilibrium with the solid (liquid solidifies) under given pressure.

Freezing time: The freezing time is the time during which the majority of the ice is formed n the product body.

French fry: A strip of potato fried in deep fat until light brown (Plate 296).

Fresh food: Food is raw and has never been frozen or heated and contains no preservatives. Some times blanched / frozen fresh / freshly frozen can also be termed as fresh.

Fricassee: A dish of cut up pieces of meat (as chicken) or vegetables stewed in stock and served in a white sauce (Plate 297).

Frozen concentrate: A juice product from which a substantial portion of water has been removed by high vacuum and then frozen and packaged.

Frozen foods: The foods which are maintained at temperature below -18°C, to retain their quality, wholesomeness, safety and shelf life (Plate 298).

Plate 296: French fries

Plate 297: Fricassee

Plate 298 : Frozen storage of carrot

Fructose: It is a crystalline sugar $C_6H_{12}O_6$ sweeter and more soluble than glucose or the very sweet levorotatory D-form of fructose that occurs in fruit juices and honey. It is also called fruit sugar levulose.

Fruit brandy: A brandy prepared from fruits other than grape.

Fruit butter: A product prepared by boiling the screened fruit pulp with or without addition of sugar, fruit juices, and spices to a semi solid mass of homogenous consistency which spiced heavily and prepared without the addition of sugar. The mixture contains not less than 5 parts by weight of the fruit ingredients to each two parts of the sugar.

Fruit cake: A cake containing nuts, dried or candied fruits, and spices is called fruit cake.

Fruit grader: The fruit graders are used for grading of the fruits and vegetables by mechanical means. Mechanical fruits graders grade the fruits / vegetables based on size, colour or weight. The grader may be roller type or conveyor type (Plate 299 and 300).

Fruit juice: A natural juice pressed out of a fruit, strained and kept practically unaltered in its composition during its preparations and preservation. Synonym of the fruit juice is unsweetened juice (Plate 301).

Plate 299: Conveyor type fruit grader

Plate 300: Roller type

Plate 301: Apple juice fruit grader

Fruit juice beverage: A fruit juice which is considerably altered in composition before consumption may be unfermented, fermented and synthetic.

Fruit Juice concentrate: Finished product of fruit juice obtained by the removal of water either by heat or freezing.

Fruit nector: A ready to serve beverage containing strained fruit juice or pulp to which sugar is added as per FPO specification. It should contain not less than 20% pulp juice (except pineapple orange where it is < 40%) and not less than 15% TSS.

Fruit paste: A product prepared like fruit butter but is dried in the sun or by artificial heat to a solid consistency like that of putty.

Fruit Product Order: The Government of India promulgated a Fruit Products Order in 1946. In 1955, the order was revised. FPO (1955) is administered by the Union Ministry of Food Processing Industries. This order is issued under Section-3 of Essential Commodities Act 1955. This order is implemented by Department of Food Processing Industries through the Directorate of Fruit & Vegetable Preservation at New Delhi. The Directorate has four regional offices located at Delhi, Mumbai, Calcutta and Chennai, as well as sub-offices at Lucknow and Guwahati. The FPO is enforced by the Department of Health. This order regulates the production and sale of fruit & vegetable products like jams, jellies, squashes, pickles, processed vegetable, synthetic vinegar and synthetic beverages etc.

Fruit sugar: Fructose sugar is called fruit sugar. It is a crystalline sugar $C_6H_{12}O_6$ sweeter and more soluble than glucose.

Fruit syrup: The product prepared like fruit butter but is dried in the sun or by artificial heat to a solid consistency like that of putty. Or The sweetened water extract of a fruit or the sweetened juice of a fruit (about 25%) usually with a high concentration of sugar (minimum 65%) and a small quantity of fine pulp with a low acid content.

Fruit tannins: Fruit tannins are reduced derivatives of flavones and comprised of catechins, leucoxanthins and some hydroxyl acids which are colourless but readily change to brownish pigments with metal ions. These are present in apple, peach, grape, almond, jamun etc.

Fruitarian: A person who lives on fruit or eats fruits only.

Frutox: This wax is emulsion of different waxes with 12% solids.

Fry bread: It is a quick bread cooked by deep-frying.

Fry check: It is an instrument is to check the polarity of the oil.

Fucoxanthin: A brown carotenoid pigment $C_{40}H_{60}O_6$ occurring in the chloroplasts of brown algae.

Fudge: A soft creamy candy made typically of sugar, milk, butter, and flavoring.

Fugu: These are very poisonous puffer fishes (family Tetraodontidae) that contain tetrodotoxin and are used as food in Japan after the removal of toxin containing parts.

Full slip: A harvesting index of muskmelon or cantaloupes for local market where the fruit can easily be removed (slip) with a slight pressure from the stem leaving a clean stem cavity.

Fumigant: A gaseous or readily volatilizable chemical substance used in fumigating and capable of destroying vermine, insects, bacteria and moulds e.g. carbon disulphide, methyl bromide, etc.

Fumigate: To apply smoke, vapor, or gas for the purpose of disinfecting or of destroying pests.

Fumigation: Treatment of materials in a close chamber by a fumigant to kill the storage pests (Plate 302).

Plate 302: Treatment of grains by fumigation

Functional food: A foodstuff (as a fortified food or dietary supplement) that provides health benefits in addition to its basic nutritional value. These are some times also known as Nutraceutical / Nutriceutical. These foods provide health benefits beyond traditional foods.

Functional layout: Functional layout is characterized by keeping similar machines or similar operations at one location. It is also called Process layout (Plate 303).

Fungus resistance test: All materials used in the fabrication of shipping containers should be tested for fungus resistance. In this case a spore suspension is prepared. Test item is exposed to RH of 25% for 28 days at 86⁰F. If after 28 days the growth of fungus is >4% on the exposed surface then material is considering to be fungus nutrient (susceptible) and if < 2%, then it is inert of fungus growth (resistant) (Plate 304).

Furfural: A liquid aldehyde $C_5H_4O_2$ of penetrating odor that is made from plant materials and used in making furan or phenolic resins and as a solvent. Furfural, a sugar dehydration product, formed as an intermediate in colour formation is also used as an index of stack burn.

F-value (Preservation): The equivalent time in minutes at 250⁰F (121°C) required to destroy spores or vegetative cells of a particular microorganism in a specific medium (Plate 305).

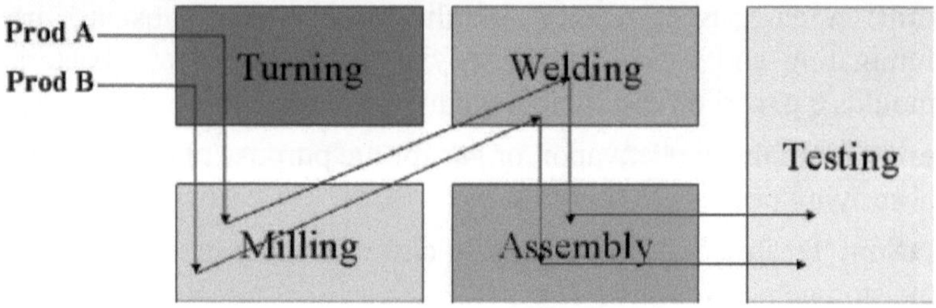

Plate 303: Functional layout

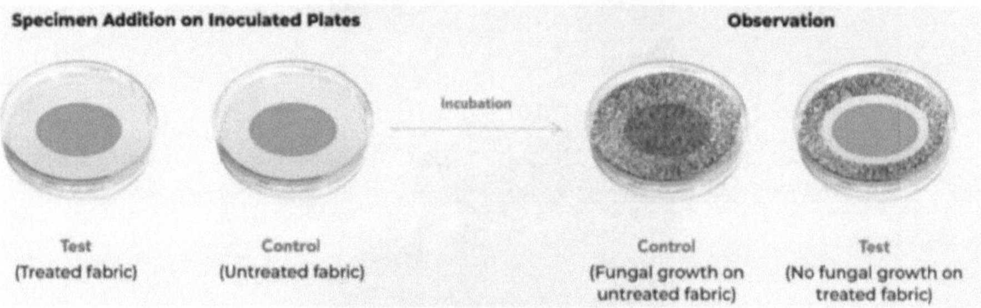

Plate 304: Fungus resistance test

$$F_0 = \Delta t \times 10^{(T - 121/Z)}$$

Where,

T= Temperature for Sterilization

Δt = 60 Sec. = 1min.

Z = 10°C

Plate 305: F - Value in Preservation

G

Galactometer: It is an instrument used for recording the specific gravity of the milk (Plate 306).

Galactose: It is an aldohexose sugar obtained by hydrolysis of lactose. Galactose is a monosaccharide and has the same chemical formula as glucose, i.e., $C_6H_{12}O_6$. It is similar to glucose in its structure, differing only in the position of one hydroxyl group. Due to this difference galactose possess different biochemical properties to glucose (Plate 307).

Galactosidase: An enzyme (actase) that hydrolyzes a galactoside. β-Galactosidase, commonly known as lactase, is an enzyme responsible to hydrolyze lactose (Plate 308).

Plate 306: Galactometer

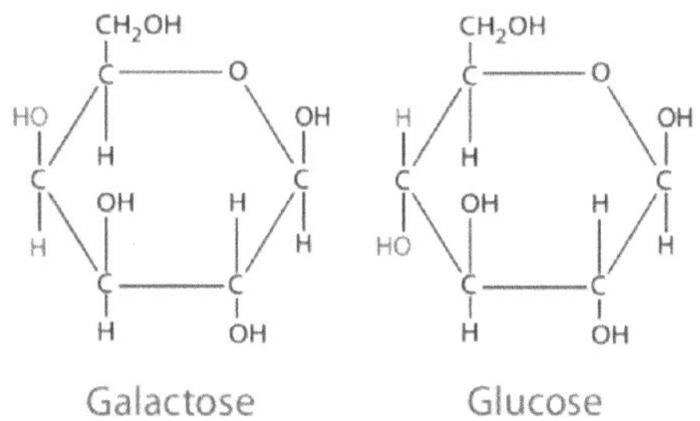

Plate 307: Galactose vs Glucose

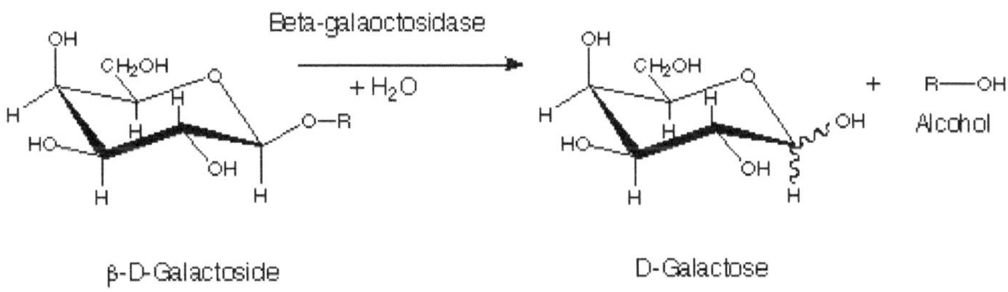

Plate 308: β-Galactosidase

Galactoside: A glycoside that yields galactose on hydrolysis. A *galactoside* is a glycoside containing galactose. The H of the OH group on carbon-1 of galactose is replaced by an organic moiety (Plate 309).

Galactosyl: A glycosyl radical $C_6H_{11}O_5$ that is derived from galactose.

Galacturonic Acid: A crystalline aldehyde acid $C_6H_{10}O_7$ that occurs in polymerized form in pectin.

Gamma ray: A photon emitted spontaneously by a radioactive substance (a photon of higher energy than that of an X ray). It is the radiant energy of very short wavelength that is capable of penetrating food but not lead (Plate 310).

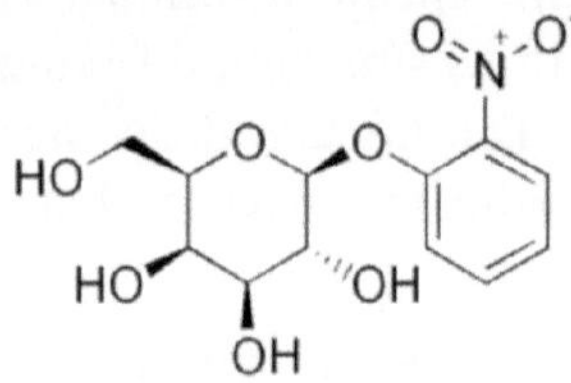

Plate 309: Galactoside

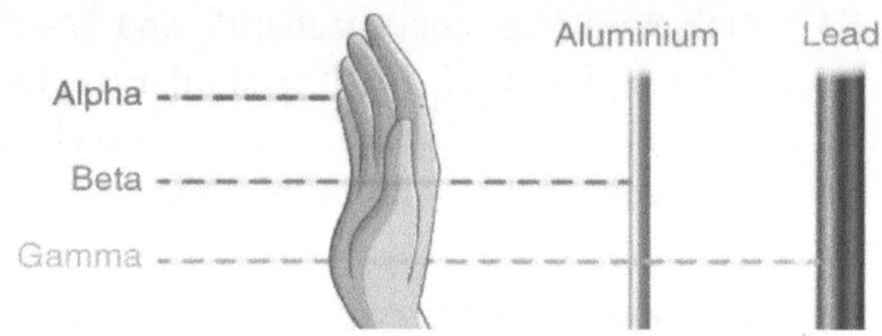

Plate 310: Penetration power of gamma rays

GAP: It stands for Good Agricultural Practices. GAP includes practices that have to be followed at the primary production, in order to ensure a safe and wholesome product at the same time minimizing the negative impact of those practices on the environment and on workers' health. One of the GAP systems that have taken off within the European community is EUREPGAP.

Garcinol: It is yellow pigment isolated from the rind of the kokum (Plate 311). Garcinol, normally found in the fruit rind of Garcinia indica. It is a potent HAT (*histone acetyltransferase*) inhibitor with high cytotoxicity towards cancer cells. Garcinol shows strong antioxidant activity since it contains both phenolic hydroxyl groups as well as a β-diketone moiety and resembles curcumin. Antioxidant actions of garcinol are believed to contribute to its chemopreventive activity.

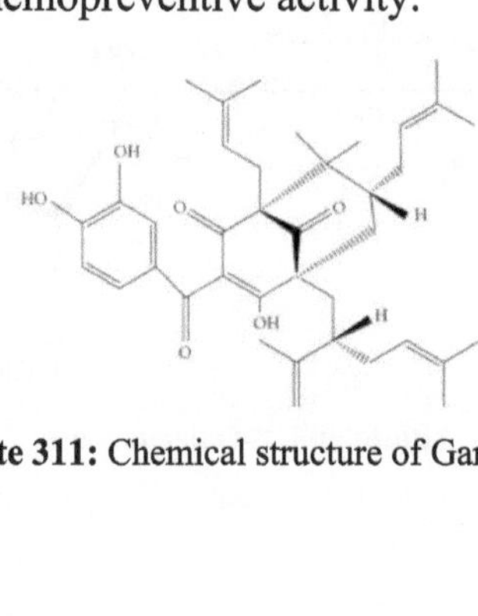

Plate 311: Chemical structure of Garcinol

Gas chromatograph: It is an instrument used for detection of the volatile compounds.

Gas chromatography: It is a separation technique which involves passage of gaseous moving phase through a column containing a fixed adsorbent phase.

Gas Liquid Chromatography (GLC): It is an apparatus used for estimation of various nutrients in the food samples (Plate 312). It involves a sample being vaporized and injected onto the head of chromatographic column. The sample is carried and transported through column by flow of inert gaseous mobile phase. The GLC comprises of gas purification chamber, injector, column and detector. Three gases like hydrogen oxygen and nitrogen are used in GLC. Nitrogen gas acts as a carrier gas while oxygen and hydrogen help in the combustion of the sample to produce flame which can be detected by the detector. Gas liquid chromatography is a chromatography in which the stationary phase is a liquid. It is a separation technique with a solid stationary phase (column) and gaseous mobile phase.

Plate 312: GLC

Gas liquor: Ammonia is known as gas liquor.

Gas purging: It is a method to remove O_2 form cans. Gas purging is removal of air from a filled can without vacuumising by more N_2 purging. In this process, few drops of liquid N_2 is dropped in a filled can, which with rapid evaporation flushes out the air, the inert gaseous N_2 replacing O_2. At this temperature liquid N_2 is of the order of -195 to -196°C. At this temperature liquid N_2 can convert to gas with a volume ratio of 300:1 and this ratio increasing upto 700 at the ambient temperature. Liquid N_2 being heavier than air reaches the bottom most position as a drop and rises all over to expel air (Plate 313).

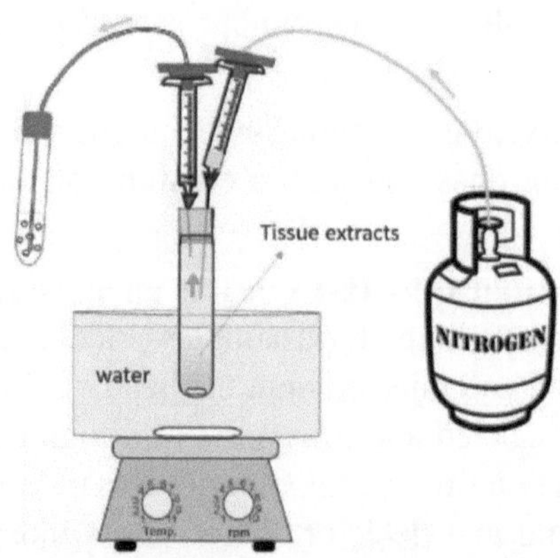

Plate 313: Illustration of gas purging system

Gas storage: Storage of products under modified or controlled atmosphere where atmospheric composition surrounding the commodity is different from that of normal air (78.08%) N_2, 20.95% O_2, 0.0, 0.3% CO_2) which usually involves reduction of oxygen and or elevation of CO_2 conc. This storage reduces ripening, senescence and respiration and thereby increases shelf life of the produce.

Gauge: It means that a measurement (linear dimension) according to some standard or system. Or The thickness of the polyethylene (packaging material) is measured in term of gauge. 4 gauge = 0.001 mm = 1 μm; 100 gauge = 0.025 mm = 25 μm, 400 gauge = 0.1mm = 100 μm. Thickness of polyethylene refers to Gauge.

Gauge pressure: It is equal to absolute pressure minus atmospheric pressure. If the pressure in the vessel is atmospheric then gauge would read "zero". It is also called gage pressure. The reading in gauge depends on absolute pressure where as it is independent of the atmospheric pressure (Plate 314).

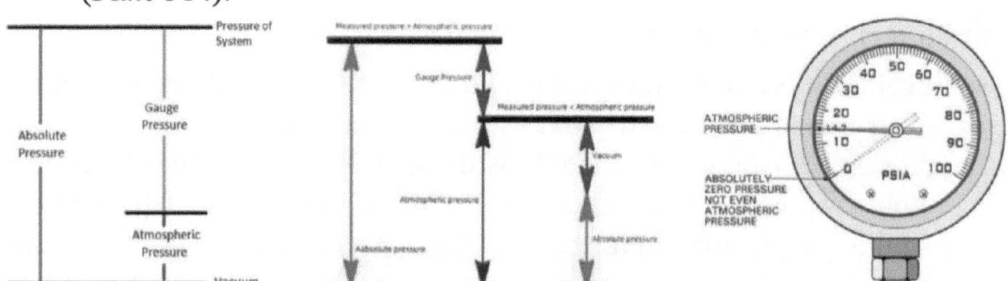

Plate 314: Gauge vs Absolute Pressure

GC-MS: It stands for Gas Chromatography Mass Spectrometry. It is a method that combines the features of gas-liquid chromatography and mass spectrometry to identify different substances within the test samples which can not be identified either by GC or by MS alone. Its application is in the field of drug detection and identification of the unknown samples.

Gel: A colloid in a more solid form than a solution i.e. jellies. It is a colloidal dispersion of a liquid dispersed in a solid.

Gelatin: Glutinous material obtained from animal tissues by boiling and is a colloidal protein used as a food, in medicine and an edible jelly made with gelatin. It is a water soluble protein made by boiling collagen with water.

Gelatinization: The process of converting a food product into a gelatinous form or into a jelly.

Gelation: The process of freezing or the formation of a gel from a sol (one in which the continuous phase is a liquid) is called gelation.

Gelato: A soft rich ice cream containing little or no air.

Gellant: A substance used to produce gelling is called gallant or gelling agent.

Gelmeter: It is a long pipette like glass apparatus used for determining the concentration of pectin present in fruit extract. It gives direct reading or indication of the amount of sugar needed for one litre of extract. Gelmeter is used for the determine of jelly grade by measuring relative viscosity of pectin.

Generator sachet: This is a sachet containing KMS is aqueous solution in a PE sachet for emission of sulphur dioxide and used for extension of shelf life of grapes. Sulphur dioxide in storage atmosphere must be between 80-300 ppm.

Genistein: It is an isoflavone $C_{15}H_{10}O_5$ found in soybeans and has antitumor activity. Genistein is *a 7-hydroxyisoflavone with additional hydroxy groups at positions 5 and 4'*. It is a phytoestrogenic isoflavone with antioxidant properties (Plate 315).

Gerber test: The test used for determination of the fat content of the milk, cream or butter is called Gerber test. The reading expressed in g/100 g (Plate 316).

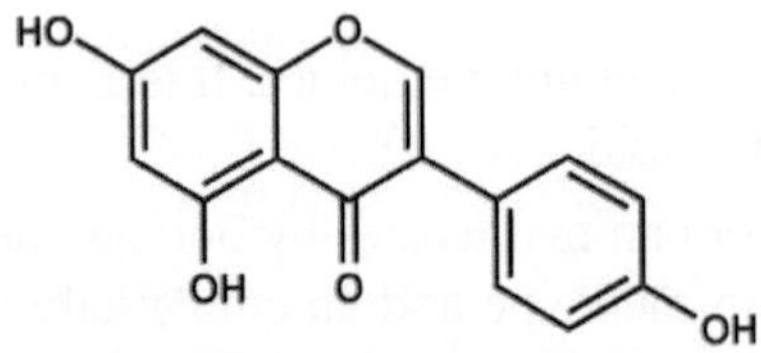

Plate 315: Genistein chemical structure

Plate 316: Gerber test by Butyrometer

Germicide: It is a substance that kills germs or pathogenic microorganisms.

Ghee: Ghee is 99.5% fat obtained by boiling off the water (removing water by boiling) from the butter. It is a clarified butter, prepared by heating and drying butter or cream, so that moisture is completely driven off.

GHP: It stands for Good Hygiene Practices. Practices oriented to ensure the safety and aptitude for use of fresh fruits and vegetables throughout the food chain, with emphasis in prevention and control of microbial hazards. GHP is a horizontal component of GAP and GMP.

Gin: A colorless alcoholic beverage made from distilled or redistilled neutral grain spirits flavored with juniper berries, orange peel, coriander, cardamom and aromatics (anise and caraway seeds) is known as Gin. It is not aged before marketing. It is a type of spirit made from grain or malt and flavoured with juniper berries. or

Gin: It is a machine separating cotton from its seeds.

Gin still: It is type of kettle which is heated by steam coils or jacket. The upper part of the kettle is provided with a gin head in which a basket holds the herbs and seeds. The vapour of alcohol passes up through these materials thereby extracting the aromatics.

Ginger ale: The ginger-flavoured non-alcoholic drink is called ginger ale.

Ginger beer: It is mildly alcoholic or non-alcoholic cloudy drink made from fermented ginger and syrup.

Ginger bread: It is ginger-flavoured treacle cake.

Ginger-nut: The ginger-flavoured biscuit is known as ginger nut.

Glace fruit: Candied fruit covered or coated with thin transparent coating of sugar which imparts to it a glossy appearance.

Glass: It is one of the amorphous materials formed from a melt by cooling to rigidity without crystallization. Or It is a transparent or translucent material consisting typically of a mixture of silicates.

Glass lug: Glass lugs are the horizontal, tapering and protruding ridges of the glass around the periphery of the finish. These help in closing of the glass jar by giving a partial turn (Plate 317).

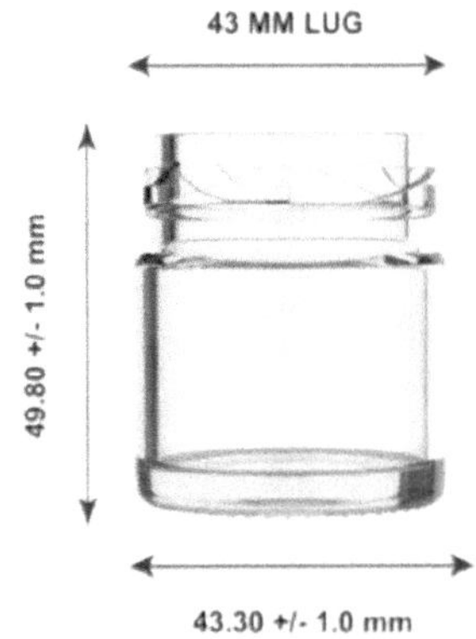

Plate 317: Glass lug

Glass temperature: The temperature at which the viscosity of material reaches $10^{13} - 10^{14}$ pa is called glass temperature.

Glass transition: The change of a material from a solid to a super cooled rubbery or viscous liquid is called glass transition.

Glass transition temperature: The temperature at which the conversion of an amorphous solid in glassy state begins to flexible rubbery state (less rigid state) is called glass transition temperature.

Glazed fruit: A candied fruit dipped for a moment in boiling syrup to cover or coat with a thin transparent coating of sugar, which imparts to it a glossy appearance is called glaced fruit.

Gliadin: A product (protein called prolamin) obtained by alcoholic extraction of gluten from wheat and rye.

GLOBAL GAP: It stands for Global Good Agricultural Practices.

GLP: It stands for good laboratory practice. It generally refers to a system of management control for laboratories and research organization. GLP helps to assure regulatory authority that the data submitted are a true reflection of the results and so can rely upon the safety assessments. It is a quality system which is concerned with the health and environmental safety during quality control in food processing organization. There are seven principles of GLP (Plate 318)

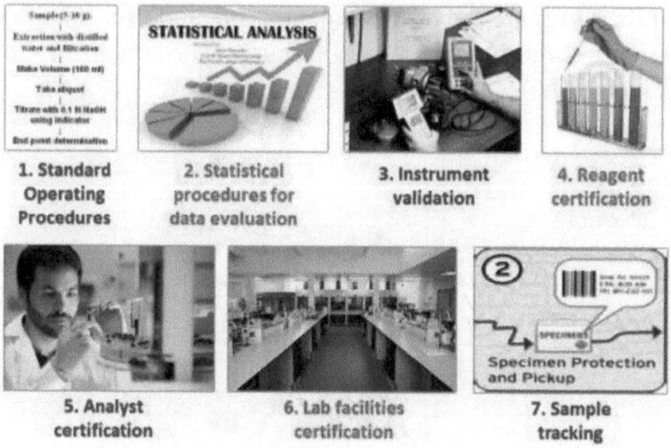

Plate 318: Principles of good laboratory practice

Glucose: A crystalline sugar $C_6H_{12}O_6$; sweet colorless soluble dextrorotatory form that occurs widely in nature and is the usual form in which carbohydrate is assimilated by animals. Light colored syrup made from cornstarch is also called glucose.

Glucose oxidase: Glucose oxidase is an enzyme that breakdown glucose and help in prevention of non enzymatic browning reactions. Glucose is breakdown to gluconic acid by Glucose oxidase

Glucosides: Nitrogenous compounds that yield sugar and some other substances on hydrolysis e.g. kolanin from the nuts of *Cola nitida.*

Gluten: It is an elastic protein substance of wheat flour and other cereals that gives cohesiveness to dough.

Glutomatic: It is an automatic gluten washing machine.

Glycemic index: An index used to check the glucose in the blood is called glycemic index.

Glyceraldehyde: A sweet crystalline compound $C_3H_6O_3$ that is formed as an intermediate in carbohydrate metabolism by the breakdown of sugars and yields glycerol on reduction.

Glyceric acid: A syrupy acid $C_3H_6O_4$ obtained by oxidation of glycerol or glyceraldehydes.

Glyceride: It is an ester of glycerol with fatty acids. These are triesters of glycerol with various long chains organic acids are known as glycerides. Common name of glycerides are oil and fat (Plate 319).

CH2OH	CH2OCOR	CH2OCOR
CHOH	CHOH	CHOH
CH2OH	CH2OH	CH2OCOR
Glycerol	Monoglycerides	Diglycerides

CH2OCOR	CH2OCOC17H35	CH2OCOC17H35
CHOCOR	CHOCORC17H35	CHOCORC15H31
CH2OCOR	CH2OCOC17H35	CH2OCOC17H35
Triglycerides	Simple Triglycerides	Mixed Triglycerides

Plate 319: Structural formula of different Glycerides

Glycerinate: To treat with or preserve in glycerin.

Glycerol / glycerin: A sweet syrupy hygroscopic trihydroxy alcohol $C_3H_8O_3$ formed by the saponification / hydrolysis of fats.

Glycine: A sweet crystalline amino acid $C_2H_5NO_2$ obtained by hydrolysis of proteins. Glycine is the simplest amino acid.

Glycogen: It is a white amorphous tasteless polysaccharide $(C_6H_{10}O_5)_x$ that is the principal form in which glucose is stored in animal tissues and especially in muscle and liver tissue. Glycogen is also called animal starch.

Glycogenolysis: It is the breakdown of the glycogen to yield energy.

Glycolysis: A series of biochemical reactions in respiratory process in which hexose sugar is converted to pyruvic acid with net gain of eight ATP molecules. or The compounds formed due to linkage of a sugar and an alcohol with a loss of one molecule of water.

Glycoside: It is one of the sugar derivatives that contain a non-sugar group bonded to an oxygen or nitrogen atom and that on hydrolysis yield a sugar (glucose) with loss of one molecule of water. Narringin and hesperidin from citrus are glycosides.

GMP: It stands for Good Manufacturing Practices. Good manufacturing practices are those practices which are in conformity with codes of practice, industry standards, regulations and laws concerning production, processing, handling, labeling and sale of foods with the intention of protecting the public from illness, product adulteration and fraud. Or GMP refers to the minimum sanitary and processing requirements necessary to ensure the production of safe and wholesome food. It includes practices oriented to the prevention and control of hazards associated to the fresh fruits and vegetables during postharvest chain, ensuring a safe and wholesome product, whilst also minimizing the negative impact of those practices on the environment and on workers health. GMP provides quality assurance without shelf testing of the produce. GMP is a more immediate and consistent way to control quality and serves as a primary vehicle for assuring quality. Thus, GMP can be termed as heart of quality management system. GMP is based on three components viz. (i) effective manufacturing operations (ii) effective food control and (iii) effective management.

Goitrogen: Thiocyanate and isothiocyanates present in soyabean, groundnut and lentils interfere with iodine uptake are known as goitrogen.

Good cholesterol: A high density lipoprotein (HDL) of blood plasma that is composed of a high proportion of protein with little triglyceride and cholesterol and that is correlated with reduced risk of atherosclerosis is called good cholesterol.

Good nutrition food: Food having 10-19 % of the daily value for a particular nutrient in a serving or per 100 gm.

Gorgonzola cheese: It is semi soft pungent blue cheese of Italian origin (Plate 320).

Gossipol: It is a phenolic pigment ($C_{30}H_{30}O$) in cotton seed that is toxic to some animals including insects.

Gouda: Gouda is a mild cheese of Dutch origin that is similar to Edam but contains more fat (Plate 321).

Plate 320: Gorgonzola cheese

Plate 321: Gouda cheese

Goulash: The stew made with meat (as beef), assorted vegetables, and paprika is called goulash.

Guarana: It is a dried paste that is made from the seeds of *Paullinia cupana* of the soapberry family that contains caffeine, tannin and is used as a stimulant

Grading: Grading refers to a voluntary program of classification of products based on certain characteristics, usually related to aesthetics. Grades usually do not pertain to safety.

Grain: It is cereal seed, single or in bulk after harvesting.

Grain grader: It is a machine used for grading of the grains.

Grain separator: It is a machine used for removing the impurities and foreign seeds etc. from the given grain material.

Grain shattering: It is the loss of the grain (mature or immature) from the ear heads, pods and spike while the crop is still standing in the field.

Grain strength: The percentage of acetic acid in vinegar is generally expressed In terms of grain strength which is ten times the percentage so acid present in it i.e. 10 grain is equal to 1% acetic acid present in vinegar.

Gram: It is metric unit of small grainy particles. weight equal to a 1000 milligram's.

Gram equivalent: The quantity of an element, group, or compound that has a mass in grams equal to the equivalent weight.

Gram molecular weight: It is the mass of one mole of a compound equal in grams to the molecular weight. It is also called gram-molecule.

Gram negative: Gram negative microbes (bacteria) do not hold the purple dye when stained by Gram's stain. Gram negative bacteria have a thin peptidoglycan layer and have an outer lipid membrane. The outer lipid membrane possessed by Gram negative bacteria means that, when referring to their physical structure, they are termed diderms. Bacteria include *Enterobacter* species, *Salmonella* species and *Pseudomonas* species

Gram positive: Gram positive microbes (bacteria) hold the purple dye when stained by Gram's stain. Bacteria include all *Staphylococci*, all *Streptococci* and some *Listeria* species. Gram positive bacteria have a thick peptidoglycan layer and no outer lipid membrane. As Gram positive bacteria lack an outer lipid membrane, when correctly referring to their structure rather than staining properties, are termed monoderms. The Gram staining technique was developed in 1884 by Danish bacteriologist Hans Christian Gram (Plate 322).

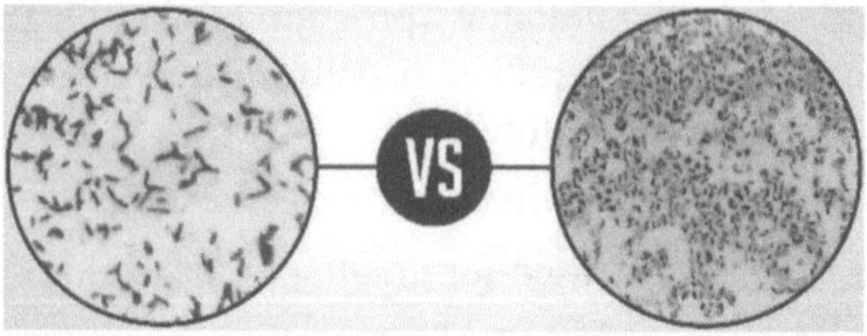

Plate 322: Gram positive vs Gram negative bacteria

Grammage: Paper quality is specified on weight per unit area which is called Grammage in the trade language.

Gram's stain: A method for the differential staining of bacteria by treatment with a watery solution of iodine and the iodide of potassium after staining with a triphenylmethane dye (crystal violet) is called Gram's stain. It is also called Gram's Method.

Grape guard: It is a paper pad containing KMS and citric acid in two different compartments. It releases sulphur dioxide in the box containing grapes when combines with the moisture and thus protect the grapes from spoilage and increases the shelf life. Therefore it is called grape guard. Grape guard suppresses fungal growth, reduce physiological loss, berry decay can be controlled, berry drop reduced, and maintain quality. Grape guard is packed in tissue paper/ butter papers.

Grappe: It is a brandy (distilled wine) prepared from pomace which has been sweetened with sugar syrup and re-fermented. It is water white in colour.

GRAS: It stands for generally regarded /recognized as safe. Any substance / additive is accorded GRAS status, if is generally recognized as safe by the panel of the experts under the condition of its intended use.

Gray G(y): It is equals to 100 rad. It absorption of one joule energy per kg of food.

Green bacon: Bacon sold after curing with brine but before being smoked is called green bacon.

Green Dot symbol on packaging: The Green Dot is the financing symbol of a European network of industry-funded systems for recycling the packaging materials of consumer goods. Packaging which can be recycled after collection carry a symbol green dot (Plate 323). It indicates that the producer of the packaging has contributed financially to the recycling of that packaging. The logo is a trademark protected and it is not a recycling logo.

Green dot symbol on label: Vegetarian food also contains green dot symbol (Plate 324).

Plate 323: Green dot symbol for recycling package

Plate 324: Green dot symbol for vegetarian food

Green extraction techniques: The green extraction techniques / methods are: pressurized liquid extraction (PLE); microwave-assisted extraction (MAE); supercritical fluid extraction (SFE); ultrasound-assisted extraction (UAE); enzyme-assisted extraction (EAE); as well as the usage of eutectic mixtures and supercritical fluids that have been applied alone, or in a combination with the above mentioned techniques. Green Extraction is based on the discovery and design of extraction processes which will reduce energy consumption, allows use of alternative solvents and renewable natural products, and ensure a safe and high quality extract/product". Pressurised hot water extraction (PHWE) or sub-critical water is one of the most promising modern green extraction techniques and methods. There are six principles of green extraction of natural products. The benefits of green extraction include enhanced efficiency, reduced solvent consumption, and the preservation of valuable bioactive compounds (Plate 325).

Plate 325: Benefits of green extraction techniques

Green solvent: Supercritical water and carbon dioxide as green solvents. The supercritical fluids, as the name suggest, are chemicals that reside in an intermediate phase across a substance that is liquid along with an air at pressures and temperatures over the corresponding critical points.

Green solvent extraction: Green solvents such as terpenes and ionic liquids have tremendous solvent properties that enable to extract the oil in eco-friendly manner. These green solvents and technologies are considered green owing to the attributes of energy reduction, eco-friendliness, non-toxicity and non-harmfulness.

Green Spotting: A disorder of citrus due to rough handling or harvesting where turgid fruits show necrosis of epidermis adjacent to oil glands. If orange is too turgid at harvest, the oil glands in the skin can be ruptured, releasing phenolic compounds and causing oleocellosis. The release of phenolic compounds and oil from the glands of citrus skin is called green spotting. It is also known as Oleocellosis or Oil green spotting.

Grending: It is the process of proper blending of flavour, enhancement of taste and smell; helpful in preparing gravy.

Grenedine: It is a product prepared from pomegranate having 60:40 ratio of juice and sugar.

Griddle: It is a flat stone or metal surface on which food is baked or fried (Plate 326).

Grilling: Cooking food in utensil of parallel bars on which food is exposed to heat (charcoal or electricity). The grill term is also used for frying or toasting.

Grist mill: It is a mill for grinding grain (Plate 327).

Plate 326: Griddle

Plate 327: Grist mill

Grist: It is a batch of grain for grinding or malt for one brewing.

Grits: Coarsely ground grains are called grits.

Grog: It is alcoholic liquor (rum) cut with water and now often served hot with lemon juice and sugar sometimes added.

Gross domestic product (GDP): It is the gross national product excluding the value of net income earned abroad. It is measure of total production and consumption of goods and services in a country. The GDP is prepared on the basis of income and expenditure. The GDP on income side is measured by adding costs of labour, capital and tax. The GDP on expenditure side is measured by adding up the expenditures on households, businesses, government and net foreign purchases. Theoretically, these two measures should be equal.

Gross energy value: The energy released as heat by a food when a given weight of the substance is completely oxidized in a bomb colorimeter. It is also known as energy of combustion (Plate 328).

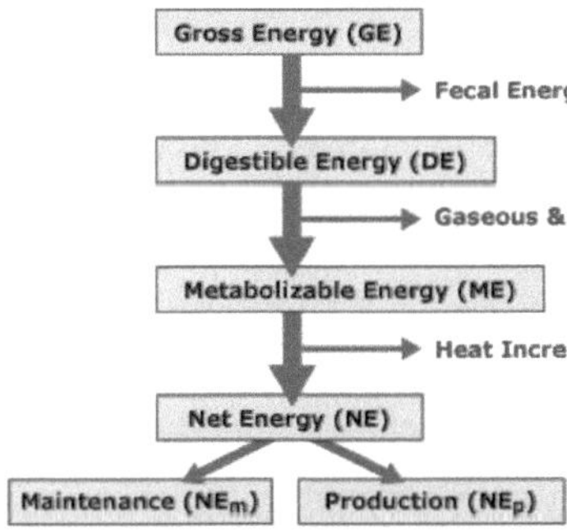

Plate 328: Illustration of Gross energy value utilization in body

Group I Nutrients: Group I nutrients are those which are added to the product and must be present at least at the levels declared on the labels.

Group II Nutrients: Group II nutrients are those that occur naturally and which must be present in at least 80 % of the level declared on the label.

Groundnut decorticator: The machine used to break the shell of the ground nut is called groundnut decorticator. It may by hand operated or machine operated (Plate 329).

Plate 329: Hand operated Groundnut decorticator

Growth: The growth is the irreversible increase in physical attributes (characteristics) of a developing plant or plant parts.

Growth curve of microbes: The growth curve of the microbe includes Lag phase, Positive accelerated phase, Logarithmic phase, Negative accelerated phase, Stationary phase, Accelerated death phase, death phase. In food preservation, it is very important to lengthen the lag phase and phase of +ve acceleration and often called as lag phase in combination. Microorganisms are very resistant in lag phase and least resistant in log phase.

Growth factor: A substance (vitamin B_{12} or interleukin) that promotes growth and cellular growth.

Growth inhibitors: A category of growth regulator which inhibits growth of shoots, blocks action of auxin, GA and cytokinin, sometimes stimulates fruit and leaf abscission and senescence, can substitute for chilling, delay bud break, and inhibit germination, e.g. Abscisic acid (ABA), abscisin, Dormin, phaseic acid, 2-transphaseic acid and theaspirone are natural growth inhibitors and synthetic inhibitor are succinic acid -2, 2-dimethyl hydrozide (SADH, B-995), B-9, Alar, (2-chloroethyl) trimethyl chloroide piperidine carboxylate (Amo-1618), Ancymidol.

Growth regulator: The synthetic or naturally occurring plant substances (auxin or gibberellin) that regulate growth.

GTR: It stands for gas transmission rate (Plate 330). The quantity of a given gas passing through a unit of the parallel surfaces of a plastic film in unit time under the conditions of test. The SI unit of GTR is 1 mol / (m^2·s).

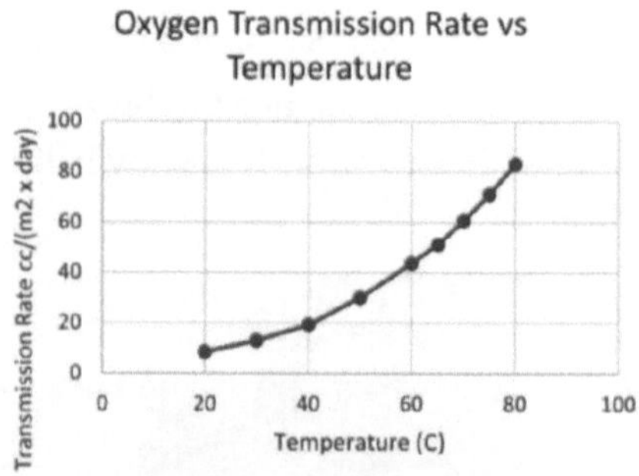

Plate 330: Gas transmiossion rate

Gulkand: It is a preserve of rose petals in sugar.

Gums: These are complex carbohydrates of plant origin usually containing galactose and atlease one other sugar or sugar derivative but excluding glucose.

Gum arabic: A water-soluble gum obtained from several acacias (Acacia, Senegal) and used in the manufacture of inks, adhesives, pharmaceuticals, and confections.

Gum resin: It is a product consisting essentially of a mixture of gum and resin obtained by making an incision in a plant and allowing the juice which exudes to solidify.

Gundruk: It is obtained from the fermentation of leaft vegetables. It is very important traditional product in Nepal. It is used as an appetizer.

Gush: A sudden flowing of a liquid in large amount from the bottle when opened.

Gushing: It is the defect of bottled beer.

H

HACCP: The Hazard Analysis and Critical Control Points (HACCP) is a scientific, rational and systematic approach to identify, assess and control hazards during production, processing, manufacturing and use of food. It ensures safety of the food. HACCP presents a good opportunity for the food industry for upgrading quality and bringing uniformity and consistency in their supplies to international market. or It is a systematic preventive approach to food safety that addresses physical, chemical and biological hazards as a mean of prevention rather than finished product inspection. It involves identification, evaluation and control of hazards which are significant for food safety.

PRINCIPLES	PROCEDURE
P 1: Conduct hazard analysis	1: Assembling of HACCP Team
P 2: Determine critical control points	2: Product description and distribution
P 3: Establishment of specification for critical limits	3: Identification of intended use and consumers
P 4: Development of monitoring and testing system to control critical point	4: Development of flow chart for description of the process
P 5: Establishment of corrective actions when particular CCP is not under control	5: On-site Verification of flow Diagram
P 6: Establishment of record for documentation	6: Applications of HACCP principles (P1 to P7)
P 7: Verification, implementation and review of HACCP system to confirm efficacy	

Plate: Principles and implementation procedures of HACCP

HACCP plan: A document prepared in accordance with the principles of HACCP to ensure control of hazards which are significant for food safety throughout the food chain under consideration.

Haemagglutinins: A group of proteins of plants that are not antibodies and do not originate in an immune system but bind specifically to carbohydrate containing receptors on cell surfaces (red blood cells). It is also known as Phytoagglutins / lectins.

Haff disease: A pyrolysis due to the excessive consumption of various fresh water fish which contain the vitamin antithiamine.

Hair hygrometer: It is an instrument used for measuring the relative humidity. The sensing element is a strand of hair (Plate 331 and 332).

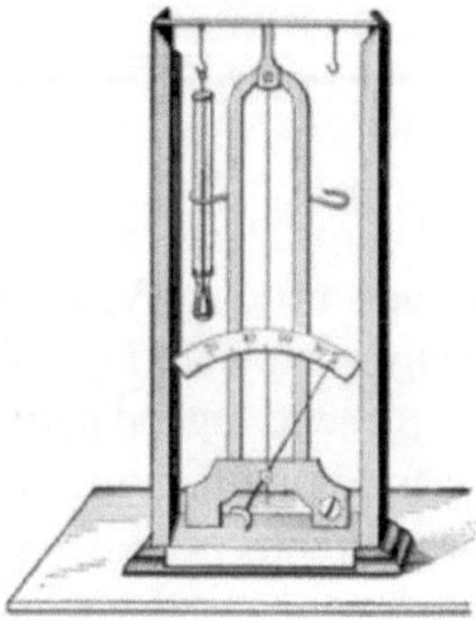

Plate 331: Schematic diagram of hair hygrometer

Plate 332: Hair hygrometer

Halal meat: The large blood vessels of the cervical region are cut and the animal is allowed to die due to loss of blood and anaemia; due to good bleeding this meat keeps well.

Halophile: An organism that flourishes in a salty environment.

Halophyte: A plant (as saltbush or sea lavender) that grows in salty soil and has a physiological resemblance to true xerophytes.

Ham: It is selected portion of the hog's flesh (pork), cured (salted and dried in smoke) for food (Plate 333).

Hamburger: Ground beef or a patty of ground beef or a sandwich consisting of a patty of amburger in a split round bun in known as hamburger.

Hand can tester: It is an apparatus used for testing the vacuum or pressure inside the "can". It is generally used in CUT OUT analysis of the "can". The measurement of the pressure inside the "can" predict where the "can" is perfect or damaged due to spoilage (Plate 334).

Plate 333: Ham

Plate 334: Hand can tester

Hand screw basket press: It consists of a wooden structure, screw basket and a tray. It is used for extraction of the juice (Plate 335).

Hard swell: A can with tightly bulged ends and often the high gas pressure distorts or buckle the ends of side seam of the cans. It is the final stage of the swell. The bulged ends cannot be pressed back to normal position and the cans ultimately burst.

Hard water: Water which does not produce lather with soap solution readily is called hard water. E.g. Sea water, river water. The hardness of the water is due to presence of bicarbonates, chlorides and sulphates of calcium and magnesium.

Hard wheat: Hard wheat (winter/spring) is used for bread making.

Harvest: It is a single deliberate action to separate a crop / food stuff with or without associated non edible material from its growth medium (A plant part from the plant or the whole plant from the soil) e.g. sweet corn (associated non edible cob portion), mango (with stalk).

Harvest index: A productive parameter depicting the economic productivity in terms of dry matter and is calculated as economic yield / biological yield x 100. It is the ratio of grain weight to the total plant weight in a cereal crop.

Harvest losses: This type of loss includes physical and mechanical damage to the commodity on qualitative or quantitative bases specifically caused by the harvesting activity.

Harvested foods: Foods removed from the medium of immediate growth are called harvested growth.

HASSOP: It stands for Hygiene and standard sanitation operating procedures. The sanitary requirements in the food processing plants should be in accordance with HASSOP. These are written procedures that an establishment develops and implements to prevent direct contamination or adulteration of product.

Haughs unit: Measurement of the height of the thick white in relation to weight of the egg white gives haughs unit (Plate 336). Good quality eggs have 72 haughs units and as it deteriorated, it comes down to 30-60. Micrometer is used to measure the height of the thick white.

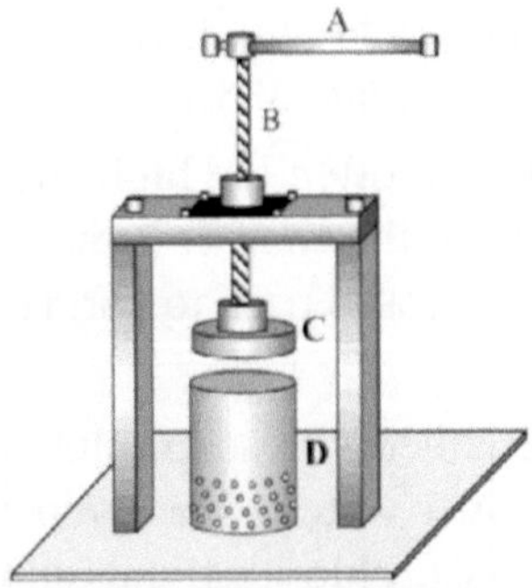

Plate 335: Hand screw basket press

Plate 336: Haugh unit tester

Hay Diet or Foolish Diet: A protein food in which there is absence of adequate carbohydrate is oxidized to provide energy and therefore not available for tissue building.

Hazard: Hazard is any factor that may be present in the food or food products and can cause harm to the consumer either through injury or illness. Hazards may be biological, chemical and physical.

Hazard analysis worksheet: It is the record of the deliberations of HACCP team during hazard analysis.

Hazard analysis: Identification of ingredients and products which might have pronounced effect on food safety.

Head space: Space between the lower surface of cap / cork (in bottles) or lid (in can) and the upper surface of liquid / other food materials which helps to accumulate CO_2, SO_2 or other gases and in the expansion and contraction of the material inside the containers during storage at different season / temperatures. The space required to be kept in the container = Coefficient of thermal expansion x temperature difference x capacity of the packaging materials. Head space analyzer scale is used to check the head space.

Health food: A food promoted as highly conducive to health is called health food.

Healthy food: A healthy food must be low in fat and saturated fat and contains limited amounts of the cholesterol (less than 60 mg) and sodium (less than 360 mg for individual foods and less than 480 mg for meal type products). In addition to this, a single item food must provide at least 10 percent or more of the Vitamin A or C, iron, calcium, protein or fibre.

Heat: It refers to transfer of the thermal energy (energy and entropy) from the hotter material to cooler material. Heat transfer may change the internal energy of the material. The heat is the added energy that causes substances to rise in temperature, fuse, evaporate, expand, or undergo any of various other related changes.

Heat capacity: It is the amount of the heat in joules required to raise the temperature of a body by one degree Celsius.

Heat engine: A mechanism (internal combustion engine) for converting heat energy into mechanical or electrical energy is called heat engine.

Heat exchanger: A device (automobile radiator) for transferring heat from one fluid to another without allowing them to mix is called heat exchanger. Or It is a device in which heat is transformed from hot fluid to a cold fluid. The different types of the heat exchangers are swept surface heat exchanger, double pipe heat exchanger, shell & tube heat exchanger and plate heat exchanger (Plate 337 and 338).

Plate 337: Heat exchanger

Plate 338: Schematic diagram showing working mechanism of heat exchanger

Heat exhaustion: It is a condition marked by weakness, nausea, dizziness, and profuse sweating that results from physical exertion in a hot environment. It is also called heat prostration.

Heat labile nutrients: These are the nutrients which readily or continually undergo chemical, physical, or biological change or breakdown and become unstable in the presence of heat are called heat labile nutrients. The vitamins are heat labile and so mostly become unstable during canning (Plate 339). Susceptibility of vitamins to factors affecting stability

Susceptibility of vitamins to factors affecting stability

Vitamin	Abbreviation	Temperature	Humidity	Light	Oxygen	Acid pH	Alkaline pH
Vitamin A	A	Very sensitive	Sensitive	Very sensitive	Very sensitive	Sensitive	Stable
Vitamin D	D	Sensitive	Sensitive	Sensitive	Very sensitive	Sensitive	Stable
Vitamin E	E	Stable	Stable	Sensitive	Sensitive	Sensitive	Sensitive
Vitamin K	K	Sensitive	Very sensitive	Stable	Sensitive	Very sensitive	Stable
Riboflavin	B_2	Stable	Sensitive	Sensitive	Stable	Stable	Stable
Niacin	B_3	Stable	Stable	Stable	Stable	Stable	Stable
Pantothenic acid	B_5	Sensitive	Sensitive	Stable	Stable	Stable	Stable
Vitamin B_{12}	B_{12}	Very sensitive	Sensitive	Sensitive	Sensitive	Stable	Stable
Pyridoxine	B_6	Very sensitive	Sensitive	Sensitive	Stable	Sensitive	Stable
Biotin	H	Sensitive	Stable	Stable	Stable	Stable	Stable
Folic acid	B_c	Very sensitive	Sensitive	Very sensitive	Stable	Very sensitive	Stable

Adapted from DSM Vitamin Nutrition Compendium and Shurson et al. (2011).

Plate 339: Factors affecting stability of vitamins

Heat of fusion: Heat released when a liquid transformed into a solid (80 calories per gram of water) is called heat of solidification or fusion.

Heat of solidification: Heat released when a liquid transformed into a solid (80 calories per gram of water) is called heat of solidification or fusion.

Heat of vaporization: Heat energy absorbed in the conversion of water into steam (540 calories per gram of the water) is called heat of vapourization.

Heat penetration: It is the process by which heat is transferred into a product. The calculation of the thermal process is possible only if heat penetration is known.

Heat processing: The fundamental principle of preserving food by application of heat is called heat processing. It is a heat treatment to which the foods are subjected after hermetic sealing in containers is called heat processing.

Heat pump: An apparatus for heating or cooling (building) by transferring heat by mechanical means from or to an external reservoir (ground, water, or outside air) is called heat pump. It operates between hot or cold body temperature and atmospheric temperature. In refrigeration system, the heat pump extracts the heat from the cold body and delivers it to a hot body.

Heat sink: A substance or device that absorbs or dissipates unwanted heat from a process or an electronic device.

Heat treatment: Any food which is treated by heating and cooling in such a way to produce desired properties / quality product.

Heavy syrup: A prepared syrup contain 1 litre water/ kg sugar.

Heavy tomato puree: It is a tomato puree having TSS of 12 ^{0}B.

Heavy water: The water composed of deuterium and oxygen (enriched in deuterium) is called heavy water. It is also called deuterium oxide (D_2O).

Hedonic rating test: A hedonic rating test is used to measure the degree of acceptance of the food product on 9 point hedonic scale. It is a scale mentioned for organoleptic evaluation of any food product i.e. extremely liked (9), like very much (8), like moderately (7), like slightly (6), neither like nor dislike (5), dislike slightly (4), dislike moderately (3), dislike very much (2), to dislike extremely (1).

Heel: The heel is the curved portion between the bottom and the beginning of the straight area of the side wall of the glass jar.

Hemicellulose: It is indigestible polysaccharide which forms the cell wall of plants.

Hermetically seal container: It is an absolute airtight container seal, which prevents entry of air or microorganisms into packed foods.

Herring: It is one of the large families (Clupeidae) of soft-finned bony fishes (herrings, shads, sardines, and menhadens) that have a laterally compressed body and a forked tail and preserved by smoking or salting.

Hertz: It is the measure of frequency and defined as one cycle per second.

Hesperidin: It is a crystalline glycoside $C_{28}H_{34}O_{15}$ found in most citrus fruits and in orange peel.

Heterofermentative: Bacteria (*Leuconostoc, Pediococcus*) which not only produce lactic acid but also produce acids alcohol and ester are called heterofermentatice bacteria.

Heterotrophs: The organisms which are unable to prepare their own food from basic nutrient like carbon dioxide, water and solar energy are called heterotrophs.

Hexokinase: It catalyses the first irreversible step of glycolysis. It is a group of enzymes that accelerate the phosphorylation of hexoses (formation of glucose-6-phosphate from glucose and ATP) in carbohydrate metabolism.

Hexosan: It is a polysaccharide yielding only hexoses on hydrolysis.

Hidden characteristic of food: It is the quality characteristics of a food. These characteristic can not be evaluated by the consumers on the basis of his sense. The characteristics includes nutritive value of food, toxicity etc.

High acid foods: The foods having pH below 3.7 are known as high acid foods. E.g. cherries and sauerkraut.

High Density Polyethylene (HDPE): HDPE isles waxy in appearance and is harder and more resistant to flexing than LDPE. It is stronger, thicker, less flexible and more brittle.

High fructose corn syrup (HFCS): HFCS is a natural sweetener. It is used as a substitute for high cost sugar in preparation of the soft drinks. HFCS is produced by converting a portion of naturally occurring glucose in starch of corn to fructose. HFCS-55 means it contain 55% fructose.

High hydrostatic pressure (HPP) processing: It is the technique of the processing of the liquid food under high pressure (Plate 340 and 341). In general, pressure over 400 Mpa at 20-25°C is used to kill the vegetative microorganisms while a higher pressure of about 900 MPa combined with elevated temperature is needed to eliminated the spores. The HPP do not affect the quality of the food. It stabilizes the colour and maintain sensory characteristic along with microbial safety of food.

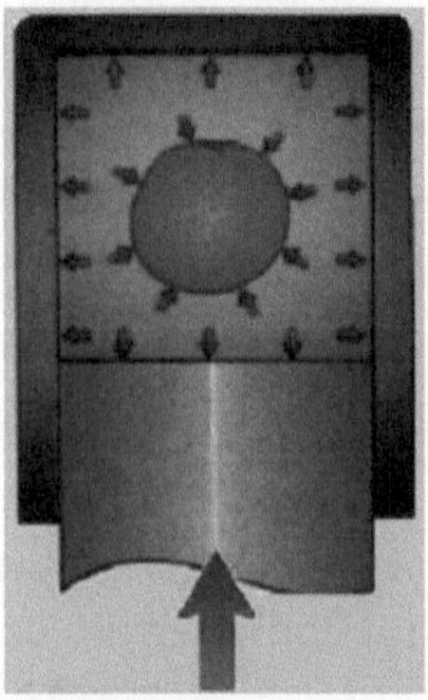

Plate 340: Mechanisms of HPP

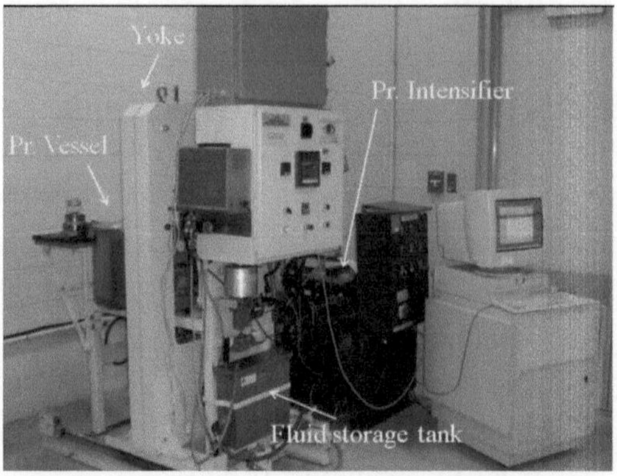

Plate 341: Batch type setup of HPP

High methoxyl pectin: Pectin molecule with a methoxyl content of more than 7% is called high methoxyl pectin (HMP). It form gel with more than 60% upto 80% soluble sugars and form gel in the pH range of 2.5 to 3.2. It is used in the preparation of jam, jellies or marmalade. Gel formation does not require divalent and polyvalent cations. The degree of methylation is 50-80%.

High nutrition food: Food having 20 % or more of the daily value for a particular nutrient in a serving or per 100 gm.

High Performance Liquid Chromatography (HPLC): Column chromatography was the first chromatographic technique. By using steel column and high pressure, the column chromatography could be fast, reproducible and flexible. It also called as high-pressure liquid chromatography (HPLC). The basic instrument consists of a solvent reservoir, pump, gradient chamber, injection port, column, detector, fraction collector and recorder (Plate 342). It permits both qualitative and quantitative examination of a wide range of nutrients. The capillary column filled with a high-performance ion-exchange packing make it possible to examine a wide variety of nutrients 1-10n mole range.

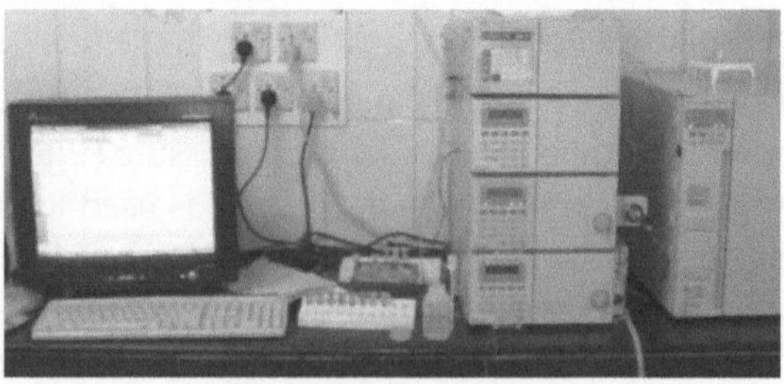

Plate 342: HPLC setup

High sugar based products: The products having sugar concentration of >50% sugar or low is called high sugar based products.

High sugar high acid foods: The dried or concentrated fruit/ vegetable products are called as high sugar high acid foods or high value low volume foods. The basic principle in the process of drying or dehydration is the removal of sufficient moisture to protect the product form spoilage.

High value low volume foods: The dried or concentrated fruit/ vegetable products are called as high sugar high acid foods or high value low volume foods. The basic principle in the process of drying or dehydration is the removal of sufficient moisture to protect the product form spoilage.

High value products: These are the agricultural products that have high value due to processing. High value products (HVP) are of three types like semi-processed products (fresh and frozen meat, flour, oils, refined sugars, roasted coffee), highly processed products (RTE /RTS like milk, cheese, wine, breakfast cereals) and high value unprocessed products (fresh fruits, eggs and nuts).

Histidine: A crystalline essential amino acid $C_6H_9N_3O_2$ formed by the hydrolysis of most proteins.

Hoist type tester: This type of tester consist of some form of release mechanism from which the package is suspended by means of a sling which allow it to drop in any position on any type of floor for any selected height (Plate 343). This drop test has ability to measure the protection provided to its contents and the ability of the container to with stand rough handling.

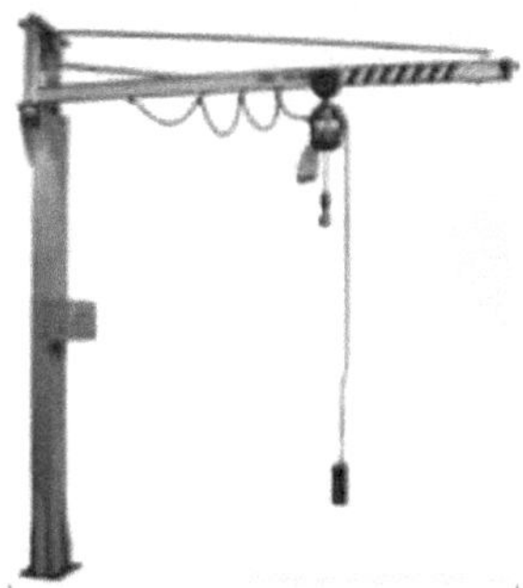

Plate 343: Hoist type tester

Holland gin: A gin made in the Netherlands is called Holland gin. It is also called Hollands. It is a gin made from alcohol by fermenting a mash of barley, malt, rye and corn.

Hollandaise: A rich sauce made of butter, egg yolks, and lemon juice or vinegar is called hollandaise.

Hollow fibre technology: This technology is used to generate Nitrogen gas. Other N_2 generators are 'Press Swing Adsorption' and Carbon molecular sieve (CMC) developed in China.

Holoenzyme: The complete active enzyme consisting of both protein part and non-protein part (co-factor) is known as Holoenzyme.

Homofermentative: The bacteria (*Lactobacillus plantarum*) which produce only lactic acid via fermentation are called homofermentative.

Homogenization: It is a process for making stable emulsion of juice, milk fat or milk serum by mechanical treatment. It is a process of reducing fat globules of milk to smaller particles of nearly the same size.

Homogenized milk: It is milk with its cream contents distributed evenly throughout so that it does not settle at the top. The milk is pumped through a valve at high pressure to break down the fat globules and then pasteurized.

Homogenizer: It is a device used for making stable emulsion of juice, milk fat or milk serum by mechanical treatment (Plate 344 and 345).

Plate 344: Homogenizer (Simple electrically operated)

Plate 345: Heavy duty homogenizer (semi-automatic)

Honey: A sweet viscid, glutinous or sticky material elaborated out of the nectar of flowers in the honey sac of various bees.

Honey comb shaped plastic bag: Honey comb shaped plastic bag is square shaped, self standing and collapsible. Therefore, these can be easily filled and emptied in a close system. An inner ring absorbs stress generated by the product during handling and transportation, so eliminate bulging. These bags are used for bulk distribution of liquid food which is placed in corrugated board containers. It is used for fruit juice concentrate, wine, vinegar, cooking oil and ice cream.

Hookean food: A food which obeys Hooks law is called Hookean food.

Hooks law: According to this law, stress applied on a food is directly proportional to strain (deformation).

Hop: Perennial hardy herb (*Himulus limulus*) with long twining stem of family Cannabinacieace cultivated for the yellow lupulin found as glands on the perianth of the female flower which is used in the manufacture of beer, essential oils and resins of the female inflorescence impart characteristic flavor of beer. The hops used in the brewery industry are the dried female flower clusters (cones) (Plate 346). Hops impart a mellow bitterness and delicate aroma to brewed beverages and aid in their preservation. The Japanese hop (*H. japonicus*) is a quick-growing annual species used as a screening vine.

Plate 346: Hops flower clusters

HOPCOMS: Horticultural producer's cooperative marketing and processing society Ltd, Lal Bagh, Bangalore. Earlier the name of HOPCOMS was Bangalore Grapes Growers marketing and processing cooperative society (1959)

Horse Power: It is the power of machine working at the rate of 550 ft lbs per second or 33000 ft lbs per minute.

Horticulture: It is art and science of cultivating and maintaining garden, growing fruits vegetables and ornamental crops along with their post harvest management till consumption by the consumer.

Horticultural maturity: It is the stage of development when a plant or plant part possesses the prerequisites for utilization by the consumer for a particular purpose. e.g. green fruit required for mango pickle making, ripe fruit required for table purpose.

Hot dog: It is a kind of frankfurter that is heated and served in a long split roll.

Hot pack concentrate: A juice product from which a substantial portion of the moisture has been removed by low vacuum concentration.

Hot pack or hot fill: It is filling of previously pasteurized or sterilized food, while still hot into clean but not necessarily sterile containers under clean but not necessarily aseptic conditions.

HPLC: HPLC stands for high-performance/pressure liquid chromatography. It is a separation technique with a solid stationary phase and a liquid mobile phase.

HPTLC: It stands for High performance Thin layer Chromatography. It is a sophisticated and automatic form of TLC.

HTST process: It stands for High temperature short time process. HTST process is the preservation of sterile products with sterile packages under sterile environment in a continuous and closed system without refrigeration with the retention of high nutritional and sensory quality. Here high temperature for short time (HTST) or ultra high temperature (UHT) is used to sterile the products. It is the continuous process in which milk is heated to 72°C for 15 seconds.

Hue: Red, blue, yellow. Green, orange and violet colour differentiation / identification is called hue.

Humectants: Compounds / substances / agents that promotes retention of moisture or absorb moisture from the air or a substance that promotes retention of moisture. They improve rehydration of dehydrated food and solubilization of flavouring agents. Propylene, glycols, glycerols, sorbitol and mannitol are water soluble and are used as humectants in foods. These function to control viscosity, texture, and bulking, retention of moisture, reduction of water activity, inhibition of crystallization etc. These substances are added to spray solutions to delay evaporation of the water carrier.

Humid heat: The energy required for raising temperature of 1 kg of dry air and it accompanying water vapour through 1°C is called humid heat.

Humid ratio: It is ratio of kg of water vapour to that of kg of dry air at a given temperature and pressure. It is also called absolute humidity.

Humid volume: It is volume of 1 kg of the dry air and its vapour at a given temperature and pressure.

Humidification: The addition of the moisture to the air, with out change in its dry bulb temperature is called humidification.

Hurdle technology: Hurdle technology is used for the preservation of the foods by applications of different combination of processing techniques at identified hurdle points (Plate 347). Hurdle technology is used because one method of preserving food is not sufficient to obtain the desired quality of the food. So, different methods like temperature, water activity, acidity, redox potential, preservatives, pH and competitive microorganisms are used to extend the shelf life and safety of the processed foods.

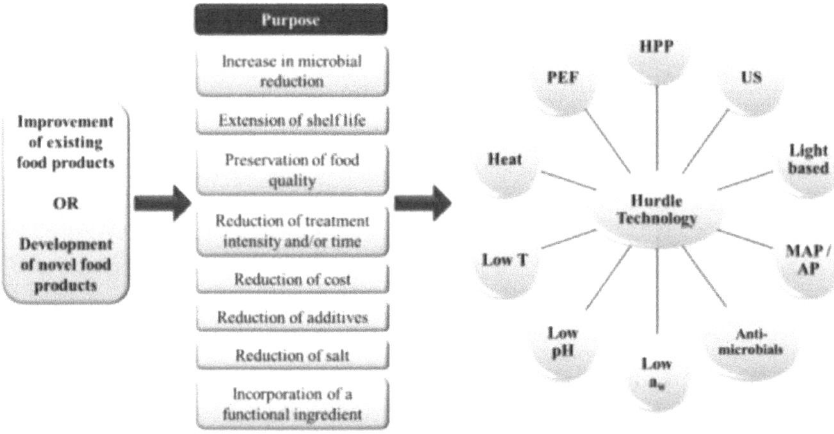

Plate 347: Hurdle Technologies

Hydraulic press: It is used for extraction of the juice. Hydraulic pressure (15psi) is used for extraction of the juice. It consists of strong steel structure, hydraulic pump, pressure gauge, basket and tray. The tray is covered by stainless steel (Plate 348 and 349).

Plate 348: Hydraulic Press

Plate 349: Extraction of juice using hydraulic press

Hydrocolloid: A substance that yields a gel with water is called hydrocolloid.

Hydrocyanic acid: An aqueous solution of hydrogen cyanide HCN that is a poisonous weak acid.

Hydro-distillation: It is one of the oldest methods of distillation. It involves heating of the plant material with water and the steam is generated with in still. It can be used for extraction of the essential oils.

Hydrodynamics: A branch of science involving principles of hydrodynamics. It is a branch of physics that deals with the motion of fluids and the forces acting on solid bodies immersed in fluids and in motion relative to them is called hydrodynamics.

Hydrogen swell: This type of bulging is due to the hydrogen gas produced by the action of food acids on the metal of the can (Plate 350).

Hydrogenation: Hydrogenation is conversion of liquid oil to solid oil by bubbling H-gas in presence of catalyst Nickel. Melting point of fat by hydrogenation increases. Hydrogenation used for soyabean oil, cotton and palm oil. It is addition of the hydrogen to an unsaturated fatty acids or fat to convert them into saturated fats. Adding hydrogen at the double bond solidifies soft fats or oils (Plate 351).

Plate 350: Hydrogen swell

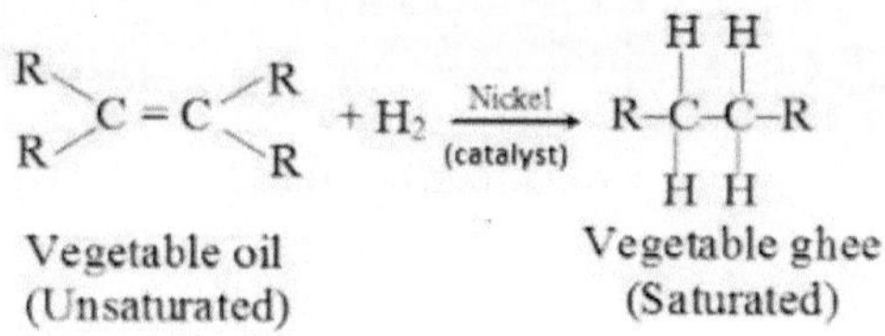

Plate 351: Hydrogenation

Hydrolysis: A chemical process of decomposition involving the splitting of a bond and the addition of the hydrogen cation and the hydroxide anion of water. It is the process of breakdown of a substance into simpler forms.

Hydrometer: It is an instrument used for measuring specific gravity of liquids or water. It is also calibrated to determine the degree Brix. It requires > 200ml the sample and not applicable to determine the TSS of carbonated drink due to buoyancy effect.

Hydrophilic: The substances which are attracted towards water are called hydrophilic (Plate 352).

Hydrophillic Balance: It is 20-point scale which is used to indicate the affinity of an emulsifying agent for oil verses water. It is also called Lipophillic balance.

Hydrophobic: The substances which are repulsive towards water are called hydrophobic (Plate 352).

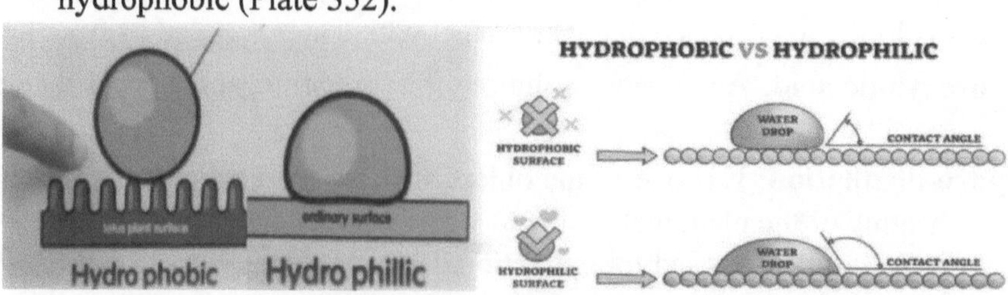

Plate 352: Hydrophilic vs hydrophobic illustration with contact angle

Hydrostatics: A branch of physics that deals with the characteristics of fluids at rest and with the pressure in a fluid or exerted by a fluid on an immersed body is called hydrostatics.

Hydrostatic sterilizer: These are continuous retort (agitating type) cans carried by conveyer through heating tower with steam by hydrostatic legs of water for heating / cooling (Plate 353).

Hydro-steam distillation: The material is kept on the screen or perforated grid inserted at some distance above the bottom of the still, lower part of the still contain water upto the level just below the grid. It is also used for extraction of essence.

Hydrovac cooler: It is a special vacuum cooler which have an in-built water spray.

Hygrometer: An instrument for measuring the RH of the atmosphere. The operation of hygrometer depends upon the expansion or contraction of certain hygroscopic substances with change in relative humidity of the surrounding atmosphere (Plate 354, 355 and 356).

Plate 353: Hydrostatic sterilizer

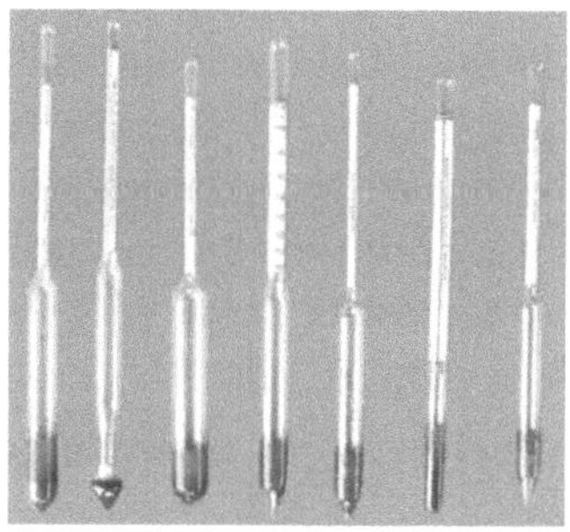

Plate 354: Different type of hydrometers

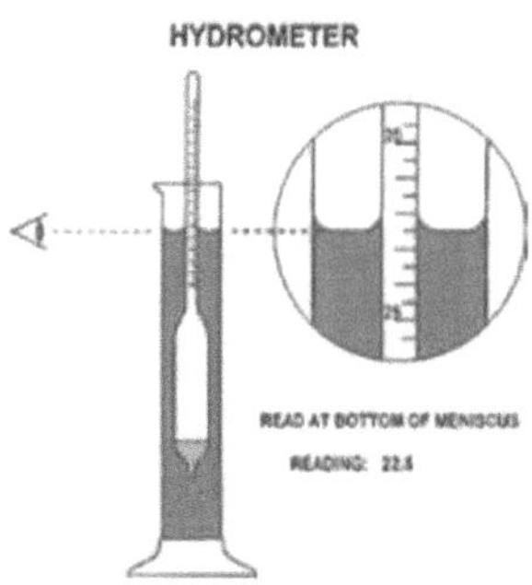

Plate 355: Use of hydrometer for measurement of specific gravity

Plate 356: Hygrometer for measuring RH

Hygroscopic: Any material which readily takes up and retains moisture under certain conditions of humidity and temperature is called hygroscopic.

Hygroscopicity: It is the ability of a substance to attract and hold water. Sugars possess varying degree of hygroscopicity.

Hyperbaric storage: Preservation of foods under high pressure (more than atmosphere, 14.7 lb/inch^2) is known as *hyperbaric storage*. Generally liquid foods like fruit juices are stored by this way under pressure and so, due to the every possibility of tissue break down of fruit and vegetable tissue; they are not stored by this method. Pressure due to inclusion of inert gas (CO_2, N_2 etc.) helps to retain the natural flavor of juice and check micro-organisms deterioration as done for carbonated soft drinks. The level of O_2 and CO_2 is not controlled in hyperbaric storage.

Hypertonic solution: If the water potential of the surrounding solution is more (positive) than that of the cell sap.

Hypobaric storage: Preservation of food (usually fresh fruit and veg.) under low pressure is known as *hypobaric storage*. In this method the commodities are kept in vacuum tight and refrigerated container (having strength to protect the negative pressure as outside pressure is more). Low pressure inside the container reduces the O_2 tension and facilitative the removal of ethylene and volatile compounds produced by the products which results in retardation of ripening and senescence by decreasing respiration. The level of O_2 and CO_2 is not controlled in hypobaric storage.

Hypoglycemia: The abnormal decrease of sugar in the blood is called hypoglycemia.

Hypotonic solution: If the water potential of the surrounding solution is less (negative) than that of the cell sap.

Hypsometer: It is an instrument used for measuring the altitude (Plate 357 and 358).

Plate 357: Physical hypsometer

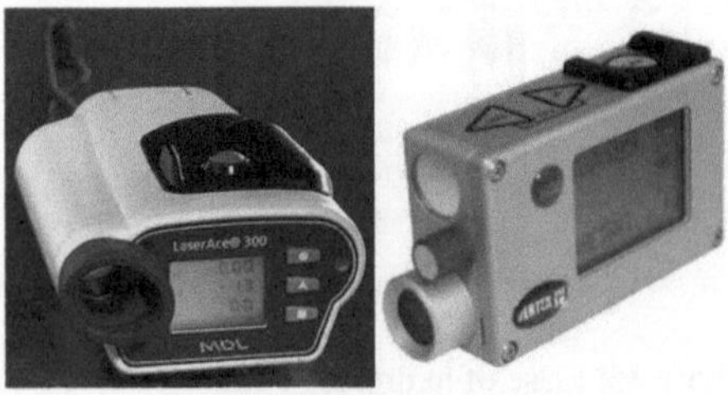

Plate 358: Digital hypsometer

I

Ice bank cooling: Storing of fruit and vegetable in storage chamber with positive ventilation and system countries where ice formation is allowed during night time (as electricity cost of night is less as compared to the day time). Melting of ice occurs during day time and with proper air circulation by means of ventilation process, approximately 0.5-1^0C temperature with 90% RH in the chamber is maintained. Ice Bank Cooler / Air forced cooler leads to desiccation of the produce (Plate 359). So use forced air high humidity cooler and these are known as Ice Bank Cooler. Ice bank coolers are useful for strawberry.

Plate 359: Ice bank cooling system

Ice cream: Ice cream may be define as frozen dairy product made by suitable blending and processing of cream and other milk products together with sugar and flavour.

Ice cream cone: A thin crisp edible cone for holding ice cream and filled with ice cream.

Ice point: It is the temperature of equilibrium between ice and water under normal atmospheric pressure (Plate 360).

Ichthylogy: It is the branch of the zoology that deals with the study of fishes (Pisces).

Icicle: It is small tapering piece of the ice formed by freezing of dropping water (Plate 361).

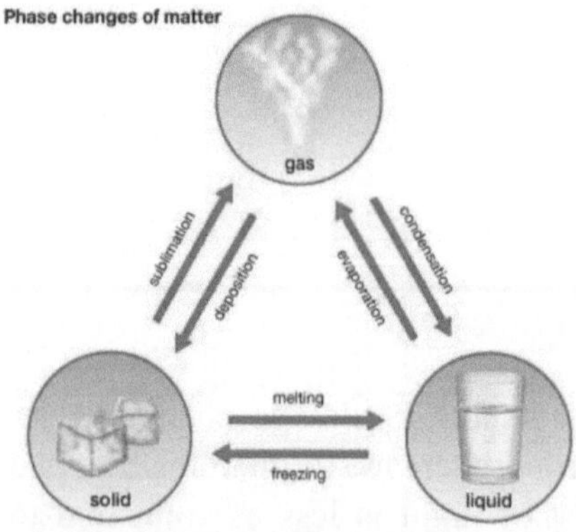

Plate 360: Phase change illustration

Plate 361: Icicle

Ideal solution: Ideal solution contains 1 mole of solute in one litre. Ideal solution obeys Raoult's law.

IFT: It stands for Institute of Food Technology.

Immobilized enzymes: Immobilized enzymes are defined as "enzymes physically confined or localized in a certain defined region of space with retention of their activity and which can be used repeatedly and continuously."

Incineration: The process of converting a substance by burning to ash is called incineration.

Incinerator: An apparatus which is used for incineration especially a furnace. It is also called as muffle furnace. It is used for estimation of the ash contents of the food samples. The sample is ashed at a temperature of about 500-550°C.

Inclined impact test: This type of test helps to study extent of crushing, breaking, cracking and distortion during handling, storage and transport. These are mainly due to shunting shock. Used to determination the ability of container to withstand impact stress particularly testing large odd shaped or heavily loaded containers that is difficult to test by the other method. It is also called contour test (Plate 362).

Plate 362: Inclined impact test

Index of stack burn: Furfural, a sugar dehydration product, formed as an intermediate in colour formation is used as an index of stack burn.

Index to volume: It is indirect method of comparing the volumes by measuring the circumference of a cross section of the product.

Indicators of blanching: Peroxidase and catalase are the two more heat resistant enzymes and is used blanching indicator. The catalase gets destroyed at much lower temperature than peroxidase so the activity of peroxidase is normally used to evaluate the effectives of blanching treatments. If this enzyme is inactivated it can be assumed that the significant enzymes are also inactivated. In beets, polyphenol oxidase is found indicator enzymes.

Indicators of pasteurization: Pasteurization is a heat treatment that kills part but not all the vegetable micro-organisms present in the food. Phosphatase activity has been widely accepted as an indicator for testing the efficiency of pasteurization of milk. Destruction of this enzyme in milk ensures safety. Pectin methyl esterase (PME) is used as an index of pasteurization efficiency for citrus juice.

Indicators of sterilization: *B. stearothermophilus* is indicator of processing /sterilization in low acid food as it is more heat resistance than *Cl. botulinum*. In tomato and tomato products, destruction of *B. coagulans* is the basis of thermal processing. Thiamin is considered as chemical index of sterilization in thermal processing. Plastic rods called biological indicator units (BIUs) containing bacterial spores. These are introduced into the container along the central vertical axis to the cold point. In the test some of the containers are filled with BIUs and some are filled with thermocouples.

Indirect heating: It can be achieved by using suitable heat exchangers (Plate, tubular, or scraped surface heat exchanger or hot water). In this type of heating, the air is in contact with a hot surface such as pipes or fins heated by steam, flame or electricity.

Individual Quick blanching (IQB): It is a two stage blanching process. In the first stage the food is heated in a single layer at a sufficient high temperature to inactivate the enzymes. In the second stage also termed as adiabatic holding, a deep bed of food is held for sufficient time to allow the temperature at the centre of each piece to increase the enzyme inactivation at the centre. There is reduction in heating time in IQB process as compared to conventional process.

Inductive Heating: It is defined as a process where electric coils and the magnets are placed near the food product which generates oscillating electromagnetic field that sends electric current through food with the primary purpose of heating them (Plate 363).

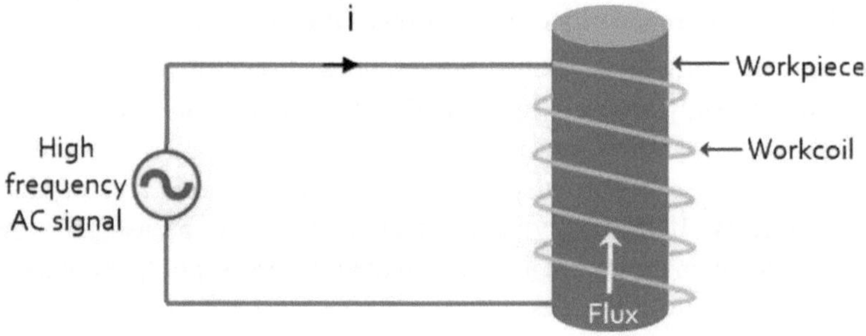

Plate 363: Setup of Inductive heating

Industrial effluent: Something that flows out waste material (as smoke, liquid industrial refuse, or sewage) into a main stream or lake and discharged into the environment serving as a pollutant.

Industry standards for quality: These standards are established by an organizational group to maintain the quality of the given commodity. These standards become effective by pressure where other legal standards are not involved.

Infrared: The radiation having a wavelength between about 760 nanometers and 1 millimeter and the radiations are situated outside the visible spectrum at red end.

Ingredients: Ingredient means any substance including a food additive used in the manufacture or preparation of a food and present in the final product although possibly in a modified form.

In-house testing: Evaluation conducted within a food company prior to field testing and test marketing.

Ink blot: In food research, it is an impression made on paper after first pressing a cross section of the sample on to an ink pad.

In-package desiccants: Desiccating chemicals used in the packages to provide partial dehydration in the storage environment; which prolong the storage period of the dehydrated vegetables. The CaO and silica gel are used as In-package desiccants

In-plant sanitation: In – plant sanitation involves quality control and storage of raw products, the proportion of a good water supply, prevention of the contamination of the foods at all stages. The waste material has to be treated to meet the standard limits laid down by the regulatory agencies.

Insignificant risk / Zero risk in food safety: Safe food does not mean zero risk, because in reality absolute safety is impossible to achieve. Therefore, these is need to shift our thinking away from the concept of "Zero risk" to a more realistic concept "Insignificant risk", So, insignificant risk is achievable while zero risk is not.

Inspection: Inspection is usually a mandatory process done by government or other agencies to ensure a product's wholesomeness, safety, and adherence to regulations. Inspection is the examination of food or systems for control of food, raw materials, processing and distribution including in-process and finished product testing, in order to verify that they conform to requirements. Inspection of the finished product is done to know or determine to extent the specification that has been followed to produce a desirable quality product. The inspection can be done for physical, chemical, microbiological and sensory attributes.

Installed capacity: The capacity of plant and machinery to produce maximum quantity of product in a day/month/year.

Instant foods: These are the foods which are prepared and packed in the powdered form and require only the addition of liquid (water or milk) for quick final preparation. They are easy to cook. E.g. Instant Halwa powder, noodles etc.

Instrumentation: The use or application of instruments for observation, measurement, and quality control is called instrumentation.

Insulator: A material that is a poor conductor (as of electricity or heat) or a device made of an electrical insulating material and used for separating or supporting conductors is called insulator.

Intentional (direct) additives: These substances are deliberately added to perform specific functions and for some technological reasons like artificial sweetener which is added to low calorie soft drinks.

Intermediary moisture Food (IMF): These are semi moist foods (SMF). IMFs contain moderate level of moisture of the order of 20-50% by weight corresponding to water activity of 0.7 to 0.9 which is less than is normally present in natural fruit and vegetables and more moisture than is left in dehydrated products. Principle of IMF is lowering of a_w by adding solute e.g. glycerol, sucrose, glucose or salt and retarding microbial growth by adding anti microbial agents e.g. Propylene glycol and sorbic acid. Examples of IMF are jam, jelly, dried figs, and dates, some type of saugage, prunes and dried apricot. These foods have non refrigerated shelf stability.

Internal combustion engine: A heat engine in which the combustion that generates the heat takes place inside the engine instead of in a furnace.

Internal respiration: It is exchange of gases between the cells of the body and the blood by way of the fluid bathing the cells.

International System of Units: A system of units based on the metric system and developed and refined by international convention for scientific work.

International unit: A quantity of a biologically active substance (vitamin) that produces a particular biological effect agreed upon as an international standard. This unit is used to measure the activity of vitamin A and D. 1 IU of vitamin A = 0.3 mcg retinol = 0.6 mcg beta-carotene; 1 IU of vitamin D = 0.25 mcg of vitamin D_3.

Intrinsic factor: A substance (mucoprotein) produced (with in body) by normal gastrointestinal mucosa that facilitates absorption of vitamin B_{12}. The physical properties of food such as flavour, texture, appearance, shelf life and nutritive value are also the intrinsic factors. These properties are directly measurable and objective. The lack of this factor leads to pernicious anaemia.

Inulin: It is a white plant polysaccharide used to improve the flavor and texture of low fat and low sugar processed foods. It is storage polysaccharide ($C_6 H_{10} O_5$) found in the roots of Jerusalem artichoke, dahlia and few other plants.

Inversion: The conversion of dextrorotatory sucrose into a levorotatory mixture of glucose and fructose. Sucrose + acid → glucose + fructose.

Invert emulsion: It is an emulsion in which water is dispersed in oil instead of oil in water.

Invert sugar: A mixture of dextrose and laevulose found in fruits or produced artificially by the inversion of sucrose.

Invertase: An enzyme that catalyzes the hydrolysis of sucrose is called invertase.

Invisible technology: Some time food products get spoiled if exposed to abuse conditions. So, the incorporation of the additional hurdles as safeguard is required for chilled foods. This is called invisible technology.

In-vitro meat: The meat grown in laboratory from the stem cells of cattle, paving the way for eating meat without animal being slaughtered. So in this way meat will be slaughter free. It can be used for preparation of hamburger etc. Netherland Maastricht University scientist claim first test tube burger.

Iodine number: It is the measure of the degree of unsaturation of the fat due to the double bonds present in the fatty acid chain. It is measured by estimating the amount of iodine absorbed in gram per 100 gram of the fat. It is also called as Iodine Value.

Iodine Value: It is the measure of the degree of instauration of the fat due to the double bonds present in the fatty acid chain. It is measured by estimating the amount of iodine absorbed in gram per 100 gram of the fat. It is also called as Iodine number.

Iodized salt: It is the table salt to which potassium or sodium iodide and a small amount of magnesium carbonate is added. It prevents from the problem of goiter.

Iodophors: It is a complex of iodine and a surface-active agent that releases iodine gradually and serves as a disinfectant. It is used for teat dipping for prevention of the spread of mastitis in dairy cow.

Ion exchange chromatography: The ion exchange chromatography apparatus contains a water jacket and a means to prevent the column from going dry. In modern ion exchange chromatography components with different charges are separated when a gradient of increasing ionic strength is used as the eluant. Sample components with negative charges will be absorbed and during desorption the negatively charges sample components are exchanged by negative ions from the salt gradient. Each sample component is then desorbed at a specific ionic strength and continuously eluted from the column. At low concentration and ordinary temperature; the extent of exchange increases with increasing valency of exchanging ions. At high concentration the differences in the exchange potential of ions of different valency diminishes and in some cases reverses. Ion-exchange chromatography is suitable for fractionating mixtures of the lowest members of homologous ionic compounds e.g. sugar phosphates from sugars.

IQF technology: It stands for Individual quick freezing (slow freezing).

Iridescent films: These films are comparatively new and unique. These are decorative plastic films exhibiting brilliant display of colour without use of pigment or dyes (Plate 364).

Plate 364: Iridescent films

Irish moss: The dried and bleached plants of a red alga (*Chondrus crispus*) used as an agent for thickening or emulsifying or demulcent (soothing effect) and as a source of carrageenan. The red alga which is a source of Irish moss is also called carrageen.

Irish whiskey: It is prepared from a mash containing malted and unmalted grains such as rye, barley, wheat and oat. The mash is distilled in pot stills.

IRS: It stands for infrared reflectance spectroscopy.

Irradiated milk: Milk in which the vitamin D content is increased by irradiation by UV rays is called irradiated milk.

Irradiation: The exposure of biological materials (food commodities) to radiation, more particularly to mutagenic radiations is known as Irradiation.

Isabgol: It is an annual herbaceous plant *Plantago ovate* and seeds of it are used in the treatment of dysentery, diarrhea and chronic constipation.

ISO: It stands for international organization for standardization and has its head quarter at Geneva. It provides guidelines for establishment and management of quality system.

Isoelectric pH: It is a pH having or representing zero difference of electric potential or it is the pH at which the electrolyte will not migrate in an electric field. The proteins have isoelectric pH. It is pH of a solution where the electric charge of an amino acid is neutral.

Isoenzyme: Any of two or more chemically distinct but functionally similar enzymes are called isoenzyme.

Isoflavone: A bioactive ketone $C_{15}H_{10}O_2$ having numerous derivatives and found in plants (soybean). These compounds / derivatives (daidzein or genistein) have antioxidant and estrogenic activity.

Isoform: Two or more functionally similar proteins that have a similar but not an identical amino acid sequence.

Isolag Light bag: The Isolag light bag is a re-useable bag made from metalized polyester (which reflects 95% of radiant heat) laminated with an extremely thin co-extruded HDPE film. The bag is packaged provides insulation effect. The bag is recyclable. This bag is used to prevent deep frozen foods from thawing out and hot foods from requiring reheating after the journey from the point of purchase.

Isoleucine: A crystalline essential amino acid $C_6H_{13}NO_2$ isomeric with leucine

Isomerization: It is a process to become changed into an isomeric form.

Isosmotic: A solution exhibiting equal osmotic pressure.

Isotherm: A line on a map or chart of the earth's surface connecting points having the same temperature at a given time or a line on a chart representing changes of volume or pressure under conditions of constant temperature is called isotherm.

Isotonic: These solutions exhibit equal osmotic pressure and no movement of solution take place. This condition is called isotonicity.

J

Jaggery: Raw brown / unrefined brown sugar in lump made from palm sap (Plate 365).

Jam: A preserved product prepared on boiling pectin contained in fruit pulp with required amount of sugar, water and acid to a reasonably thick consistency, firm enough to hold the fruit tissues in position (Plate 366).

Jam pack: The packing of the commodity tightly or in excess is called Jam pack.

Jamaica rum: Heavy bodied rum made by slow fermentation and marked by a pungent bouquet.

Jamboline: A type of glucose found in the jamun fruit help in cure of the diabetes in the patients.

Jelly: A product prepared by boiling clear extract of fruits with sugar. It should be transparent, well set, and tender enough to quiver but not flower and when cut; it should retain its shape and show a clean cut surface (Plate 367).

Plate 365: Jaggery

Plate 366: Jam

Plate 367: Jelly

Jelly grade: It is defined as grams of the sugar required for forming a gel of 65°Brix with one gram of the pectin under standard test conditions. E.g. If 1 gram of the pectin forming a gel with 150 gram of sugar under standard test conditions, then it is called 150 grade pectin. Micrometer and ridgelimeter are used for estimation of the jelly grade.

Jelly seed: Jelly seed is a physiological disorder of mango during storage and marketing. Tommy atkin is susceptible for jelly seed.

Jelmeter: An apparatus used for pectin test during the preparation of jam and jelly. It directly indicates the amount of sugar to be added per liter of extract.

Jelmeter test: It is used to identify the viscosity of the jam and jelly.

Jeotkal: The salted and fermented sea foods are known as jeotkal (Plate 368).

Plate 368: Jeotkal

Jetting: It is a process of removal of the air from the head space from the bottle under pressure.

Jhatka meat: It is the serving the head of the animal with one big blow of an axe or sword etc; the animal dies very soon and the carcase bleeds less.

Joule: It is a unit of work or energy equal to the work done by a force of one Newton (N) acting through a distance of one meter. 1 Kilocalorie = 4.184 kilojoules (KJ).

Juice: The extractable fluid content from cells or tissues of fruits is called juice. It may be sweetened or natural / unsweetened juice.

Juice extractor: A machine used for extracting juice form fruit by pressure or sieving through fine screens. The unique spiral design (helicoidal) of the machine ensure high yield (Plate 369).

Juice line: It is a part of the fruit processing plant comprising of juice extractor, blender, homogeniser, pasteuriser, juice collection tank, bottle filling machine and sterilizer. The line may be automatic or semi-automatic (Plate 370).

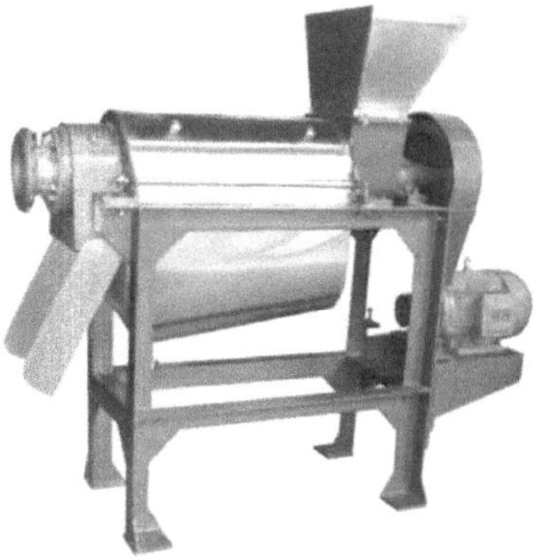

Plate 369: Helicoidal juice extractor

Plate 370: Juice line with juice extractor, blender, homogeniser, pasteuriser, juice collection tank, bottle filling machine

Juice oil: The oil recovery from single strength juice in deoiling operation is called juice oil. Single strength juice contains 0.005% of oil.

Juice sacs: Juice sacs refer to the coarse membrane material that is screened from the juice during juice finishing operations.

K

K ration: It is a lightweight packaged ration of emergency foods developed for the U.S. armed forces in World War II (Plate 371).

Kaiser rolls: A round crusty roll often used for sandwiches (Plate 372).

Kalamata /calamata: It is brine cured black olive grown in Greece.

Kamucha: Tea cider or fermented tea beverages is known as kamucha. The fermentation process take place by the presence of yeast and bacteria

Kanji: Clean, black coloured, crisp and sound product of characteristic flavor prepared by full fermentation (chiefly lactic acid fermentation) of black carrot with 2.5% salt, 1% powder mustard and some amount of water. It is fermented to make ready to serve drink which is known as Kanji. Each kilogram of the grated carrots yields just 7-8 litres of kanji.

Karl Fischer Titration: Karl Fischer Titration is used for estimation of moisture content of foods (low moisture products).

Kasseri cheese: The simple unpasteurized sheep's milk cheese (sometimes mixed with goat's milk) is called kasseri. It is popular in Greece. It is made in the shape of wheel, fairly firm and pale yellow (Plate 373). The word kasseri comes from the Turkish *kasar* which means that it has ability to hold its shape when heated.

Plate 371: K ration packing

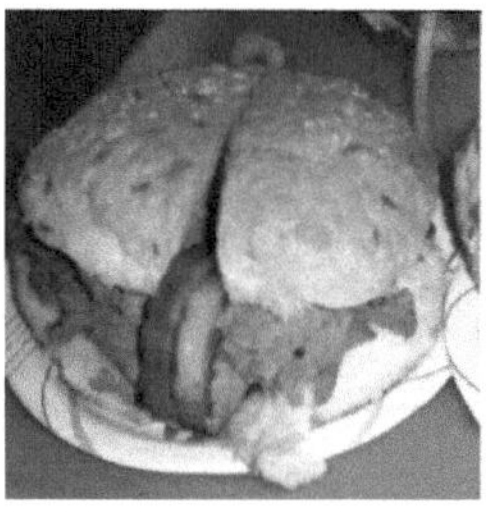

Plate 372: Kaiser roll

Plate 373: Kasseri cheese

Katcha Arhatia: Katcha arhatia advances money to the cultivators and village banias on the condition that the produce will be disposed off through him alone and hence charges a very nominal rate of interest on the money advanced. Katcha arhatia charges commission for services rendered by him. He plays an important link between the village cultivator or traders on the one hand.

Kebab / kebob / kabob: Cubes of meat (as lamb or beef) marinated and cooked with vegetables on a skewer (Plate 374).

Plate 374: Kabob

Ketchup: A concentrated juice or pulp without seed and skin in which spices, salt, sugar, vinegar etc are added to extent that it contains $<$ 12% fruit and vegetable solids and 25% total solids.

Ketoglutaric acid: It is crystalline keto derivative of glutaric acid ($C_5H_6O_5$) and is alpha keto isomer formed in metabolic processes of Krebs cycle.

Ketone bodies: These are the substances which are synthesized by the liver in the process of combustion of fats. There are three compounds like acetoacetic acid, acetone, and beta-hydroxybutyric acid which are normal intermediates in lipid metabolism and accumulate in the blood and urine. The abnormal amounts of these ketone bodies results in impaired metabolism (diabetes mellitus).

Ketose: A sugar (fructose) containing in its acyclic form one ketone group per molecule.

Ketosis: The accumulation of large quantities of ketone bodies in the body tissues or an abnormal increase of ketone bodies in the body fluids is called ketosis. This results from the incomplete oxidation of fatty acids by the liver.

Khalpi: It is a cucumber pickle prepared by lactic acid fermentation and is popular in Nepal during summer.

Khoa: It is a heat desiccated milk product.

Killing out percentage: The weight of the dressed carcase as a percentage of the live weight of the animal is called killing out percentage. It is also called dead weight. It gives the indication of the amount of meat on a carcase after the removal of the head (except pigs), limbs, hide, blood and offal. For beef cattle the percentage of dead weight is 50-55%.

Kilocalorie: It is the amount of heat required to raise the temperature of 1 kg of water from 15 to 16 °C. It is a unit of heat used in nutrition.

Kilorad (Krad): It is equal to 1000 rad.

Kimchi: A vegetable pickle seasoned with garlic, red pepper, and ginger that is the national dish of Korea is known as kimchi (Plate 375). It is also known as kimchee. Chinese cabbage or radish is generally used as a source of vegetable for kimchi. It is similar to sauerkraut.

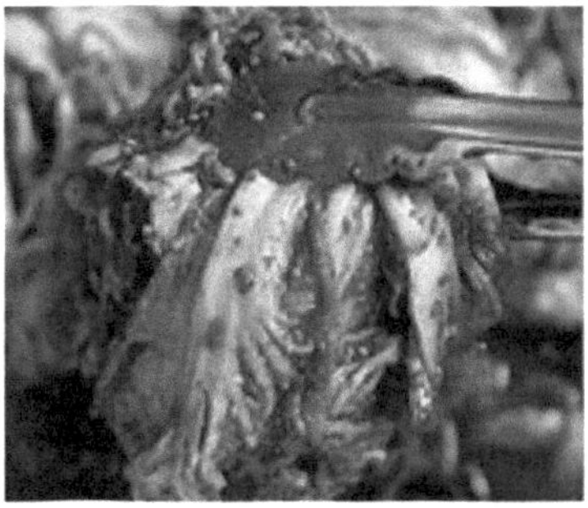

Plate 375: Kimchi

Kinetics: A branch of science that deals with the effects of forces upon the motions of material bodies or with changes in a physical or chemical system.

Kinetic energy: The energy associated with motion is called kinetic energy.

Kinetic theory: There are two theories in physics based on the vigorous motion of the minute particles of a substance. First theory is the theory at which temperature of a substance increases with an increase in either the average kinetic energy of the particles or the average potential energy of separation (as in fusion) of the particles or in both when heat is added it is called kinetic theory of heat. Second theory is the theory at which the particles of a gas move in straight lines with high average velocity, continually encounter one another and thus change their individual velocities and directions, and cause pressure by their impact against the walls of a container and is called kinetic theory of gases.

Kinetin: A cytokinin $C_{10}H_9N_5O$ used to stimulate cell division in plant tissue culture.

Knacker: An animal slaughterer is called Knacker.

Kneading: It is a process of making dough / paste by pummeling for preparation of bread. Kneading is also done during extrusion cooking process in the kneading section with the help of kneader (Plate 376).

Kocho: False banana is fermented in a pit to produce a pulp known as kocho. Food preserved in pits can last for years without deterioration.

Koji: This is a product prepared during manufacturing of sake/ rice beer. Koji is prepared when steamed, cooled rice are inoculated with spores of *Aspergillus oryzae* at 65-70°F until entire mould mycelium penetrate.

Kola nut / cola nut: The bitter caffeine containing chestnut sized seed of a kola tree used as a masticatory and in beverages (Plate 377).

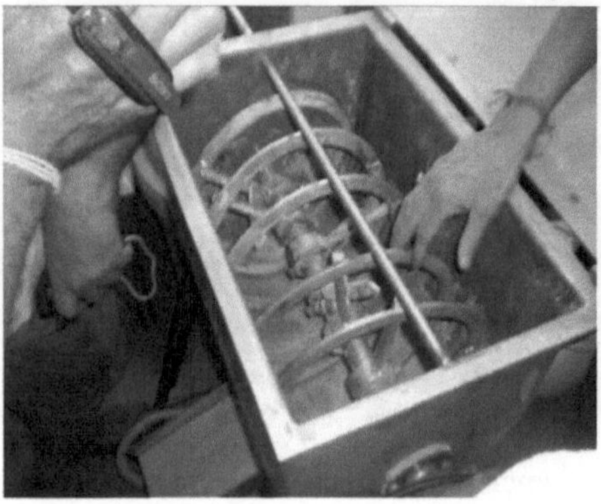

Plate 376: Kneading unit

Plate 377: Kola nut

Kola tree /cola tree: An African tree (genus *Cola,* esp. *C. nitida* and *C. acuminata* of the family Sterculiaceae) cultivated in various tropical areas for its kola nuts.

KPag: It stands for kilopascal gauge

K-plate: It is electrolytic tin plate (ETP) and it is more corrosion resistant tinplates (Plate 378).

Kraft: It is Germen word for strength and used for paper strength (Plate 379).

Kramer shear press: It is widely used to measure the Tensile strength of chapatti for quality attributes (Plate 380).

Krausing stage: It is a stage after dead period in case of wort. At this stage yeast form a ring around the top of the tub.

Krebs cycle: A sequence of reactions in the living organism in which oxidation of acetic acid or acetyl equivalent provides energy for storage in phosphate bonds (ATP). It is also called citric acid cycle or tricarboxylic acid cycle (Plate 381).

Kurumba: An immature coconut containing a refreshing clear liquid (Plate 382).

Kwashiorkor: It is severe malnutrition in infants and children especially of impoverished regions caused by a diet low in protein (Plate 383).

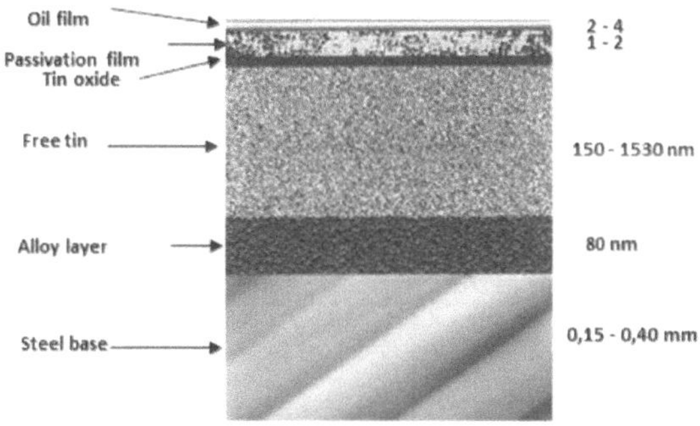

Plate 378: Electrolytic tin plate

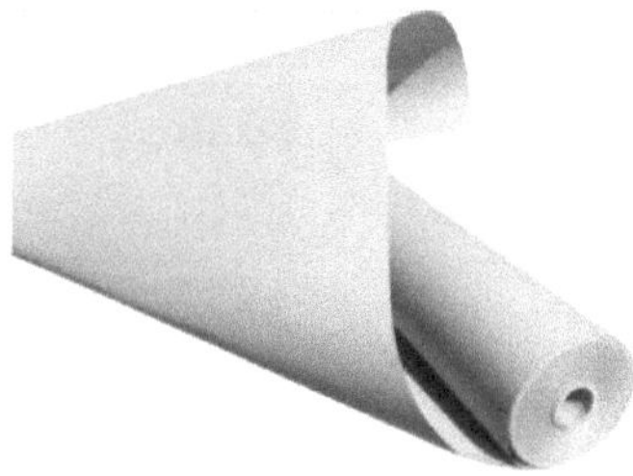

Plate 379: Kraft paper

Plate 380: Kramer shear press

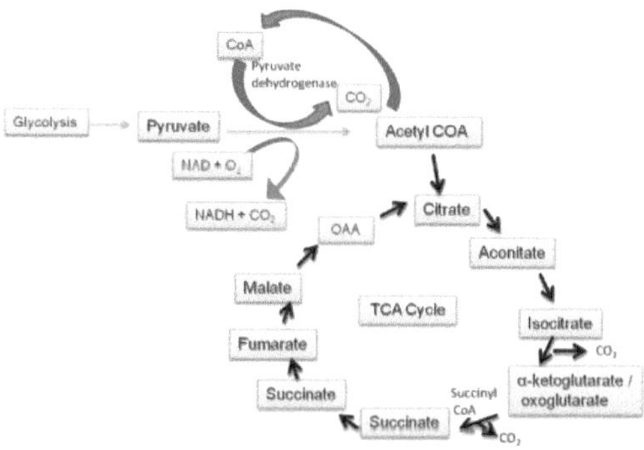

Plate 381: Krebs cycle

Plate 382: Immature coconut used for making Kurumba

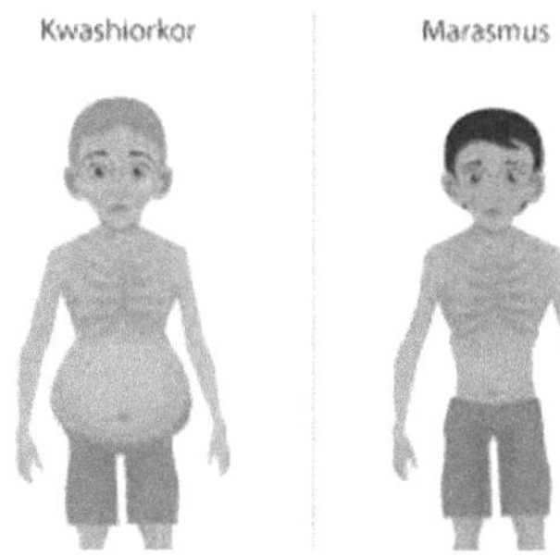

Plate 383: Kwashiorkor vs Marasmus

L

Lab coat: A loose white coat with deep pockets that is worn in a laboratory or medical office.

Labelling: It means any written, printed or graphic matter that is present on the label, accompanies the food or is displayed near the food, including that for the purpose of promoting its sale or disposal. After packaging, labels are required on the finished products intended for distribution and sale. Labelling can reflect on the quality of the product. It must be clear and informative. Information includes Name of the Manufacturer, Date of Manufacture, Net Weight, Gross Weight, Best before etc. Proper storage is needed for maintenance of the quality. Storage must be done only at recommended temperature, relative humidity, light and air / oxygen and carbon dioxide conditions etc. to maintain the quality of the food product. Label may include tag, brand, and pictorial or descriptive matter, written, printed, embossed, graphic, perforated, stencilled or stamped impressions. It is attached to container, cover, lid, or crown of a food package (Plate 384).

Plate 384: Mandatory informations required on label

Laboratory: A place equipped for experimental study in a science or for testing and analysis. Broadly, it is a place providing opportunity for experimentation, observation, or practice in a field of study.

Lacquer: It is a colloidal dispersion of a solution of cellulose derivative, resin and plasticizer in solvent and diluents, where in all these constituents dried in air by evaporation of solvents, thus yielding a transparent, hard and water proof film.

Lacquering: It is a process of providing a resinous material inside can to prevent discolouration of food due to formation of black stains of the iron sulphide. Lacquering is of two types. Acid resistant or R enamel cans are used for acid food while sulphur resistant or C enamel cans are used for non-acid foods.

Lacquering of can: It is used for providing a lining of a resinous material inside the can to prevent discoloration of the food due to formation of black stains of iron sulphide or corrosion on the tin plate of can and outside the can to prevent rusting. Lacquering are of two types (a) acid resistant used for acid foods and this lacquered cans are called R-enamel cans (b) sulphur resistant sued for non acid foods and this lacquered can are called C-enamel or S.R. cans.

Lactic acid: A hygroscopic organic acid $C_3H_6O_3$ present normally in muscle tissue as a product of anaerobic glycolysis, produced in carbohydrate matter by bacterial fermentation, and used in food and medicine industry.

Lactic acid bacteria: It is a group of gram positive anaerobic bacteria, and is used in food industry especially as starter cultures in cheese making.

Lacto vegetarian: A vegetarian whose diet includes dairy products, vegetables, fruits, grains, and nuts.

Lactobacillus: A genus (Lactobacillus) of bacteria that produce lactic acid.

Lactoferrina: Lactoferrina is an iron binding 80 KD glycoprotein and is one of the most important anti microbial substance present in milk.

Lactoflavin: It is the oldest and obsolute name of the riboflavin.

Lactoglobulin: A crystalline protein fraction that is obtained from the whey of milk.

Lactometer: It is a hydrometer for testing the relative density (specific gravity) of milk.

Lacto-ovo vegetarian: A vegetarian whose diet includes dairy products, eggs, vegetables, fruits, grains, and nuts is called ovo-lacto vegetarian.

Lactoprotein: It is albuminous constituent of the milk.

Lactory: It is milking parlour.

Lactoscope: It is the instrument used for measuring purity or richness of the milk (Plate 385).

Lactose: A disaccharide sugar $C_{12}H_{22}O_{11}$ that is present in milk and yields glucose and galactose upon hydrolysis and yields especially lactic acid upon fermentation. It is 12-carbon milk sugar (Plate 386).

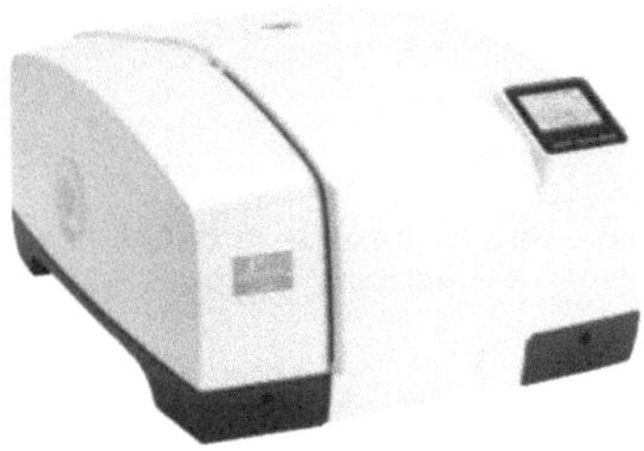

Plate 385: Lactoscope

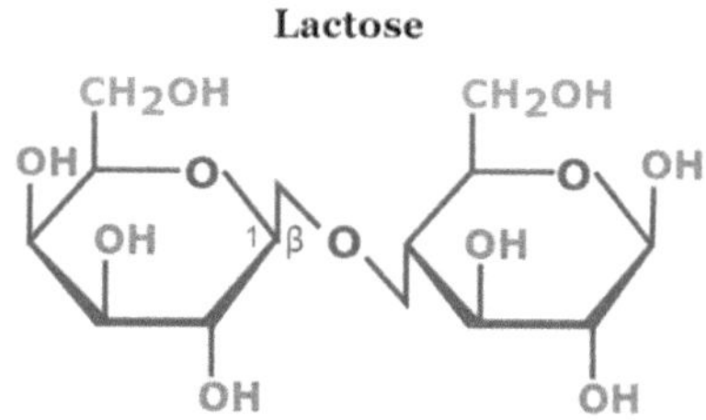

Plate 386: Lactose

Lager: A beer brewed by slow fermentation and matured under refrigeration.

Lager beers: Lager beers are bottom fermented and stored at a low temperature for several months; mostly light in colour, with high carbonation, medium hop flavor, and alcohol content of 3–5% by volume. Lager yeasts typically settle at the bottom of the fermentor kept at 8-14^{0}C while ale fermentation processed at 16-12^{0}C and yeast rises to the top.

Lagering: Storage of young or green beer at 0°C in tanks for several weeks to several months is known as lagering.

Lairage: It is a place where cattles are housed or laired especially near a market place (for slaughtering), docks (for export) or a fair (for sale).

Lamb: Flesh of young ovine animals of both sexes whose aging is 12 month or under 12 months.

Laminar air flow: Laminar air flow is referred to as a system that provides continuous air flow that is uniform in both direction and velocity. Laminar air flow cabinets facilitate unidirectional air flow in a confined space. It is used for microbial study whichuses UV light along with laminar air flow (Plate 387).

Laminar flow: The uninterrupted flow in a fluid near a solid boundary in which the direction of flow at every point remains constant (Plate 388).

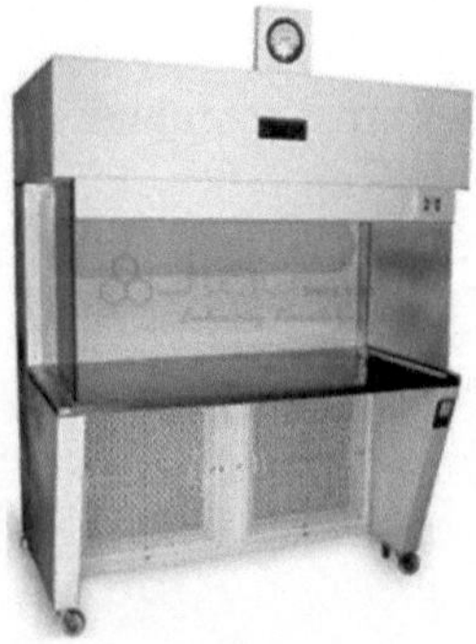

Plate 387: UV light Laminar air flow

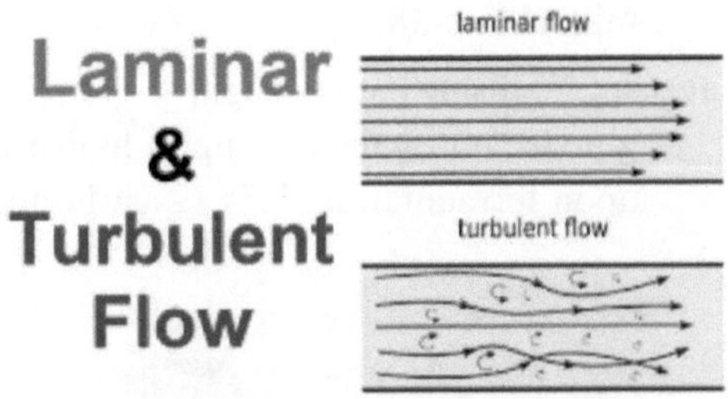

Plate 388: Laminar vs turbulent flow

Laminate: A laminate is defined as a combination of distinctly different flexible materials such as paper, cellophane, plastics and foil where in each film is generally thicker than 6 microns. Laminate films are bonded by hot melt lamination, dry bonding, wet bonding and thermal extrusion. The available combination of the laminate includes polyester + PE +Aluminium foil + PE +LLDP.

Large calorie: It is the amount of heat required to raise the temperature of one kilogram of water by one degree Celsius (1°C) that is equal to about 1000 gram calories or 3.968 BTU. It is also known as Cal / large calorie. It is an amount of food having an energy-producing value of one large calorie.

Latent heat: Energy required to change the phase of the matter per unit mass is called latent heat.

Latent heat of crystallization: It is the heat released during transition from the liquid state (higher energy) to the frozen state (lower energy).

Lathyrism: It is a nervous disease / spinal cord disease that cripples man. It is caused by the consumption of Lathyrus stivus (Kesari Dhal). Variety Pusa 24 and LSD are having higher yield along with low toxin content. Lathyrism's Stages: First stage is Non stick stage, second stage is one stick stage, 3rd stage is 2 stick stages and at final stage patient is unable to walk (Plate 389). Neurotoxin responsible for lathyrism is β-N-Oxalyl-L-α, β, diaminopropionic acid.

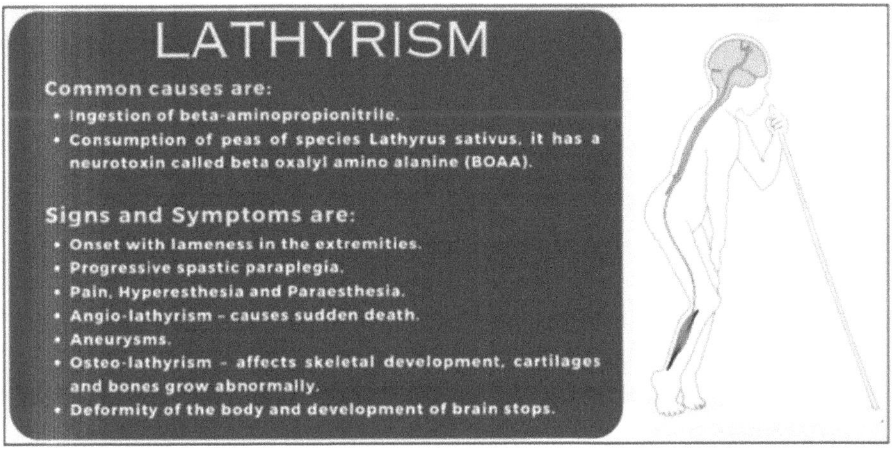

Plate 389: Lathyrism

Lauren's half shade polarimeter: It is also called saccharimeter. It is used for estimation of sugars.

Lauric acid: A crystalline fatty acid $C_{12}H_{24}O_2$ found in coconut oil and used chiefly in making soaps and esters.

Law of conservation of energy: According to this law, in a given system, energy can neither be created nor destroyed. In other words, the total energy content in an isolated system remains constant.

Law of conservation of mass: According to this law, in a given system, matter (mass) can neither be created nor destroyed. In other words, the total mass content in an isolated system remains constant.

LD_{50}: LD stands for lethal dose. It is the dose that is lethal to 50% of the group of test animals. It is expressed in milligrams per kilogram of body weight of the test animal.

LDL: It is a lipoprotein of blood plasma that is composed of a moderate proportion of protein with little triglyceride and a high proportion of cholesterol and that is associated with increased probability of developing atherosclerosis. It is also called bad cholesterol.

L–DOPA: The levorotatory form of DOPA that is obtained from broad beans or prepared synthetically and is converted to dopamine in the brain, and is used in treating Parkinson's disease. It is also called levodopa.

Leakage flow: It occurs between the clearance of the screw fright and barrel (open space). Leakage flow decreases the output of the extruder. This is also called pressure flow or pressure gradient.

Leaker: A can with visible leakage of the contents through the seam's perforations or nail holes. The leak may appear due to faulty seam, faulty lock seam or pinholes as a result of corrosion from the inside of the can or rusting from the outside. It is also called leakage.

Lean food: Food having less than 10 gm fat, 4.5 gm or less saturated fat and less than 95 mg of the cholesterol per serving and per 100 gm.

Leather: The dried fruit / vegetable pulp in layers to the thickness of 2-3 cm is called leather.

Leavening agent: A substance (yeast) used to produce fermentation in dough or a liquid. Leavening agents are the agents who create bubbles of gas in batters and dough causing then to rise in shapes e.g. baking powder / yeast.

Lecithin: A waxy hygroscopic phospholipid that are widely distributed in animals and plants and form colloidal solutions in water, and have emulsifying, wetting, and antioxidant properties. This phopholipid occurs in the nervous system and also found in egg yolk. These are enzymes that hydrolyze lecithin's or phosphatidyl ethanolamines. It is also called Phospholipase.

Legal standards for quality: These are established by federal, central, state or municipal agencies and are generally mandatory. These are set up by the law or through regulation. They generally concerned with freedom from adulteration by insects, mold, yeasts and pesticides.

Leicester: It is a hard orange colored cheese. It is similar to cheddar.

Lethal rate: It is the rate of sterilization process at a temperature of 121.1°C; $T = F_o/F_t$.

Lethality: Lethality is a measure of heat treatment or sterilization process to compare the relative sterilizing capacities of heat processes. So, it is defined as an equivalent heating of 1 min at a reference temperature of 121.1°C. Lethality (L) = F_T x lethal rate.

Leucine: A white crystalline essential amino acid $C_6H_{13}NO_2$ obtained by the hydrolysis of dietary proteins.

Levigation: It is a process to grind food commodity to a fine smooth powder while in moist condition (Plate 390).

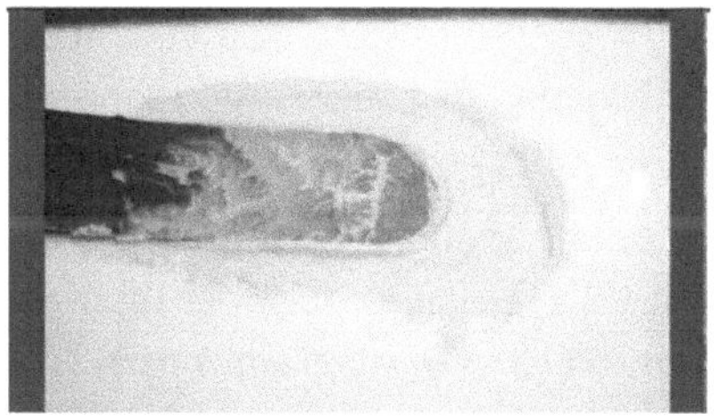

Plate 390: Grinding of moist food

Levulose: It is a crystalline sugar $C_6H_{12}O_6$ sweeter and more soluble than glucose. It is the very sweet levorotatory D-form of fructose that occurs in fruit juices and honey. It is also called fruit sugar levulose.

Leyden jar: It is an electrical capacitor consisting of a glass jar coated inside and outside with metal foil and having the inner coating connected to a conducting rod passed through an insulating stopper.

Lidding: It is the process of loosely covering the cans with lid after syruping or brining followed by exhausting. The cans after syruping and brining are covered with loose lid followed by exhausting. This process is called lidding.

Light Ale: Light ale is a type of fairly week pale beer.

Light beer: A beer having alcohol content below 3.9% (v/v) is called light beer.

Light food: Nutritionally altered food containing one third calories or half amount of fat and 50% of the sodium of the reference food.

Light syrup: A prepared syrup contains 2 litre water/ kg sugar.

Light wine: A wine having alcohol strength of 7-9% is called light wine.

Lignin: Lignin is a complex polymer of aromatic compound linked by propyl units, is as a major component of fibre. It is an amorphous polymer related to cellulose that provides rigidity and together with cellulose forms the woody cell walls of plants and the cementing material between them.

Lignocellulose: The substances constituting the essential part of woody cell walls of plants and consisting of cellulose intimately associated with lignin.

Limburger cheese: It is a pungent semisoft surface ripened cheese.

Lime Doser: It is a part of effluent treatment plant. In this unit alum, calcium, lime etc are added to maintain the pH of the effluent waste water.

Lime water: Quick lime (CaO) and water is mixed and filtered. The filtrate obtain is called lime water (Ca $(OH)_2$.

Line spread test: It is a measure of flow of a viscous liquid or semisolid food by determining the spread of a measured amount of sample in a specified length of time at 90° intervals on the template of concentric rings.

Liner: It is Kraft paper of about 80gm/m² upto 225 gm/m² and is used for liner for preparation of CFFB. The outermost liner used for a box should be of maximum grammage (No. of gramm/m²). Higher the gsm, high will be strength.

Linoleic acid: It is a liquid unsaturated fatty acid $C_{18}H_{32}O_2$ found in semidrying oils (corn oil) and essential for the nutrition of some animals.

Linolenic acid: A liquid unsaturated fatty acid $C_{18}H_{30}O_2$ found in drying oils (linseed oil) and essential for the nutrition of some animals.

Lipid: The lipid is soluble in nonpolar organic solvents (chloroform and ether) and usually insoluble in water. It along with proteins and carbohydrates constitute the principal structural components of living cells, and that include fats, waxes, phosphatides, cerebrosides, and related and derived compounds (Plate 391).

Lipolysis: The hydrolysis of fat by heat is called lipolysis (Plate 392).

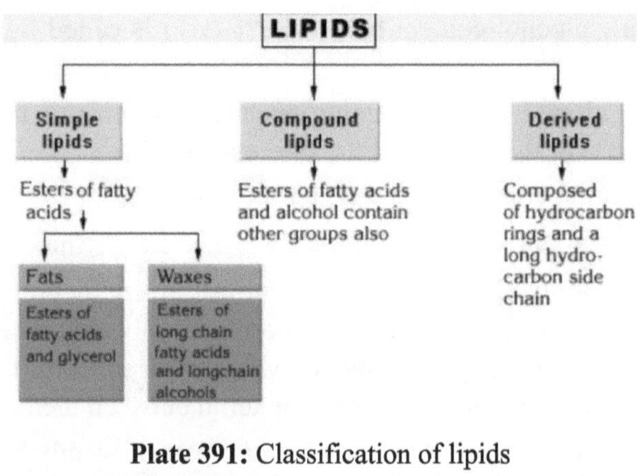

Plate 391: Classification of lipids

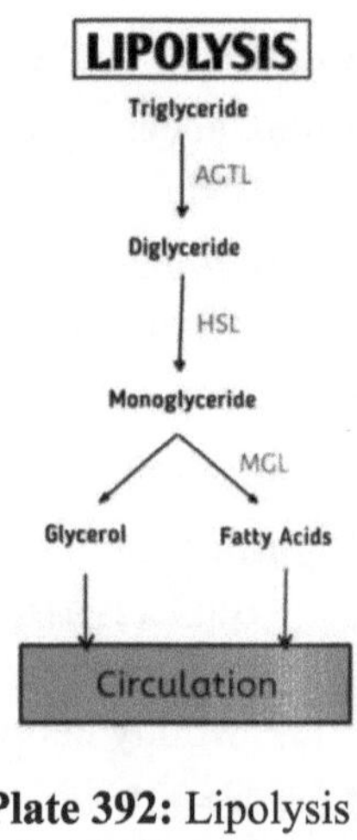

Plate 392: Lipolysis

Lipopolysaccharide: A large molecule consisting of lipids and sugars joined by chemical bonds.

Lipoprotein: A conjugated protein that is a complex of protein and lipid.

Lipotropic: A dietary factor that is promoting the physiological utilization of fat.

Liquefaction: It is the process of making or becoming liquid or conversion of solid into a fluid like mass during extrusion process.

Liqueurs: These are alcoholic beverages prepared from ethyl alcohol, sugar and flavouring essences. These contain about 33% alcohol and 0 to 50% sugar. These are also called cordials.

Liquid fruit: Any juice from the solid fruit is called liquid fruit.

Litchi nut: Dried litchi is commonly known as litchi nut which is very popular among the Chinese.

Lithographed cans: These cans are decorated cans and these cans gives good appearance.

Lithography: It is the process of coating and printing both inner and outer surfaces of the cap of container (glass jar cap or crown cork) on which the image to be printed with ink. It help to protect the metal cap from moisture attach, adhere gasket material and decorate the closure.

Loaf: A shaped or molded often symmetrical mass of bread / food.

Logarithmic order of death: It is the percentage of the bacteria and bacterial spores killed per minute at a constant temperature.

Loin: It is a cut of meat comprising one or both sides of a carcass with the adjoining half of the vertebrae included without the flank. It is a cut / piece of meat between upperband lower abdominal region and the region above the hip (pubic region and reproductive organs).

Long ton: One long ton is equal to 2,240 lbs. There are three types of ton for measurement of mass (weight). The short ton is US ton which is 2,000/ lbs. The long ton isImperial (British) ton which is 2240 lbs. The third ton is the metric tonne which is, equal to 1000 kilograms, or approximately 2204 pounds (Plate 393).

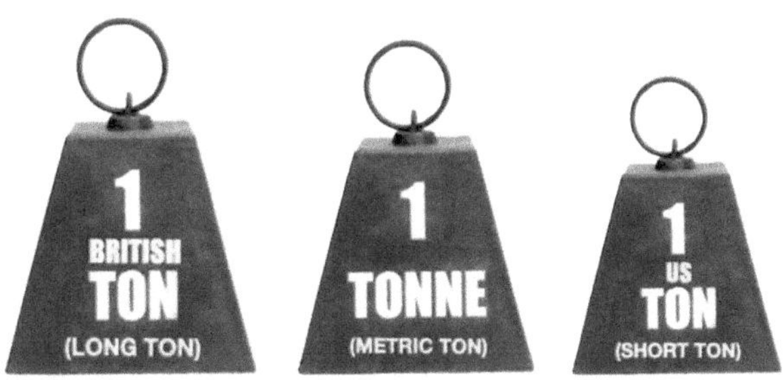

Plate 393: Diferent types of ton

Loss: Loss mean any change in the physical, chemical or sensory properties of the commodity that directly or indirectly affect the quality and / or quantity of the commodity and make the commodity undesirable or unuseable for human consumption.

Loss Assessment: This assessment is used to denote approximately a rough quantitative food loss.

Loss measurement: Loss measurement is more précised and objective processes by which quantitative fact about a loss situation are calculated.

Lot inspection: It means the inspection and grading of specific lots of processed fruits and vegetables which are located in plant warehouses, commercial storage, railway cars, trucks, or any other storage place. Generally, lot inspector does not have knowledge of conditions and practices under which the product is packed and his grading is limited to examination of the finished product only.

Low acid foods: These are the foods having pH > 4.6 and processed above 100°C. The acidity in these foods is not sufficient to prevent the growth of the bacterium *Clostridium botulinum.*

Low caloric beer: It is a beer in which enzymes breakdown dextrin's during the fermentation and results in production of a beer with normal alcohol strength, but of a lower calorie value.

Low calorie food: Food having 40 calorie or less sodium per serving or per 100 gm.

Low cholesterol food: Food having 20 mg or less cholesterol and 2 gm or less of the saturated fat per serving or per 100 gm.

Low fat food: Food having 3 gm or less fat per serving or per 100 gm.

Low methoxyl pectin: Pectin molecule with a methoxyl content of less than 7% is called low methoxyl pectin (LMP). It forms gel with less than 50% soluble sugars and form gel in the pH range of 2.5 to 6. It is used in the preparation of low sugar jam, jellies or sugar free jellies. It is galacturonic acid polymer in which an eighth, 6th and a fourth of acid radical have been esterified with methanol; form a pectic substances found in fruit that is just begin to ripen. Methylation of $COOCH_3$ means that at 6th position methyl group instead of COOH group. Gel formation requires divalent and polyvalent cations. The degree of methylation is 25-50%.

Low Nutrition food: Food eaten without exceeding dietary guidelines for one or more food components like fat, saturated fat, cholesterol, sodium and calories.

Low oxygen injury: A storage injury of apple and pear caused by the accumulation of ethyl alcohol in the fruit which reaches toxic level under the anaerobic conditions and symptoms appear as dark brown water soaked lesion in the skin.

Low saturated fat food: Food having 1 gm or less saturated fat per serving.

Low setting pectin: Pectin containing 7-10% methyl ester groups and by using this pectin, more than 25 minutes is required to set the gel, used mainly for jelly preparation.

Low sodium food: Food having 140 mg or less sodium content per serving or per 100 gm.

Low sugar based products: The products having sugar concentration of 30% or low is called low sugar based products.

Lozenge: A small sweetened and flavored medicated material that is designed to be held in the mouth for slow dissolution and have soothing effect on sore throat. It is also called pastille troche.

LTLT process: It stands for low temperature long time process. It is batch process in which milk is heated to 63°C for 30 min.

Lug Cap sealing machine: It is machine used for sealing a glass jar with a lug cap (Plate 394).

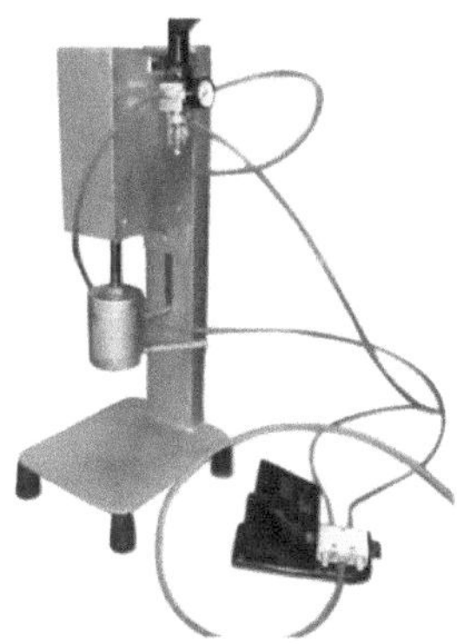

Plate 394: Lug Cap Sealing machine

Lutein: An orange xanthophyll $C_{40}H_{56}O_2$ occurring in plants, animal fat, egg yolk, and the corpus luteum.

Lycopene: It is a carotenoid pigment $C_{40}H_{56}$ that have red coloring matter. Red pigment found in ripe tomato which is a straight chain derivatives of carotene with no vitamin A activity. Its chemical composition (C_{40} H_{56}) is same to that of carotene.

Lye peeling: Peeling of fruits (peach, apricot, orange etc) and vegetable (carrot, sweet potato etc) by dipping them in boiling lye solution (caustic soda) of 0.5-2% strength for short period (<30 second - 2 minutes) depending on the maturity and nature of fruits and vegetables to loose the skin from the flesh underneath to facilitate easy peeling.

Lyophilization: A method of preserving the microorganisms where the organisms are subjected to extreme dehydration in the frozen state and then sealed in a vacuum. A desicated (Lyophilized) culture of microorganisms remains viable for many years. Lyophilizing is the term used to drying from the frozen state (Plate 395).

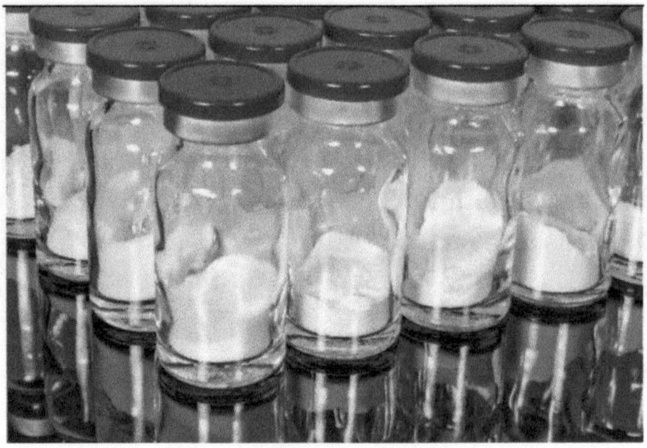

Plate 395: Lyophilized culture

Lysin: It is an antibody (bacteriophage) capable of causing breakdown of the cells.

Lysine: A crystalline essential amino acid $C_6H_{14}N_2O_2$ obtained from the hydrolysis of various proteins.

Lysis: It is the degradation of bacterial cells by the action of lysin.

Lysozyme: It is a bacteriolytic enzyme present in certain plants, animals and secretions like egg white, tears etc.

Liquid Death Water: The name "Liquid Death" refers to the idea of "murdering your thirst," as well as "death to plastic," according to founder and CEO Mike Cessario. Liquid Death beverages are packaged in aluminium, which is more recyclable than plastic

M

Macaroni: Pasta made from semolina and shaped in the form of slender tubes (Plate **396**).

Macaroni wheat: It is tetraploid wheat and is preferred for making macaroni, vermicelli etc.

Macaroon: A small cookie composed chiefly of egg whites, sugar, and ground almonds or coconut (Plate **397**).

Mace: An aromatic spice consisting of the dried external fibrous covering of a nutmeg (Plate **398**).

Plate 396: Macaroni

Plate 397: Macaroon

Plate 398: Mace and Nutmeg

Mach number: It is a number representing the ratio of the speed of a body (flow velocity) to the speed of sound (sonic velocity) in a surrounding medium (as air).

Macroglobulin: A highly polymerized globulin of high molecular weight.

Macronutrient: A chemical element (as nitrogen, phosphorus, or potassium) of which relatively large quantities are essential to the growth and health of a plant or a substance (as protein or carbohydrate) essential in large amounts to the growth and health of an animal.

Magnesium silicate: It is use as glazing and polishing agents in confectionery products.

Magness - Taylor Pressure Tester: It is used to study the compression of peas, pear, kiwi and apples.

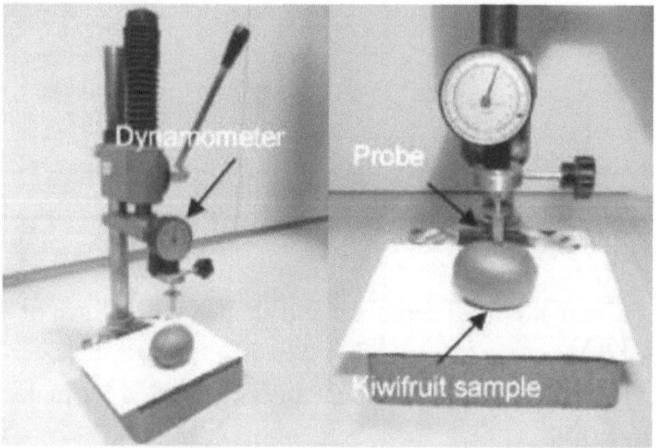

Plate 399: Magness - Taylor Pressure Tester

Magnetron: It is a vacuum tube in which the flow of electrons is controlled by an applied magnetic field to generate power at microwave frequencies. It is a part of microwave heating device (Plate 400).

Maillard reaction: The reactions between (i) nitrogenous compound and sugars (ii) nitrogenous compound and organic acids (iii) sugars and organic acids (iv) among organic acids themselves are known as maillard reaction. It may be independently causing browning of products which is of great important to food preservation techniques. It is difficult to control and also known as non enzymatic browning. Maillard reactions occur between reducing sugars, proteinaceous materials; is a major quality problem in dried and concentrated fruit products (Plate 401 and 402).

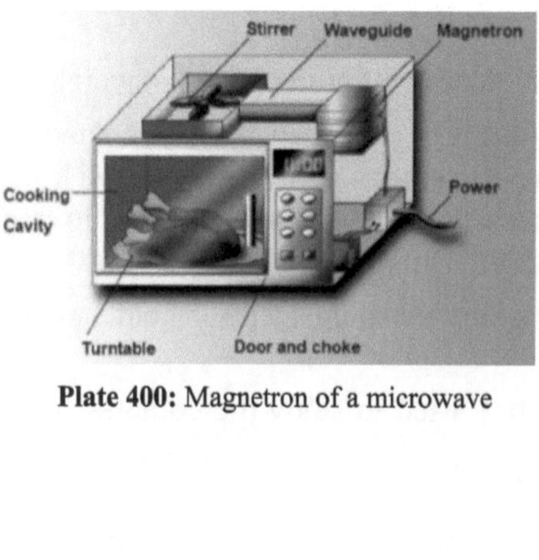

Plate 400: Magnetron of a microwave

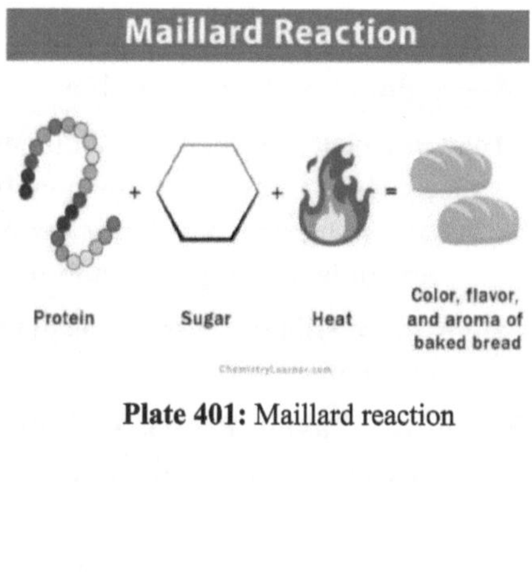

Plate 401: Maillard reaction

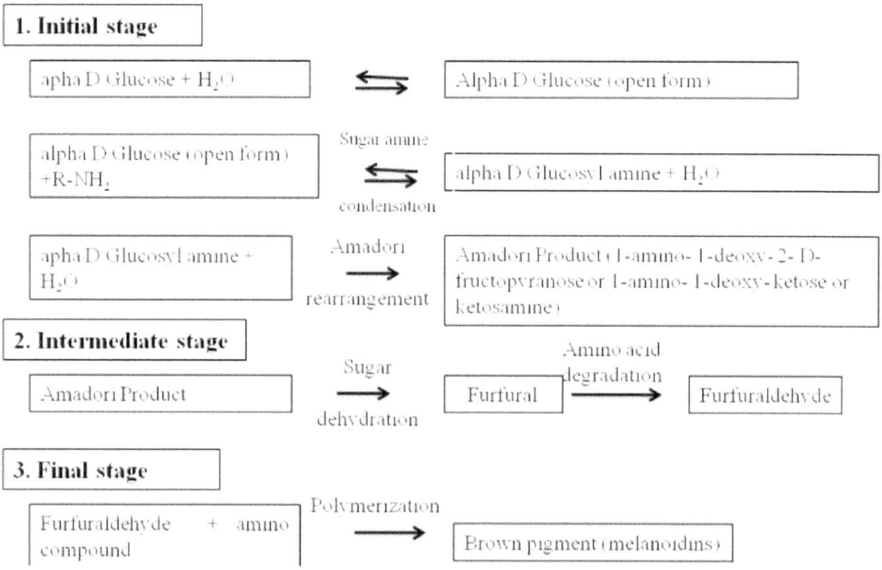

Plate 402: Principal steps involved in Maillard's browning reaction

Maize sheller: It is equipment used for removing maize grains from the cobs (Plate 403).

Malnutrition: An undesirable kind of nutrition leads to ill health and results from lack, excess or imbalance of nutrients in diet is called malnutrition (Plate 404).

Plate 403: Hand operated maize sheller

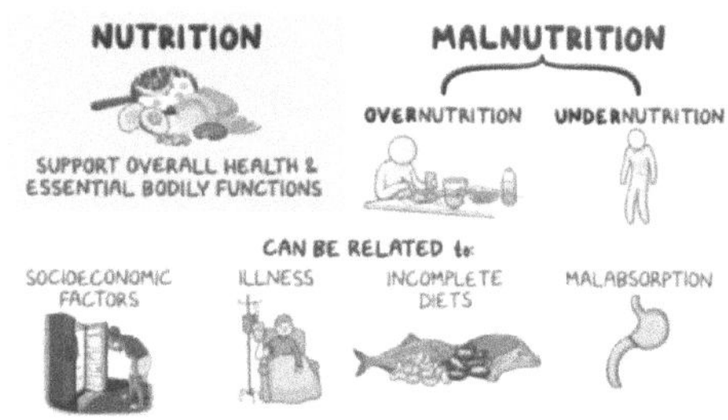

Plate 404: Malnutrtion

Malt culms: A by-product of the malting process used as a feeding stuff for diary cows, sheep and horse after thorough soaking. It is also called Coombs (Plate 405).

Malt liquor: Malt liquor contain alcohol content 6% or above.

Maltase: It is an enzyme that breaks down maltose into glucose. It assists in brewing process.

Maltodextrin: The carbohydrates derived from the partial hydrolysis of starch (corn or potatoes) and used in prepared foods as non-sweet filler and to enhance texture and flavour (Plate 406).

Maltose: It is a hard, white, crystalline dextrorotatory fermentable sugar $C_{12}H_{22}O_{11}$ formed from starch by amylase / diastase. It is also called malt sugar. It is produced as a result of seed germination. It is a reducing sugars (Plate 407).

Mannitol: It is a common sugar alcohol obtained by reduction of fructose to alcohol. It is the main soluble sugar in fungi, lichen and some brown algae (Plate 408).

Plate 405: Malt culms

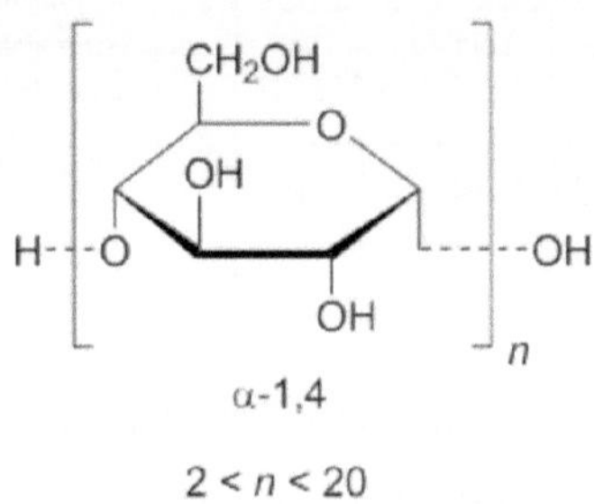

Plate 406: Maltodextrin

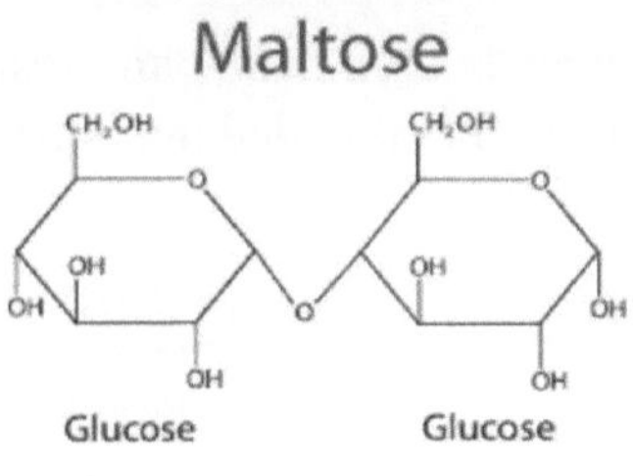

Plate 407: Maltose structure

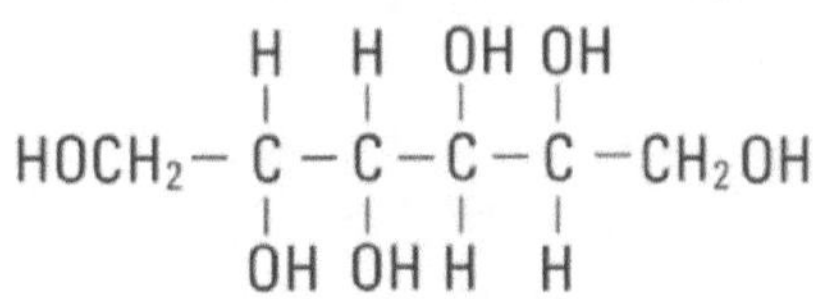

Plate 408: Mannitol

Manometric techniques: These techniques are used for study of membrane transport properties like diffusion, sorption and permeation coefficients and used to study the transport properties of CO_2, O_2, and nitrogen in the dense / asymmetric membrane (Plate 409).

Manothermo sonication: A combination of low pressure (0.3 MPa), heat treatment and ultrasonic wave treatment for preservation of the food is called manothermo sonication (Plate 410).

MAP: It stands for modified atmospheric packaging (Plate 411 and 412). Modified atmosphere packaging (MAP) is the imposition of a gas atmosphere, typically containing an inert gas, such as nitrogen combined with an antimicrobially active gas, such as carbon dioxide, upon a packaged food product to extend its shelf life.

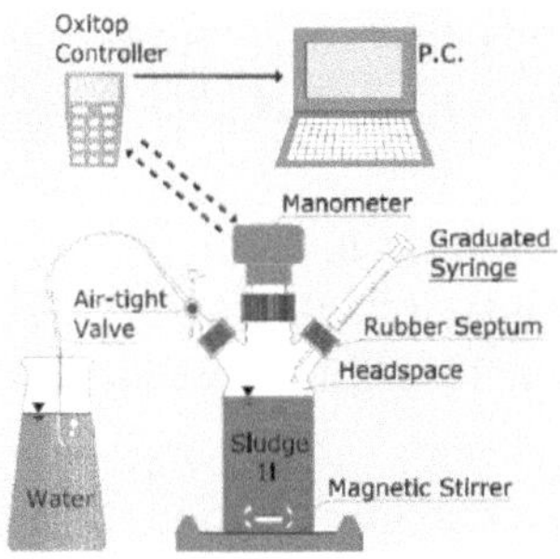

Plate 409: Manometric equipment set-up

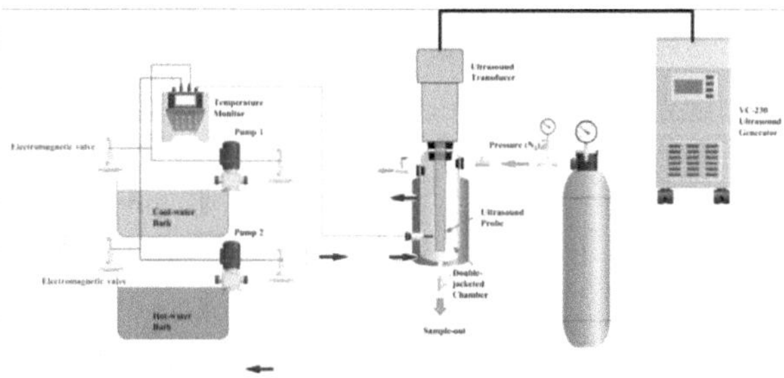

Plate 410: Laboratory Manothermo sonication

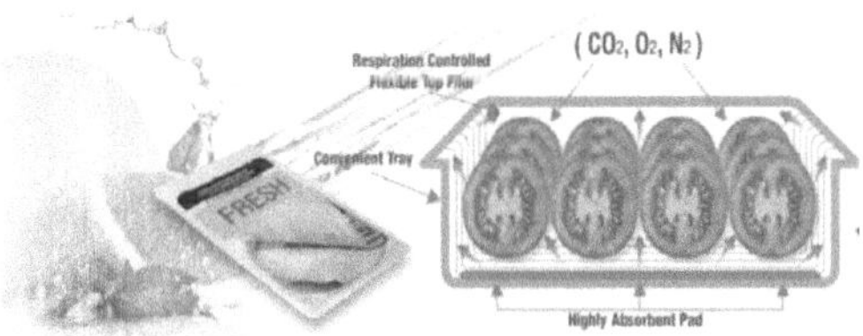

Plate 411: Modified atmospheric packaging

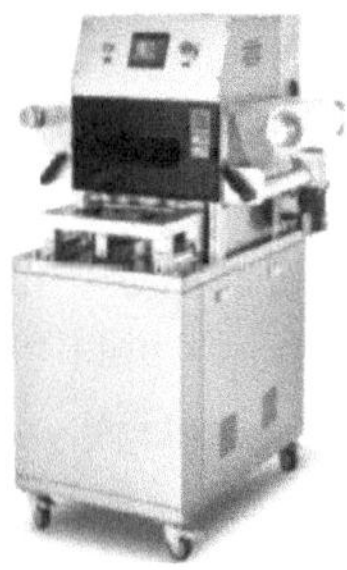

Plate 412: MAP machine

Maple: A genus (Acer of the family Aceraceae, the maple family) of chiefly deciduous trees or shrubs with opposite leaves and a fruit of two united samaras; the hard light colored close grained wood of a maple used especially for flooring and furniture.

Maple sugar: The sugar made by boiling maple syrup is called maple sugar.

Maple syrup: Syrup made by concentrating the sap of maple trees and mainly the sugar maple.

MAR resistance: The resistance of the various plastic contains to scrotching, scuffing, denting, bruising and abrasion is referred as the MAR resistance.

Maraschino: A sweet liqueur distilled from the fermented juice of a bitter wild cherry or a large cherry preserved in true or imitation maraschino.

Marasmic / marasmus: A condition of chronic undernourishment / malnutrition (protein energy malnutrition - PEM) occurring in children and caused by a diet deficient in calories and proteins.

Marble cake: A cake made with light and dark batter so as to have a mottled appearance.

Marbling: An intermixture of fat and lean mainly when it is evenly distributed in a cut of meat.

Margarine: A food product made from vegetable oils churned with ripened skim milk to a smooth emulsion and used like butter. It is also called oleomargarine.

Marinating: It is a process of soaking a food in a marinade to add flavour or to tenderize it or both. Marination is done to brinjal, onion, radish, bitter gourd, potatoes and chillies. Marination prevents browning reaction in potatoes and apple.

Market: Market is a place or building where buying or selling of goods takes place. The word 'market' is a derivative of the Latin word 'marcatus' meaning thereby merchandise, ware, trade or a place where business is conducted. The market is a place or a building where commodities are brought and sold e.g. super market, cotton market, Asian market etc. Market is an organization which provides facilities for exchange of commodities. In other words a market means any place where persons assemble for sale or purchase of commodities intended for satisfying human wants.

Market basket: A variety of consumer goods and services used to calculate a consumer price index.

Market forecasting: To calculate or predict market condition from some future event or usually as a result of study and analysis of available pertinent data.

Market garden / truck farm: A farm devoted to the production of vegetables for the market.

Market maker: An intermediary in a stock exchange who controls buy and sell orders (by purchase and resale) for a particular stock or group of stocks.

Market order: An order to buy or sell commodities immediately at the best price obtainable in the market.

Market place: An open place in a town where markets or public sales are held.

Market planning: The process of carrying out plans for establishment of goals, policies, and procedures for an economic market.

Market price: A price actually given in current market dealings is called market price.

Market research: It is research into the size, location, and makeup of a product market.

Market segmentation: The process of dividing market into segments.

Market share: The percentage of the market for a product or service that a company supplies.

Marketeer: Marketeer is a specialist in promoting or selling a product or service.

Marketer: Marketer is one that deals in a market to promote or sell a product or service.

Marketing: It is a total of functions concerned with various activities such as product planning, product development, product change, pricing and packaging, selling, sales promotion, advertising and marketing research. Marketing represents the full effort in getting the product to the market place. Selling is one of the most important aspects of the marketing job. The selling job is broken down into the domestic and international markets. Marketing is the process or technique of promoting, selling, and distributing a product or service from producer to consumer. Marketing is the act / process of selling / purchasing in a market.

Marketing agencies: The transfer of produce takes place through a chain of middlemen or functionaries.

Marketing research: It is a research for the purpose of promotion, selling, and distribution of a product or service.

Marketization: It is the process of entering into, participating in, or introducing a free market economy.

Marmalade: A fruit jelly generally prepared form citrus fruit like oranges and lemons in which the slices of fruits or peels are suspended.

Martin aseptic canning process: HTST is known as Martin aseptic canning process (1950) and involve four separate operations.

Marzipan: It is a confection of crushed almonds or almond paste, sugar, and egg whites that is often shaped into various forms (Plate 413).

Plate 413. Marzipan

Mash: It is a crushed malt or grain meal steeped and stirred in hot water to produce wort.

Mashing: It is a process to reduce to a soft pulpy state by beating or pressure. Or It is the process to subject crushed malt to the action of water with heating and stirring in preparing wort.

Mass spectroscopy: The basic idea of mass spectroscopy is to produce ions by bombarding organic molecules with high energy electrons, then accelerating these ions in a defined direction so that they can be separated according to their mass or velocity. Each spectrometer has three major components like ion source, analyzer and detector.

Mass transfer: The movement of the material in fluid system (gases and liquids) under the influence of concentration gradient is called mass transfer.

Masticometer: It is a machine that measure comparative tenderness of meat and other foods by stimulating chewing action.

Matting: It is the process that involves fusion of curd pieces to form continuous rubbery slab.

Maturation: The process of becoming fully developed which is the final stage of fruit development prior to ripening. Or The stage of development leading to the attainment of physiological or horticultural maturity is called maturation.

Matured mutton: Meat obtained from >20 month aged sheep (Plate 414).

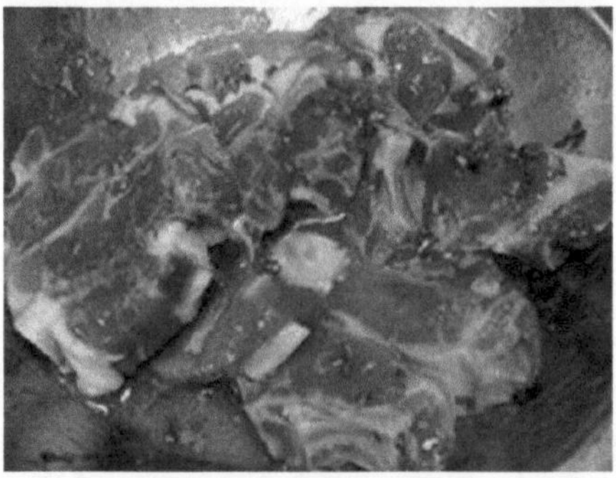

Plate 414: Marinated mutton

Maturing: It is simply keeping of the wine or vinegar in oak wood barrels for 6-12 months. During this period, it's raw and harsh flavor changes to a more pleasant aroma due to oxidation of the wine/ vinegar by air entering through the pores in the wood and formation of esters having a good fruit flavor.

Maturity: Main harvesting index of fruit and vegetable or any other food commodity can be attributed to a distinct stage of development when the commodity has completed its natural growth and development and attained desired physical and chemical attributes prior to ripening. The climacteric fruits are to be harvested at this stage but non climacteric fruits to be after this stage as they do not undergo ripening after harvesting.

Maturity index: The critical factor / index which is used for determining the harvesting time of fruit and vegetables is known as maturity index. It can be determined according to consumers' purpose, type of commodity etc. and can be judged by the (a) visual means (colour, size, shape etc.) (b) Physical means (firmness, softness), (c) chemical analysis (sugar content, acid content, etc.) (d) Computation (heat unit) and physiological method (respiration).

Mayonnaise: A dressing made of egg yolks, vegetable oils, and vinegar or lemon juice is called Mayonnaise. Mayonnaise is an emulsion of oil, egg yolk, and an acid, either vinegar or lemon juice; there are many variants using additional flavorings. The color varies from near-white to pale yellow, and its texture from a light cream to a thick gel (Plate 415).

Mead: A wine prepared from honey is called mead. Distilled mead or honey brandy is called eau-de-viede Miel (Plate 416).

Plate 415: Mayonnaise

Plate 416: Mead

Meals: It is sum of one or more foods eaten at one time.

Meat: The animal flesh that is used as food especially of goat or sheep (not fish and poultry). It is an important source of nourishment and is rich in proteins. It contains roughly about 75-80 % water, about 20% proteins, 3-4 % fats and 1-1.5% minerals.

Meat ax: It is a butcher's implement for cutting animal carcasses into joints or pieces.

Meat Food Product Order (MPO), 1973: It provides means to detect and destroy meat of diseased animals. It ensure that the preparation and handling of meat and meat products be conducted in a clean and sanitary manner. It prevents the use of harmful substances in meat foods. It sees that every cut of meat is inspected before sale to ensure its wholesomeness. The order also lays down rules and conditions for procedure to be adopted for the selection of disease free animals, slaughterhouse practices. It deals with the products prepared from meat and their licensing manufacture, trade and distribution including hygiene condition. This order is operated by the Directorate of Marketing & Inspection.

Meat loaf: A dish of ground meat seasoned and baked in the form of a loaf is known as meat loaf.

Mechanical decorticator: It is a machine used for removal of the husk from seed; oilseeds are generally decorticated before extracting the oil. The mechanical decorticator is used for decortications of stones prior to extraction of oil (Plate 417). In the equipment two roller moves in inward and opposite direction and thus results in breaking of the hard seed coat. Mechanical decorticator can be used for decortications of apricot / peach stones.

Mechanical harvester: It is an apparatus used for harvesting mango, apple, pear, plum etc from the tree (Plate 418).

Plate 417: Mechanical apricot stone decorticator

Plate 418: Mechanical harvester

Median Lethal Dose (MLD): It is the dose of the radiation / chemical which is required to kill 50% of the individuals in a large group of animals or organisms with in a specified time. In toxicology, the **median lethal dose, LD_{50}** ("lethal dose, 50%"), **LC_{50}** (lethal concentration, 50%) or **LCt_{50}** is a toxic unit that measures the lethal dose of a given substance. The value of LD_{50} for a substance is the dose required to kill half the members of a tested population after specified test duration. The term LD_{50} is generally attributed to John William Trevan (1927).

Median Lethal Time: It refers to time required following administration of a specific dose of the radiation / chemical, for the death of 50% of the individuals in a large group of animals or organisms.

Medical foods: The foods which are consumed to get rid of some specific disorder or disease are called medical foods (Plate 419).

Medicinal plants: These plants are generally rich in secondary metabolites and are potential sources of drugs, alkaloids, glycosides, flavonoids and steroids etc.

Medicinal wine: A flavoured wine with herbs is called medicinal wine. Vermouth is a medicinal wine containing 15-21% alcohol and flavoured with herbs and spices.

Medium acid food: Foods having pH 5.0-4.5 which are usually spoiled by thermophillic anaerobes e.g. vegetable mixture, soups, sauces, meat etc.

Medium sugar based products: The products having sugar concentration of 30-50% sugar or low is called medium sugar based products.

Medium syrup: A prepared syrup containing 1.5 litre water/ kg sugar.

Medium tomato puree: It is a tomato puree having TSS of 9 ^{0}B.

Medium wine: A wine having alcohol strength of 9-16% is called medium wine.

Megahertz: It is the measure of frequency and defined as one million cycle per second.

Megarad (M rad): It is equal to 1 million rad.

Melanin: The black, dark brown, reddish-brown, or yellow pigment of animal or plant structures (as skin or hair) is called melanin (Plate 420).

Plate 419: Medical Food

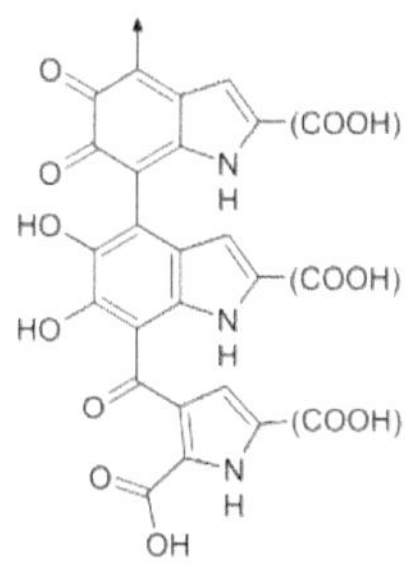

Plate 420: Melanin

Melanoids: The brown to black, amorphous unsaturated heterogenous polymers are called melanoids.

Melatonin: It is a powerful antioxidant (Plate 421). Melatonin is a natural hormone that plays an important role in sleep. Foods like pistachios, tart cherries, mushrooms, tomatoes, pineapples, oranges, kiwifruits, apples, and bananas are good sources of melatonin. Some diseases can lead to low melatonin levels.

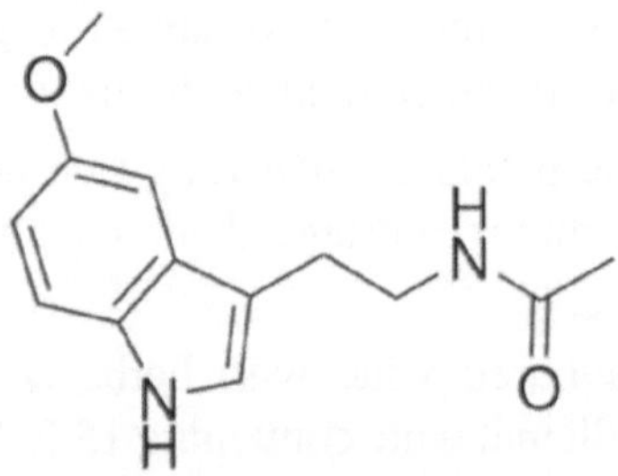

Plate 421: Melatonin structure

Melting point: It is the temperature at which a solid substance melts into a liquid.

Menadione: It is a yellow crystalline compound $C_{11}H_8O_2$ with the biological activity of natural vitamin K. It is a synthetic vitamin K.

Mescal: The distilled pulque (distilled agave wine) is called Mescal. It is produced in Mexico.

Mesophile: These organisms require a temperature below 38°C for growth (Plate 422 and 423).

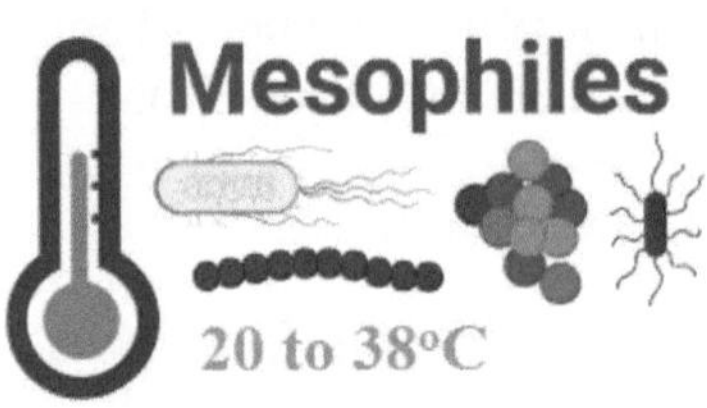

Plate 422: Mesophile

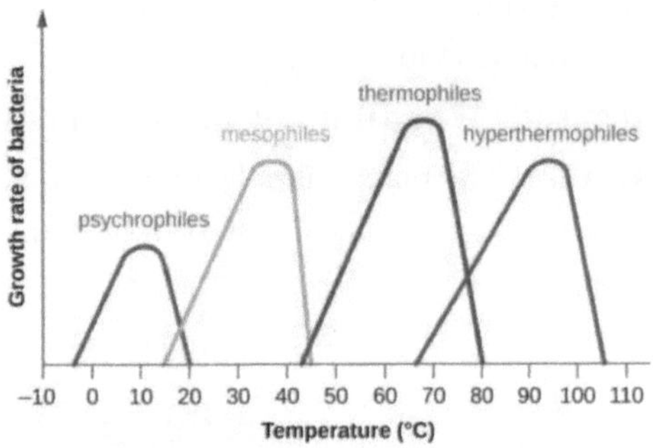

Plate 423: Mesophile vs others

Metabolism: The sum of the processes in the buildup (building) and destruction (breakdown) of protoplasm; specifically, the chemical changes in living cells by which energy is provided for vital processes and activities and new material is assimilated.

Metabolite: It is a substance essential for the metabolism of a particular organism or to a particular metabolic process and is a product of metabolism process.

Metal Cans: Deep drawn or 2 or 3 piece cans are found very suitable for food and beverages when these were coated electrophoretically from inner side with an aqueous solution of 0.5 – 3% by weight (solid basis) of a wax with melting point of 60-100 °C. Two piece aluminium cans were found to be very suitable for canning of wide variety of Indian foods. Micro-seam is found suitable for sealing of cans to increase the shelf life of dried and heat processed food as compared to double seam.

Metcryotic liquid: During storage the unfrozen concentrated solution of sugars salts etc may ooze out from package of fruits or concentrates as viscous materials called metcryotic liquid. Unfrozen concentrated solution of sugars, salts etc. May ooze out from fruits or concentrates during storage as a viscous material called 'metacryotic liquid'. Fluctuation in storage temperature results in an increase in the size of ice crystals resulting in physical damage to the food.

Methionine: It is crystalline sulfur containing essential amino acid $C_5H_{11}NO_2S$ that occurs in the levorotatory form as a constituent of many proteins (Plate 424). Methionine is an aliphatic, sulfur-containing, essential amino acid, and a precursor of succinyl-CoA, homocysteine, cysteine, creatine, and carnitine. It is also called as 2-amino-4-(methylthio) butanoic acid. Recent research has demonstrated that methionine can regulate metabolic processes, the innate immune system, and digestive functioning in mammals. It is precursor of ripening in fruits. Foods with a particularly high percentage include eggs (31 mg/g protein), cod (30 mg/g), and chicken (28 mg/g). Intermediate content is in beef (26 mg/g), pork (26 mg/g), milk (25 mg/g), and rice (24 mg/g). Grains and other plant-derived protein sources tend to contain a lower percentage.

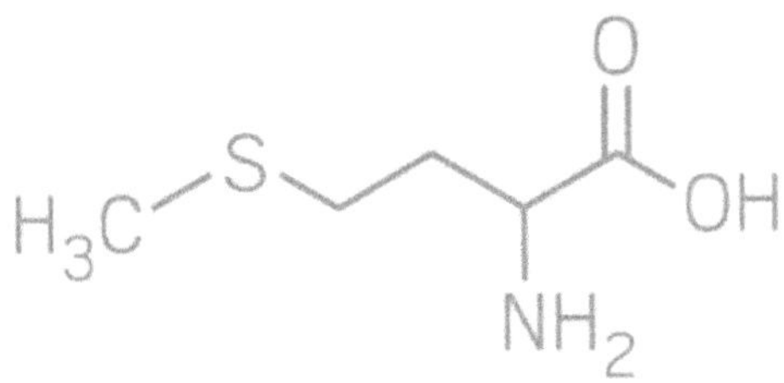

Plate 424: Methionine structure

Methyl bromide: It is poisonous gaseous compound CH_3Br used chiefly as a fumigant against rodents, worms, and insects. It is used to control pests in post harvest storage of fruits, vegetables and grains before export.

Methyl orange: It is an alkaline dye used as a chemical indicator. It is used for acid and base.

Methyl parathion: It is a potent synthetic organophosphate insecticide $C_8H_{10}NO_5PS$ that is more toxic than parathion.

Methylene blue: It is a basic thiazine dye $C_{16}H_{18}ClN_3S{\cdot}3H_2O$ used as a biological stain, an antidote in cyanide poisoning, and an oxidation-reduction indicator. It is used as an indicator for reducing sugars test @1%.

Metric ton: It is equal to 2,204.62 pounds or 1000 kg. It is usually abbreviated by MT.

MeV: It is million electron volts. It is a measure of the intensity of irradiation.

Microbalance: A balance designed to measure very small weights.

Microbarograph: A barograph for recording small and rapid changes in atmospheric pressure 9 Plate 425).

Plate 425: Microbarograph device

Microbicidal: Microbicidal methods of preservation are canning, making jams, jellies, preserves and irradiation.

Microbiology: A branch of biology dealing with microscopic forms of life.

Microbistatic: Antimicrobials that kill microbes are called microbicidal; those that merely inhibit their growth are called microbiostatic / microbistatic. Microbistatic methods of preservation are drying, freezing, desiccation, low temperature storage, pickling, brining, salting, smoking, use of antibiotics and other preservatives such as organic acids.

Microbrew: A beer produced by a microbrewery is called microbrew.

Microbrewery: A small brewery making specialty beer in limited quantities.

Microcalorimeter: An instrument for measuring very small quantities of heat.

Microcapsule: A tiny capsule containing material (adhesive or medicine) that is released when the capsule is broken, melted, or dissolved.

Micrococcus: A small spherical gram positive bacterium of genus (Micrococcus) chiefly harmless and typically occur in irregular clusters (Plate 426).

Microcurie: A unit of quantity or of radioactivity equal to one millionth of a curie.

Microencapsulation: It is the process to enclose a substance in a microcapsule e.g. microencapsulated aspirin (Plate 427).

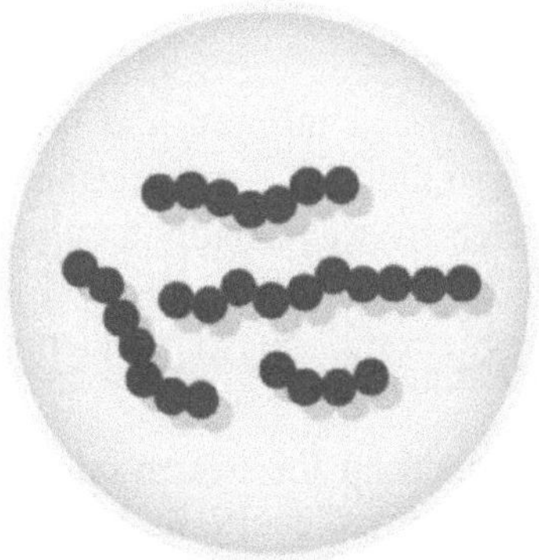

Plate 426: Micrococcus

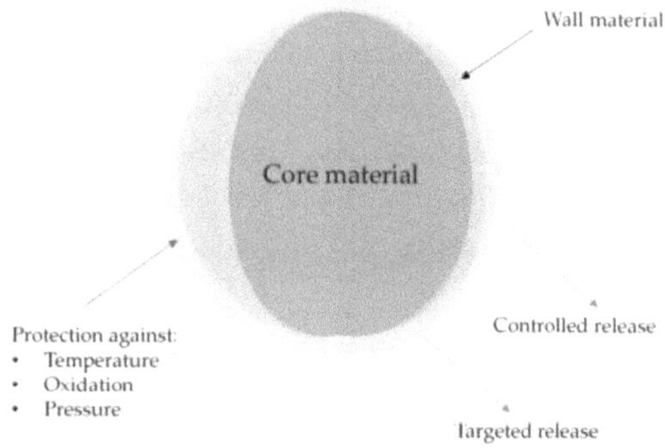

Plate 427: Micro-encapsulation

Microfiltration (MF): Microfiltration (MF) is membrane process which is mainly used for clarification purposes due to large membrane pore size (Plate 428).

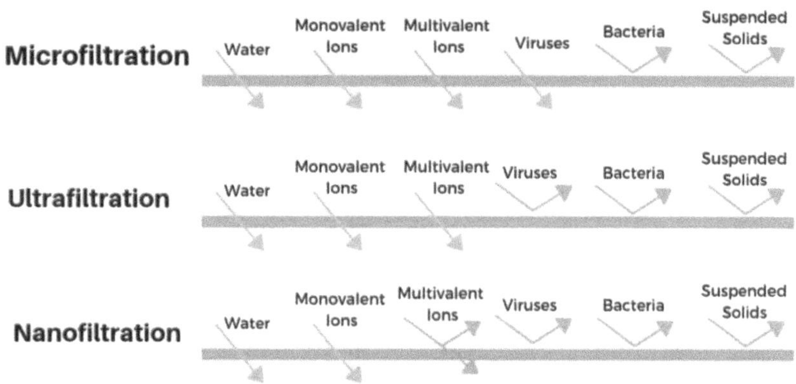

Plate 428: Characteristics of different membranes for filtration

Microgard: It is prepared from grade A skim milk that has been fermented by *Propionibacterium frendenreichii subsp shermani*. It is used for preservation of cheese in the US.

Micronutrient: An organic compound (vitamin) essential in minute amounts to the growth and health of an animal.

Microorganism: An organism (bacterium or protozoan) of microscopic or ultramicroscopic size. These include bacteria, yeast and mold. When they are alive in a suitable environment, they grow rapidly and may divide or reproduce at every 10 to 30 minutes.

Microscope: An optical instrument consisting of a lens or combination of lenses for making enlarged images of minute objects.

Microwave: These are electromagnetic waves between infrared and radio waves. Microwave is comparatively short non ionizing electromagnetic wave having one millimeter to one meter in wavelength and has frequency between 300 MHz and 300GHz. These waves are used to cook or heat food in a microwave oven. Magnetron results in production of the microwaves.

Microwave oven: An oven in which food is cooked by the heat produced by the absorption of microwave energy by water molecules in the food (Plate 429).

Plate 429: Microwave oven

Milk: A fluid secreted by the mammary glands of females for the nourishment of their young is called milk. Cow's milk is used as a food by humans. Liquid resembling milk in appearance is also called milk e.g. the latex of a plant, the juice of a coconut composed of liquid endosperm or the contents of an unripe kernel of grain. Or Milk is a white or yellowish liquid consisting of small fat globules suspended in a solution, secreted by mammary glands for nutrition of the new born. It is rich sources of proteins, vitamins, minerals and fats. It is lacking vitamin C and iron. It is very complex food and often called as complete food.

Milk & Milk Products Order: Milk & Milk Products Order (MMPO), 1992 is exercised under the Essential Commodities Act and is regulated by the

Ministry of Agriculture through the Department of Animal Husbandry and Dairying. According to this order, it is essential for a dairy plant to process more than 10,000 liters of milk per day or handle more than 500 tones of milk solids per annum.

Milk adulteration: The milk is commonly adulterated by adding water which lowers the proportion of its constituents and also its specific gravity (Plate 430). The detection can be avoided by raising specific gravity by adding skim milk, sugar or other soluble substances. Buffalo milk which contains more fat is often adulterated and sold as cow's milk, which contains less fat.

Plate 430: Milk Adulteration

Milk chocolate: The chocolate made with milk solids is called milk chocolate.

Milk fat: The natural fat of milk and chief constituent of butter consisting essentially of a mixture of glycerides as those derived from butyric, capric, caproic, and caprylic acids.

Milk house: A building for the cooling, handling, or bottling of milk is called milk house.

Milk powder: It is fine white powder obtained by drying milk at low temperature, so that it can be reconstituted into fluid milk, used in the preparation of infant foods and bakery products. It is also used in preparation of tea and coffee drink at homes and hotels.

Milk punch: A mixed drink of alcoholic liquor / wine, milk, and sugar is called milk punch.

Milk shake: A thoroughly shaken or blended drink made of milk, a flavoring syrup, and often ice cream.

Milk sugar: Disaccharide sugar lactose ($C_{12}H_{22}O_{11}$) is hard, gritty and crystalline and is called milk sugar. Women, cow and buffalo milk contain about 7%, 5 and 5.5% sugar respectively.

Milk viscosity: The viscosity of the whole milk at 25°C is about 2.0 cp.

Milk bottle caps: Red caps indicate that milk is skimmed whole milk. Pink or light blue caps indicate skim milk. Blue caps /tops indicate 2% milk. Yellow or purple caps indicate that milk is 1%. Gold colour indicate Channel Island milk. Blue indicate Homogenized whole milk. Green means Semi-skimmed milk.

Milky stage: It is the stage in the development of grains, where the contents of the caryopsis when squeezed out look like milky white fluid.

Mince meat: A finely chopped mixture of raisins, apples, and spices sometimes with meat that is often used as pie filling.

Mince: Small chopped bits of food are called mince.

Mincing: The process to cut or chop into very small pieces is called mincing.

Mineral: An inorganic substance as in the ash of calcined tissue is called mineral.

Mineral oil: An oil of mineral origin especially refined petroleum oil and is used as a laxative.

Mineral water: Water naturally or artificially infused with mineral salts or gases (carbon dioxide).

Minimal processing methods: This method causes the least possible change in food quality but at the same time provide the food with enough useful life to transport it from production site to the consumer. The processing done is at a minimal level so it is called minimal processing.

Minimally processed foods: The foods which have fresh like characteristics and safety atleast partially are called minimally processed foods. The fruits and vegetables are sealed pack after preparation like trimming and cutting and preserved without causing significant loss or change in the fresh like properties. These foods have a demand like that of fresh commodity and possess high quality.

Minimum Support Price: This is the price fixed by the Government to protect the producer – farmers against excessive fall in price during bumper production years. In case, the market price for the commodity falls below the announced minimum price due to bumper production and glut in the market, government agencies purchase the entire quantity offered by the farmers at the announced minimum price. Minimum support prices for different agricultural crops viz., Food grains, oilseeds, fibre crops, sugarcane and tobacco are announced by the govt. of India before the start of the sowing season of the crop.

MIR: It stands for Mid Infra red spectroscopy.

Mixing: It is the process of mixing of two or more separate phases such as two fluids or a fluid and a powdered solid and causing them to be randomly distributed through one another.

Modern dairy: A modern dairy is a complex, complete with a commercial processing facilities and distribution of milk is done in bottles or pouches. Some dairies may be purely milk-bottling plants whilst a few may confined to making milk products.

Modified Atmosphere Packaging (MAP): MAP has been defined as the enclosing of food products inside gas barrier materials or in a gaseous environment which has been modified to slow down respiration rate, microbial growth and reduced enzymatic degradation all with the intention of extending the shelf life. The films used in MAP restrict the transmission of the respiratory gases (Plate 431). Such modifications are usually associated with reduced oxygen content in the headspace atmosphere, but increased carbon dioxide and nitrogen levels. Carbon dioxide is the most important constituent in the gas mixture because of its microbial inhibitory effect, which is influenced by many factors including microbial load, gas concentration, temperature and packaging film permeability. A concentration of 20-60% of carbon dioxide in the package headspace is effective against aerobic spoilage microorganisms. The success or failure of MAP for foods depends on the impermeability of the packaging films to both oxygen and carbon dioxide. A key feature of the MAP for fruits and vegetables is the controlled permeation of oxygen and carbon dioxide as well as scavenging of ethylene gas or the controlled release of other compounds which could have preservative effect on foods. It is the storage of commodity in plastic films in controlled environment. Air box is the example of MAP.

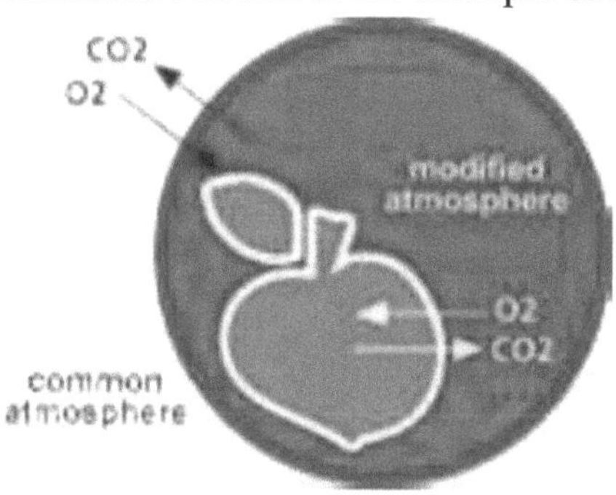

Plate 431: Mechanisms of MAP

Modified starch: These are normal, natural starches that have been altered chemically or physically to have greater viscosity, to assist in the food processing industry. They are used in canning, instant puddings, frozen foods etc. The use of the label is usually owned by the standard setting body. The label is usually used in communication with the end consumers.

Modulus: It is a constant or coefficient that expresses numerically the degree to which a food substance possesses a particular property e.g. elasticity (Plate 432).

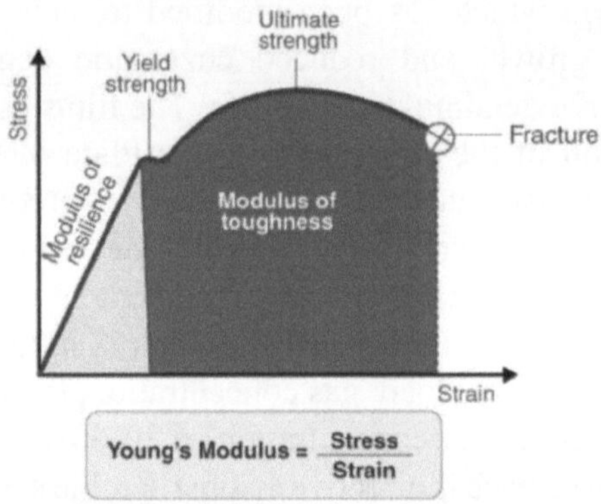

Plate 432: Modulus of elasticity

Moist air: It is a mixture of dry air and water vapour. The amount of the water vapour present in air depends upon absolute pressure and temperature of mixture.

Moisture absorber: The substances which have capability to absorb moisture from the product are called moisture absorber. These are also called desiccants. E.g. poly vinyl alcohol, silica gel and clay base material.

Moisture content: It is moisture / water present in a fresh / dried material. It is expressed on wet basis.

Molality (m): A solution in which one gram of moles (molecular weight) of the solute dissolved in 1000 gm of the solvent is known as molality.

Molar solution (M): A solution in which one gram mol weight of the substance is dissolved in the solvent and volume is made up to 1 litre.

Molasses: It is the thick dark to light brown syrup that is separated from raw sugarcane during sugar manufacture. Syrup made from boiling down sweet vegetable or fruit juice is also called molasses. It contains 20-30% moisture or water.

Mold: A superficial often woolly growth produced on damp / decaying organic matter or on living organisms by a fungus of the order Mucorales. It is a fungus type microorganism whose growth is visible on food and is colourful. It can grow on different foods including low acid foods like jams, jelly and canned foods.

Mole fraction: It is the ratio of number of moles of one component (solute or solvent) to the total number of moles present in the solution.

Monoethanol amine: It is used for absorption of the carbon dioxide in the atmosphere.

Monopoly market: It is a market situation, wherein there is only one seller of a commodity. Such competition occurs in Imperfect markets.

Monosaccharides: A sugar that is not decomposable into simpler sugars by hydrolysis and is classed as either an aldose or ketose, and contains one or more hydroxyl groups per molecule. These are also called simple sugar. The unbranched group of carbohydrate (CHO) compounds also known as simple or reducing sugars with the formula $(CH_2O)n$ in where n=3 or a large number (Hexoses are by most abundant) e.g. glucose, fructose, galactose, mannose, L-Rhambose, L-sorbose (Plate 433). Most of monosaccharides are hexose sugars which are glucose, fructose, galactose (xylose, arabinose, ribose).

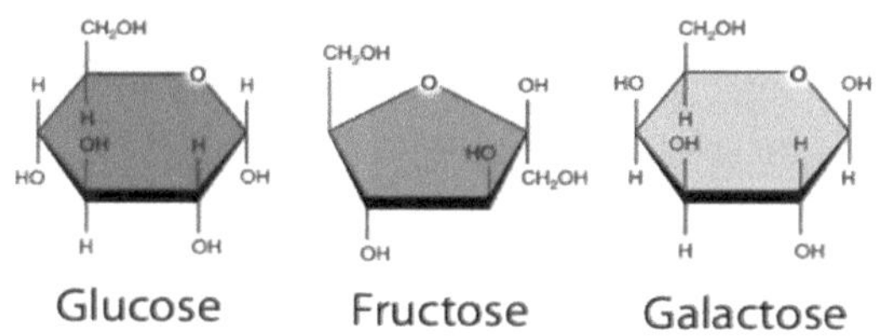

Plate 433: Examples of Monosaccharides

Monosodium glutamate (MSG): MSG can be produced by *Corynebacterium, Artherobacter, Brevibacterium* or *Micrococcus glutamicus* (*C. glutamicus*) because these organisms lack enzyme α – ketoglutamate dehydrogenase in Kreb's cycle and therefore convert α – ketoglutamate to glutamate instead of succinyl CoA. It is crystalline sodium salt $C_5H_8NO_4Na$ derived from glutamic acid and used to enhance the flavor of food. MSG (Monosodium glutamate) is one of the best known and widely used flavor enhancer.

Morchella: The morels comprise the genus commonly referred to as Gucchi in India which is an edible fungus (Plate 434).

Plate 434: Morchella / Gucchi

Mordant : It is a chemical that fixes a dye in or on a substance by combining with the dye to form an insoluble compound.

Moto: It is a mixture of koji along with a thick liquid consisting of steamed rice and water. Moto in incubated at 90-113°F for conversion of starch into sugar by *Aspergillus oryzae*. It is used in preparation of rice beer/ sake.

Mouth coating: Type and degree of coating in the mouth after mastication (for example, fat/oil).

Mouth feel: The textural quality of the food perceived by the mouth is called mouth feel. It is a product's physical and chemical interaction in the mouth.

MPF: It stands for multipurpose food. Indian MPF contains 75% low fat peanut flour, 25% Bengal gram flour plum vitamins and calcium. American MPF is based on soya.

MRI: It stands for magnetic resonance imaging. It is the procedure in which magnetic resonance imaging is used to check the quality.

MRL: It stands for maximum residue limit.

Mucilage: It is a complex glutinous carbohydrate (CHO) secreted by certain plants (Okra, Basella, Isabgol) which absorbs water freely.

Muenster cheese: It is a semisoft cheese that may be bland or sharp in flavor.

Muffin: It is a quick bread made of batter containing egg and baked in a pan having cuplike molds.

Muffle furnace: An apparatus which is used for incineration and estimation of the ash contents of the food samples. The sample is ashed at a temperature of about 500-550°C (Plate 435).

Multi effect evaporator: Evaporator operates at multi stages and has more than one heat exchanger (Plate 436).

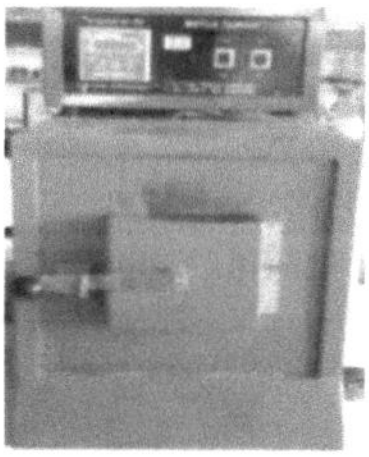

Plate 435: Muffle furnace

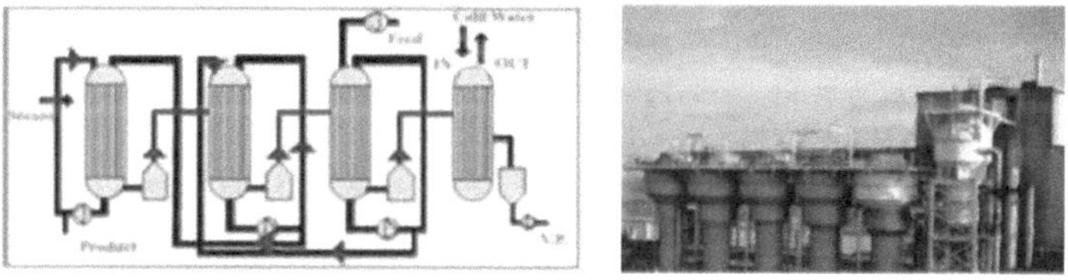

Plate 436: Multi-effect evaporator

Multiple Sample Difference / Rating Tests: These tests are used to test more than two samples of same food product having some differences in their qualities. The differences may be due to variation in the treatments etc. One sample is a known standard and so one test sample is a duplicate of standard. The difference of the test samples from standard sample can be evaluated on the basis of difference scale like none, slight, moderate, and large.

Munacca: Seeded grape in dried condition is known as munacca.

Munsell system: It is system of identifying colour on the basis of hue, value and chroma using a numerical scale.

Murphy's Law: It is a good rule of thumb regarding physical or chemical contamination.

MWA- 9002: It is hollowness detector in melon.

Mycotoxins: Mycotoxins are produced as secondary metabolites of certain moulds / fungi and can cause long term carcinogenic effects at ongoing low levels or short term acute toxic effects at high level of exposure (Plate 437). Mycotoxins are being considered here under biological hazards because they are products of microbial growth although they may also be considered as chemical hazards. The mycotoxins of concern in food production are aflatoxins, patulin, vomitoxin and fumanisins which may be consumed by humans via two routes, firstly, direct consumption of contaminated grain or fruit and secondly by indirect consumption via animal products such as milk.

Plate 437: Groundnut containing mycotoxins

Myristic acid: A crystalline fatty acid $C_{14}H_{28}O_2$ occurring in the form of glycerides in most fats.

Myxoxanthin: It is a carotenoids pigment found in algae and has vitamin A activity.

N

Naïve panel: It refers to the sensory evaluation panel that has not been trained specifically regarding the product evaluation being undertaken in the study. It is untrained panel of sensory evaluation.

Nano technology: It is the art of manipulating materials on an atomic or molecular scale to build microscopic devices. The word "*nano'* meaning *'dwarf" in* Greek language refers to dimension / magnitude of the material / particle in the order of 10^{-9}. Nanotechnology is nothing but manipulating of particles into very small particles called nano-particles. These particles have dimensions in the scale of nano-meter and used to design / build materials and devices. Nanotechnology has been provisionally defined as relating to materials, systems and processes which operate at a scale of 100 nanometers (nm) or less. It will allow many things to be manufactured at low cost and with no pollution. The concept of nanotechnology was first given by Nobel Laureate physicist Richard P. Feynman in Southern California in 1959.

$NaNO_2$: Pink colour of cured meat is preserved by $NaNO_2$. It is effective against *Cl. Botulinosm*

Nanogram: It is one billionth of a gram.

Nano-materials: Nano material have been defined as having one or more dimensions measuring 100 nm or less, or having at least one dimension at this scale which affects the materials behavior and properties (Plate 438).

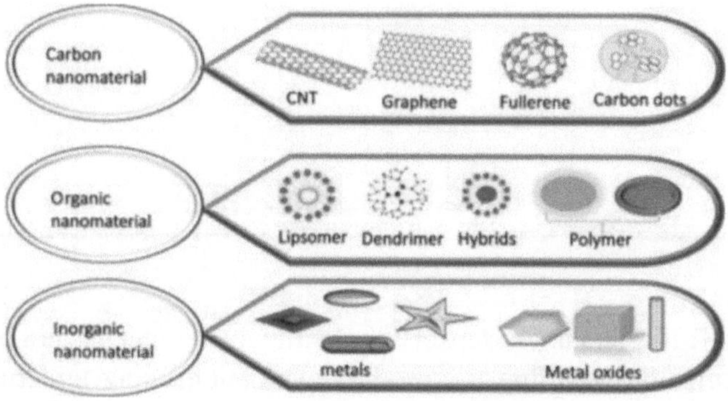

Plate 438: Different nano-materials

Nanometer: One nanometer (nm) is one thousandth of a micrometer (μm), one millionth of a millimeter (mm) and one billionth of a meter (m).

Nanoparticle: A microscopic particle whose size is measured in nanometers.

National canners association (NCA): National canners association (NCA) lab was established in 1907 by W.D Bigelow and Catheart, to study effect of acidity in lowering the processing requirement of canned foods. NCA is now known as national food processors association (NFPA).

Natural foods: The foods which are free from artificial flavouring, colouring, preservative etc are called natural foods.

Natural juice: It is pure juice that is extracted from the ripe fruits and contains only natural sugars.

Natural sparkling wine: If CO_2 occurs naturally as a result of fermentation it is called natural sparkling wine and if it is added artificially then it is carbonated wines.

NBS: It stands for National Bureau of Standards. The National Institute of Standards and Technology has had several different names. Founded as the National Bureau of Standards in 1901. It was renamed Bureau of Standards in 1903. In 1934, the word "national" was affixed again to its name. For more than 50 years it remained the National Bureau of Standards, or NBS. Now, it became the National Institute of Standards and Technology, or NIST, since 1988. Samuel W. Stratton (1901 – 1922) was the first director.

NCL: It stands for national chemical laboratory. National Chemical Laboratory (CSIR-NCL), Pune, established in 1950, is a constituent laboratory of the Council of Scientific and Industrial Research (CSIR). CSIR-NCL is a science and knowledge-based research, development and consulting organization.

Near infrared: It is relating to the shorter wavelengths of radiation in the infrared spectrum and. to those between 0.7 and 2.5 micrometers.

Near infrared reflectance (NIR): Near infrared reflectance (NIR) can be used for measuring moisture content using light emitting diode. NIR is a spectroscopic method that uses the NIR region of the electromagnetic spectrum (800-2500 nm). The typical application includes pharmaceutical, medical diagnosis, food and agrochemicals quality control (Plate 439).

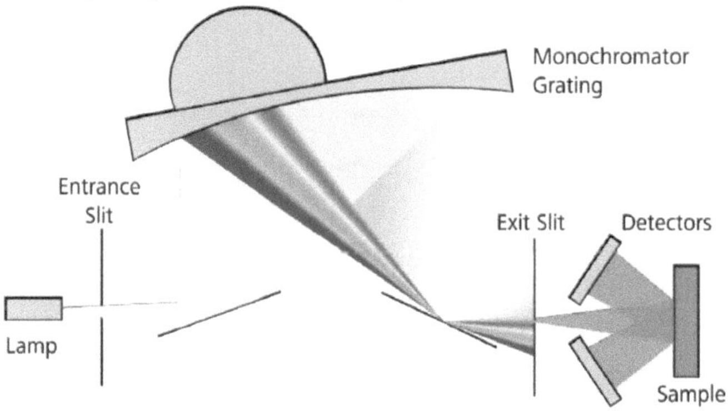

Plate 439: Near infrared reflectance / spectroscopy

Nectar: Fruit beverage contains at least 20% fruit juice and 15% TSS.

Neglected nutrient: The dietary fibre / roughage present in food are often regarded as the neglected nutrient.

Net's web: It is a part of the net that traps the fishes.

Neufchatel cheese: A soft unripened cheese similar to cream cheese but containing less fat and more moisture.

Neurotoxin: It is a poisonous complex especially of protein that acts on the nervous system. An endotoxin produced by *Cl. botulinum* which affects the peripheral nervous system by adversely disturbing the production of acetyl cholin necessary for conduction of nerve impulses.

Neutral spirits: Ethyl alcohol of 190 or higher proof used for blending other alcoholic liquors is known as neutral spirit.

Newtonian fluids: The fluids which obey the Newton's law of viscosity are called Newtonian fluids. Newton's law of viscosity is given by $\tau = \mu\, du/dy$. In such fluids, the flow rate is not affected by the shear rate e.g. water and sugar syrups. Newtonian fluids are chemically pure homogenous fluids whose viscosity or resistance to deformation does not change with shear rate (Plate 440).

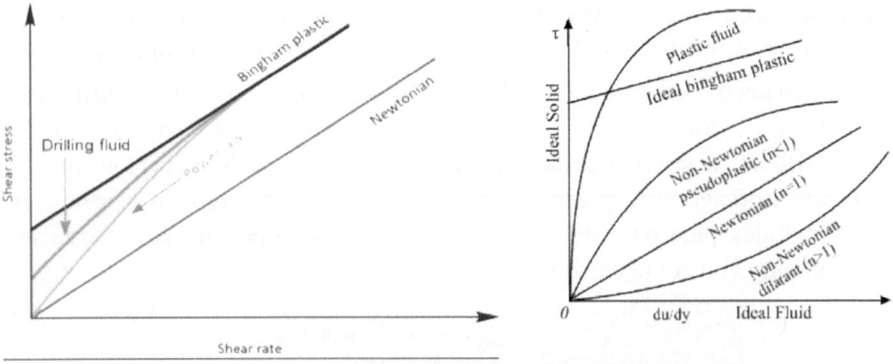

Plate 440: Newtonian fluids

Niacin: An acid $C_6H_5NO_2$ of the vitamin B complex found widely in animals and plants and used against pellagra. It is also called nicotinamide / nicotinic acid/ PP factor / antipellagera factor. Niacin is very stable to heat, light, oxidation, but as it is water soluble and so may leach out form food.

Niacinamide: It is a compound $C_6H_6N_2O$ of the vitamin B complex found as a constituent of coenzymes and used similarly to niacin.

Nickel: In hydrogenation powdered nickel is used as catalyst. Alternative to hydrogenation is inter-esterification. In this case no hydrogen atoms are added as in case of hydrogenation. Hydrogenation reaction is accomplished by bubbling of stirring H gas into a batch of pressurized hot oil contain finely powdered nickel.

Night blindness: Reduced visual capacity in faint light (at night).

NIN: It stands for National Institute of Nutrition Hyderabad. The National Institute of Nutrition (NIN) is an Indian public health, nutrition and translational research centre located in Hyderabad, India. It is under ICMR. National Institute of Nutrition (NIN) was founded by Sir Robert McCarrison in the year 1918 as 'Beri-Beri' Enquiry Unit. Originally, NIN was known as "Nutrition Research Laboratories". This Institute was established in Coonoor in 1918 and was located in the Pasteur Institute of South India. The Institute was shifted to Hyderabad in 1959. It was renamed as National Institute of Nutrition (NIN) during its Golden Jubilee held in 1978, and Platinum Jubilee in 1993 (Plate 441).

Ninhydrin reactions: It is a blue colour reaction used in the chromatographic identification of amino acids (Plate 442). Ninhydrin is the most widely used chemical reagent for the detection of latent fingermarks on porous surfaces such as paper and cardboard. The compound reacts with the amino acid (eccrine) component of the fingerprint deposit to give a dark purple product known as Ruhemann's purple.

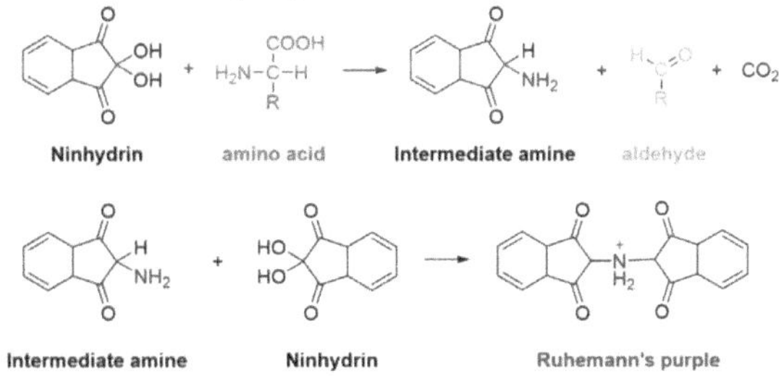

Plate 441: Logo of National Institute of Nutrition (NIN)

Plate 442: Ninhydrin reactions

NIR: It stands for Near Infra Red Spectroscopy. NIR is a spectroscopic method that uses the NIR region of the electromagnetic spectrum having wavelength of 800 to 2500 nm. The typical applications of the NIR includes in pharmaceutical, medical diagnosis, food and agrochemical quality control.

Nira: Alcoholic drink prepared from the juice of the palm tree is called Nira.

Nisin: It is an antibiotic obtained from *Streptococcus lactic*, used in the preservation of acid foods in which it is more stable. It is a polypeptide bacteriocin and is produced by *Lactococcus lactis.* It is a bacteriocidal rather than bacteriostatic.

Nitrite: A salt or ester of nitrous acid is called nitrite.

NMR: It stands for nuclear magnetic resonance.

No / Free /Zero Nutrition food: Foods contain no /zero or trivial (of little worth) level of nutrients like fat, saturated fat, cholesterol, sodium, sugar or calorie.

Nomograph: A nomograph is defined as a graph, usually containing three parallel scales graduated for different variables so that when a straight line connects values of any two, the related value may be read directly from the third vertical line at the point intersected by the line. Ex. Nomographs use three scales to predict boiling point. A nomograph shows a boiling point temperature scale, a boiling point temperature at sea level pressure scale and a general pressure scale. To use the nomograph, connect two known values using a ruler and read the unknown value on the third scale. The graphic representation used for measurement is called nomograph.

Non - essential amino acids: These are the amino acids which can be produce / synthesis by the human bodies itself and need not required to be taken from the out side sources (Plate 443).

Essential AA	Nonessential AA	Conditionally essential AA
Arginine	Alanine	Cysteine
Histidine	Asparagine	Glutamine
Isoleucine	Aspartate	Hydroxyproline
Leucine	Glutamate	Proline
Lysine	Glycine	Taurine
Methionine	Serine	
Phenylalanine	Tyrosine	
Threonine		
Tryptophan		
Valine		

Plate 443: Essential amino acids and Non-essential amino acids

Non climacteric fruits: These fruits are ripened on the fruits and not capable of continuing their ripening process once removed from the plant (once harvested). These fruits produce very small quantity of ethylene and do not respond to ethylene treatment except degreening in citrus and pineapple.

Non-enzymatic browning: Non-enzymatic browning reactions are responsible for the colour and flavor of foods (Plate 444). Roasting of potatoes, toasting of biscuits, baking of breads and cakes produces a golden brown colour. Sometimes this reaction produces a desirable flavor (the chocolate flavor of cocoa beans) at other times the reaction is responsible for browning in foods. On heating the sugars undergoes ring opening, enolisation dehydration, and fragmentation into unsaturated carbonyl compounds. Heat induced browning reactions can be divided into two groups maillard reaction and carmalizaton. The Maillard reaction occur between (i) nitrogenous compound and sugars (ii) nitrogenous compound and organic acids (iii) sugars and organic acids (iv) among organic acids themselves are known as maillard reaction. It may be independently causing browning of products which is of great important to food preservation techniques. It is difficult to control.

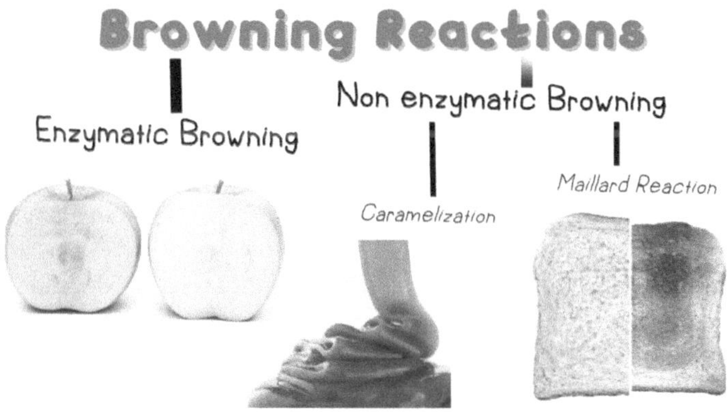

Plate 444: Enzymatic vs Non-enzymatic browning

Non perishable foods: These are the foods in which the spoilage is relatively slow.

Non-caloric: A food / drink free from or very low in calories are called non-caloric.

Non-Hookean food: A food which do not obeys Hooks law is called Non-Hookean food.

Non-Newtonian fluids: The fluids whose flow rate is influenced by the shear rare are called non-Newtonian fluids. E.g. chocolate and emulsions. Non Newtonian fluids have different apparent viscosities at different shear rate and are heterogenous solution such as citrus juice and concentrates. Citrus juice is primarily of non Newtonian nature due to pulp and cloud material with in the juice. Removal of these material results in serum of Newtonian nature. The term consistency and apparent viscosity is generally used for non-Newtonian fluids while viscosity is generally used with Newtonian fluids (Plate 445).

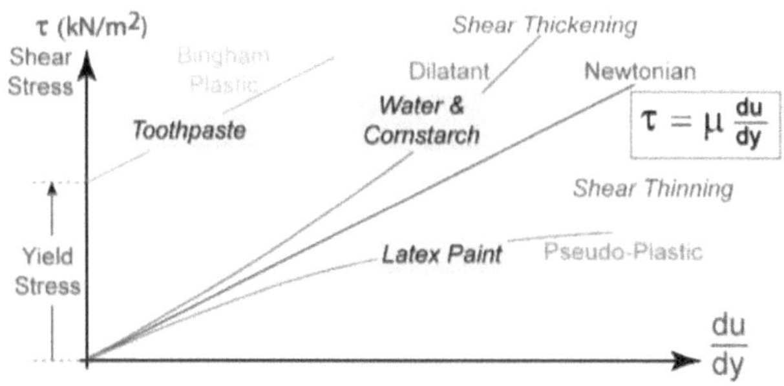

Plate 445: Non-Newtonian fluids

Non-nutritive/ low calories/ special dietary sweeteners: These are the sweetener which do not possess any nutritive value in the form of calori and provide all the benefits of the sweet taste as provided by the sugars like "sucrose, glucose or fructose" which is also a source of the energy. Various sweetening agents have been used as substitute for sucrose, which are 10-3000 times sweet as sucrose. Non-nutritive sweeteners are mainly used to manufacture low calories soft drinks, low calorie liquid foods, canned fruit, salad dressings, gelatine desserts and some baked products. Non-sugar sweeteners help to control weight and made it possible for diabetics to enjoy foods. Present sucrose alternatives viz; saccharine, cyclamate, aspartame fall short of expectations in delivering complete sucrose taste experience. Saccharin the first synthetic sweetener, which is 300 times sweeter than sucrose. Cyclamates are 15-30 times as sweet as sucrose, but FDA of USA has banned their use on basis of report that they cause bladder cancer in animals.

Non-reducing sugars: Reducing sugars which are locked up in a glycosidic linkage are said to be non reducing e.g. sucrose, maltose. Or When two or more monosaccharides are linked together through their aldehyde or ketone group so that their reducing groups are not free.

Noodles: A food paste made usually with or without egg or something other substances and shaped typically in ribbon form is called noodles (Plate 446). Very hard durum wheat is used for making noodles. Manual operated extruder or electrically operated single or twin screw cold extruders are employed in the preparation of noodles.

Plate 446: Wheat flour noodles

Normal solution: One normal solution of a substance contains one equivalent or one gram equivalent weight of the substance in one litre of solution i.e. mol. weight divided by the hydrogen equivalent of the substance. Or A solution having a concentration of one gram equivalent of solute per liter. It is also called as normality.

Normal starch: The starch which contains 40-70% of the amylose is called normal starch. Normal starch has 1:3 ratios of amylose and amylopectin content. Normal starch have smooth surface of the granules.

Novel processing techniques: HPP, PEF, Irradiation, ultrasonication and cold plasma which influence on consumer's health have been the major innovations in the field of processing technology. These novel techniques act by prolonging the shelf life, enhancing or maintaining the quality, and to regulate freshness of food product (Plate 447).

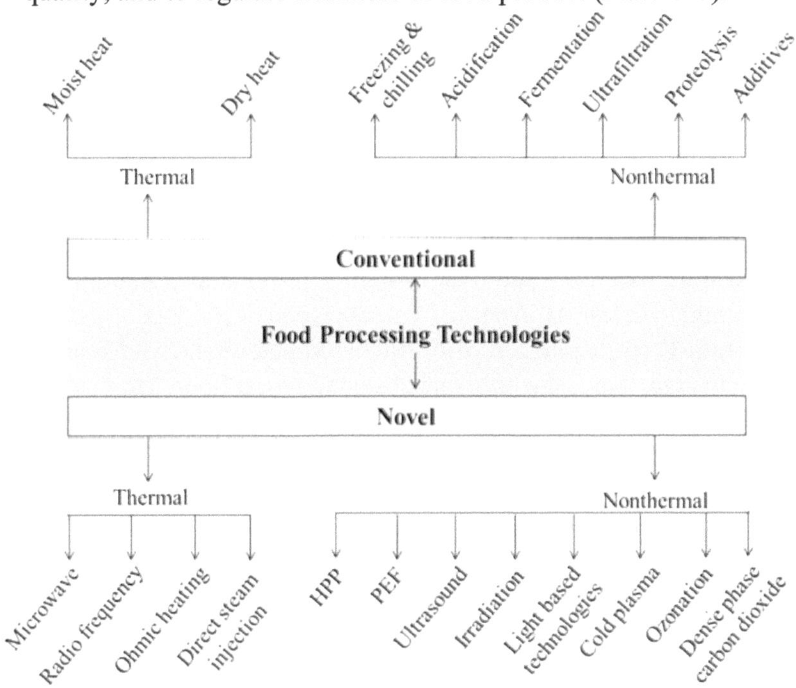

Plate 447: Novel vs conventional processing techniques

Nougat: It is a confection of nuts or fruit pieces in a sugar paste. It is high boiled syrup containing fat to which frappe is added.

NPDES: It stands for National Pollution Discharge Elimination System.

NTB's: It stands for non-tariff barriers to trade.

Nuclear magnetic resonance: The magnetic resonance of an atomic nucleus and chemical analysis that uses such resonance to study molecular structure.

Number of transfer unit (NTU): It is a method to evaluate the effectiveness of heat exchanger.

Numerical scoring / rating test: The numerical scoring test is used to evaluate particular characteristics of one or more samples. The scales like excellent, good, fair and poor are used for scoring / rating of the samples.

Nutraceutical: A foodstuff (fortified food or dietary supplement) that provides health benefits in addition to its basic nutritional value. These are naturally occurring chemical compounds that are derived from a plant, animal or marine source which impart health benefits to human. Nutraceutical is the name given to a range of food supplements which are designed to nutritionally support certain illness, improve metabolic processes and correct biochemical balance etc. This food act as blocking agents, suppressing agents or chemo preventive compounds for prevention of disease.

Nutrient supplements: During processing of foods and food products, there may be certain loss of some nutrients. Nutrient supplements help to restore this loss. These are vitamins & minerals, vitamin D (added to milk), vitamin B, iron and calcium (cereal products), iodine (salt), vitamin C (fruit juices & fruit flavoured desserts). Lysine an essential amino acid, and is the only one absent in wheat flour. Moreover, Lysine is not permitted in wheat flour and bread because sufficient lysine is available from other foods in a normal diet.

Nutrient: A substance that provides essential nourishment. It may be organic or inorganic.

Nutriment: It is something that nourishes or promotes growth, provides energy, repairs body tissues, and maintains life.

Nutrition: The act or process of nourishing or being nourished or the sum of the processes by which an animal or plant takes in and utilizes food substances. Or It is the study of composition and quantities of food materials required by living organisms for growth, maintenance and survival.

Nutritional marasmus: A condition of chronic undernourishment / malnutrition (PEM) occurring in children and caused by a diet deficient in calories and proteins (Plate 448 and 449).

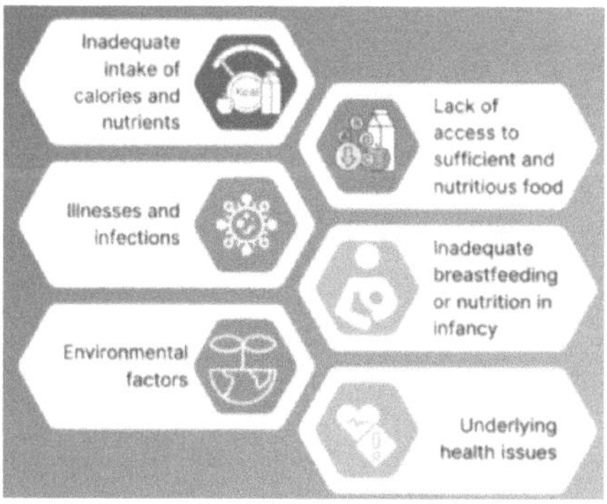

Plate 448: Cuases of Marasmus

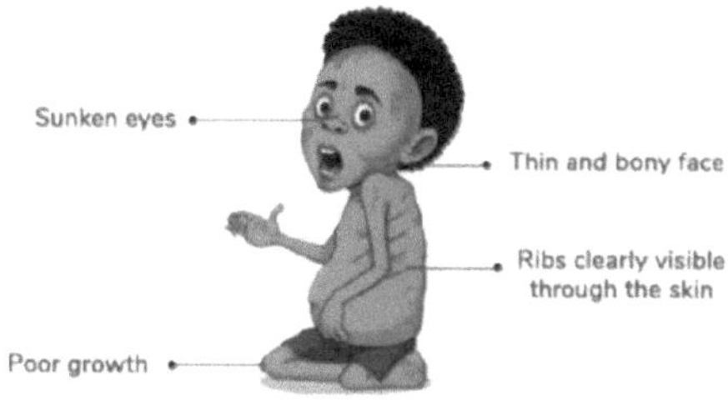

Plate 449: Symptoms of marasmus

Nutritive ratio: It is the ratio of digestible protein to other nutrients in a foodstuff or ration. It is the ratio of non-nitrogenous nutrient including fats multiplied by 2.25 to the digestible crude protein. NR = Digestible carbohydrate + digestible fibre + (digestible fat x 2.25) divided by digestible crude protein.

Nutritive value: It is characterized by the chemical composition, digestibility and nature of the digested products of the food material.

O

Objective Methods: Objective Methods are used for evaluation of food quality. These are based on recognized standard scientific tests to any sample of the product without regard to its previous history. They represent the modern idea in quality control (QC) because the human element has been excluded. This method divided into three groups like physical, chemical and microbiological. The devices are used in this evaluation.

Obligate aerobes: These microbes grow only in the presence of oxygen (Plate 450).

Obligate anaerobes: These microbes do no grow in the presence of oxygen (Plate 450).

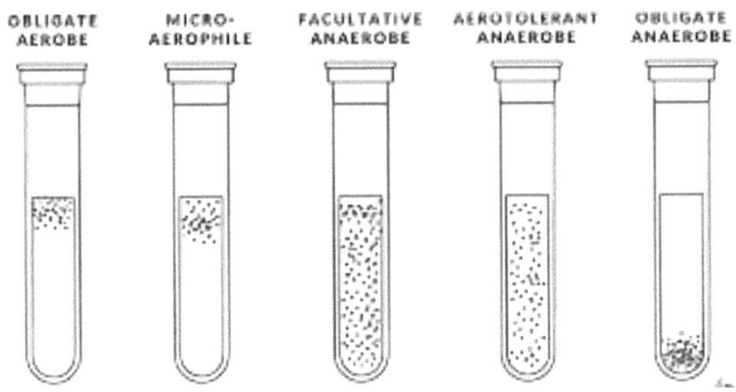

Obligate Aerobe. Strictly aerobic, gets kill under anaerobic conditions, e.g., Bacillus subtilis,

Facultative Anaerobe. Aerobe which can switch can switch over to anaerobic respiration e.g., Clostridium botulinum.

Obligate Anaerobe. Strictly anaerobic, gets killed under aerobic conditions, e.g., Costridium botulinum.

Facultative Aerobe. Anaerobe which can switch over to aerobic respiration, e.g., photosynthetic bacteria.

Plate 450: Illustration and examples of Obligate aerobes / anaerobes vs facultative

Obligate parasite: Obligate parasites is host specific and grows on living host only and attach limited range of host and are very difficult or impossible to grow on artificial culture/media.

Obligate thermophile: These microorganisms grow at a temperature between 38-80°C.

Oenology: Treatments used to prevent spoilage of wine is called oenology. It is synonym to that of enology.

Offal: It is the viscera and trimmings of a butchered animal removed in dressing of meat. It is by-product of slaughtering process.

Ohmic heating: It is novel sterilization process in which heat is generated within a food by the passage of electric current, making it possible for particles to heat at the same or faster rate than fluids (an ideal ohmic process) whereas in practical scenario, heating rates depends on the electrical and thermal properties of both particle and liquid. This technique utilizes the principle that heat can be generated throughout a material by the passage of electric current, using the inherent electrical resistance of the material. Ohmic heating is defined as a process where primary alternating electric current is passed through a food or conducing solution (salt brine) with the purpose of heating them. It is also known as Joule heating, electric resistance heating, direct electrical resistance heating, electro heating and electro conductive heating. In ohmic heating, the presence of the electrode results in heating of the foods (Plate 451, 452, 453, 454).

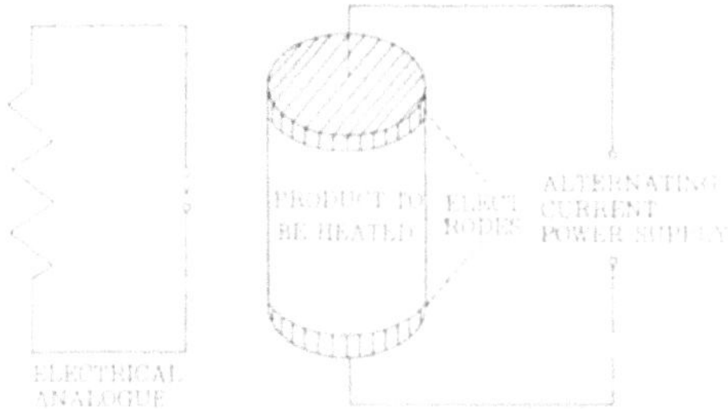

Plate 451: Mechanism of ohmic heating

Plate 452: Stainless steel electrodes used for heating cheese through ohmic heating

Plate 453: Insulation box containing heating system

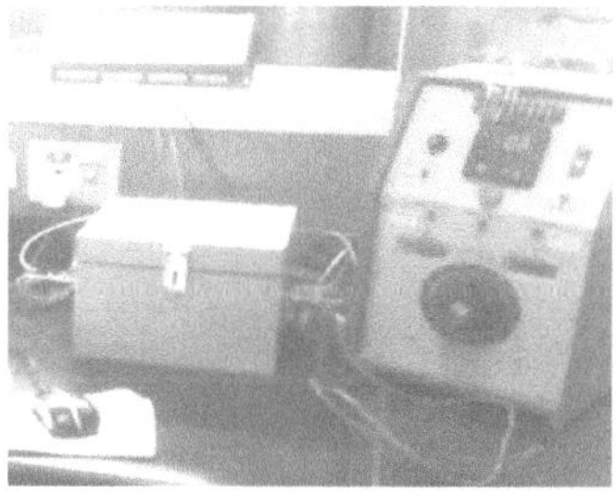

Plate 454: Complete ohmic heating system with control panel

Oil: Fats liquid at room temperature is called oil. The greater the degree of instauration in the fatty acids of fat molecule, the softer the fat is at a given temperature and lower is its melting point. Or Oil is long chain fatty acids esterified to glycerol which may be saturated or unsaturated (i.e. contain one or more double bonds) and liquid at room temperature (Plate 455).

Oil cake: It is also called oil press cake (Plate 456). The solid residue left after extracting the oil from seeds (cotton, apricot, mustard etc.). The apricot oil press cake can be used for extraction of protein isolate, extraction of essential oil and preparation of hawan samagri.

Plate 455: Apricot oil

Plate 456: Oil cake

Oil point: It is the critical temperature / point at which there is maximum oil recovery.

Oil of vitriol: The concentrated sulfuric acid is called Oil of vitriol.

Oil Spotting: A disorder of citrus due to rough handling or harvesting where turgid fruits show necrosis of epidermis adjacent to oil glands (Plate 457). If oranges are too turgid at harvest, the oil glands in the skin can be ruptured, releasing phenolic compounds and causing oleocellosis. The release of phenolic compounds and oil from the glands of citrus skin is called oil spotting. It is also known as Oleocellosis or green spotting.

Oleic acid: A monounsaturated fatty acid $C_{18}H_{34}O_2$ obtained from natural fats and oils.

Oleocellosis: A disorder of citrus due to rough handling or harvesting where turgid fruits show necrosis of epidermis adjacent to oil glands (Plate 457). If oranges are too turgid at harvest, the oil glands in the skin can be ruptured, releasing phenolic compounds and causing oleocellosis. The release of phenolic compounds and oil from the glands of citrus skin is called oleocellosis. It is damage to the rind of the citrus fruits due to the release of the oil. It is also known as oil spotting or green spotting.

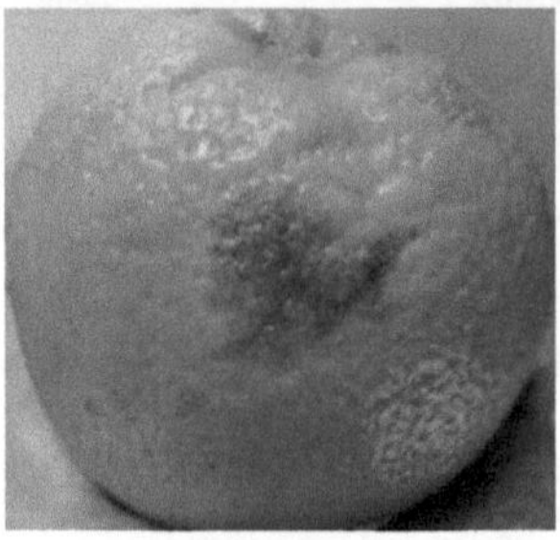

Plate 457: Oleocellosis/Greening spotting/oil spotting

Oleo-margarine: It is a fatty substance extracted from clarified animal fat (meat) and made into margarine (cooking medium) with addition of butyrin, milk etc.

Oleoresin: A natural combination of resinous substances and essential oil present in certain crop plants like chilli fruit. These are extracted by using solvents.

Oleuropein: Bitter glycoside of olive (Olea europaea) (Plate 458 and 459). Oleuropein is a phenolic compound found in olive leaves and the oil of its fruit. Oleuropein occurs in glycosylated and non-glycosylated forms that alters its solubility. The non-glycosylated form oleuropein predominates in the lipid fraction (olive oil).

Plate 458: Olive fruit

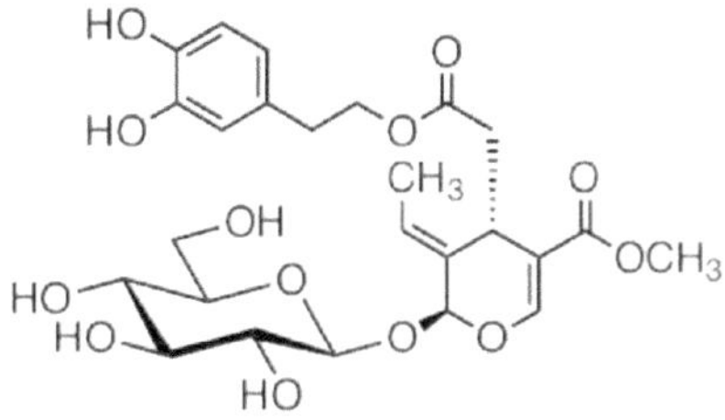

Plate 459: Oleuropein

Oligodynamic sterilization: Preservation or sterilization of liquids by metallic ions such as Ag^{++} due to their germincidal effects on microorganisms, concentration of Ag^{++} to the tune of $2x10^{-11}$ mole/l has got germicidal activity towards the microorganisms.

Oligomer: It is protein comprising of only a small number of identical polypcptidc subunits.

Oligopoly market: In this market there are more than two but still a few sellers of commodity. It is Imperfect market.

Oligosaccharides: Monosaccharide units (2 to 10) join in glycosidic linkage to form oligosaccharides the most common disaccharides include sucrose (glucose + fructose), maltose (Glucose + glucose), lactose (Glucose + galactose) in milk, cellobiose (glucose + glucose) and among trisaccharides found in nature are raffinose (Fructose + glucose + galactose) in sugar beets and melezitose (glucose+ fructose +glucose) in some conifers. These are saccharides (disaccharide) that contains a known small number of monosaccharide units (Plate 460).

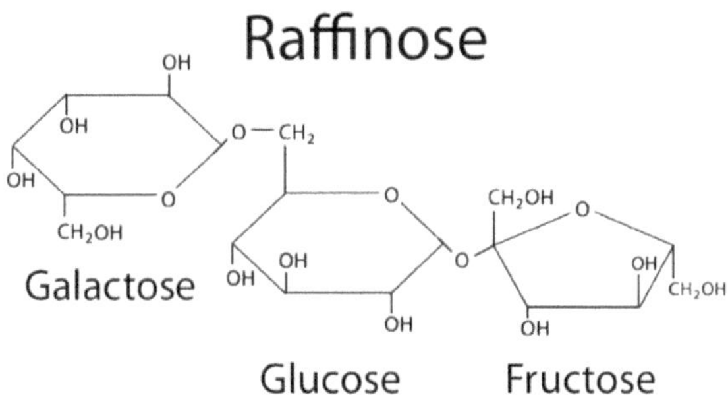

Plate 460: Raffinose an example of oligosaccharide

Omega–3: These are composed of polyunsaturated fatty acids that have the final double bond in the hydrocarbon chain between the third and fourth carbon atoms from the end of the molecule opposite that of the carboxyl group and that are found in fish, fish oils, green leafy vegetables, and some nuts and vegetable oils. Fatty acids having carboxyl group attached at one end is called omega fatty acid. The lower the ratio of omega 3 to omega 6, better is the oil.

Omega–6: These are composed of polyunsaturated fatty acids that have the final double bond in the hydrocarbon chain between the sixth and seventh carbon atoms from the end of the molecule opposite that of the carboxyl group. Omega 6 must be in large quanyity than omega 3.

Omelet or omelette: The beaten eggs cooked without stirring until set and served folded in half.

Onsite storage: It means delaying harvest until crop is required for consumption. It can be done in certain cases with root crops. It is also called in-situ storage.

One stage drying: The drying of the food at one single constant temperature is called single stage drying. One stage drying mostly result poor quality of the dehydrated product.

Oolong: Tea made from leaves that have been partially fermented before firing.

Open Market auction/bid: Under this system arhatia or broker invites bids for their produce and produce is sold to highest bidder. It is better than any other system as sale is by an open auction. The buyers and sellers know clearly the price quoted and leaves no room for being cheated. Different systems as ahead ensure fair dealing to the parties and secure a premium for superior quality.

Optical Imaging: It is an optical imaging technique. Optics usually describes the behavior of the visible, UV, IR light used in imaging. Optical imaging is divided into diffusive ana ballistic imaging system. Diffusive Optical Imaging is also known as Near Infrared Optical Tomography (NIROT).

Orange pekoe: A tea made from the smallest and youngest leaves of the shoot is called orange pekoe (Plate 461).

Plate 461: Orange pekoe tea

Ordinary Food: If vitamins and minerals in a diet are less than 50% of the RDA, then it is called as Ordinary Food.

Organic: A fertilizer of plant / animal origin or a pesticide whose active component is an organic compound or a mixture of organic compounds or a food produced by organic farming.

Organic food: Food crops (food grains, fruit, vegetable etc.) which are produced by organic approach which is an integrated living system aims to enhance and develop the biological cycle's in the system involving microorganisms, soil life, plants and animals. These are the food products which are produced or handled under organic farming, handling and manufacturing practices are stated by state organic certifying agencies.

Organic preservatives: Organic preservative are Benzoic acid, esters of p-hydroxyl benzoic acid, o- chlorobenzoic acid or salicylic acid.

Organization: The organization is characterized by complete conformity to the standards and requirements. Or It is an administrative and functional structure.

Organoleptic: This term is related to senses like taste, aroma, flavour, odour, colour and feel.

Organoleptic evaluation: Descriptive index characters of the fruits or cooked sample of vegetables for appearance, firmness, taste, mouth feel etc.

Osmol or osmole: A standard unit of osmotic pressure based on one molal concentration of an ion in a solution. It is blend of osmosis and mol.

Osmolality: The concentration of an osmotic solution when measured in osmols or milliosmols per 1000 grams of solvent.

Osmolarity: The concentration of an osmotic solution when measured in osmols or milliosmols per liter of solution.

Osmometer: An apparatus for measuring osmotic pressure (Plate 462).

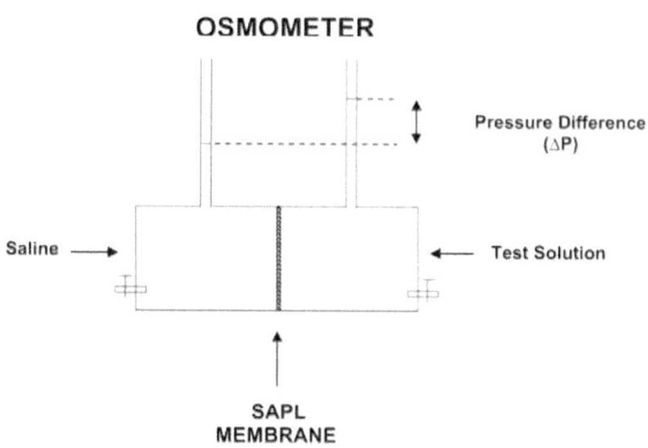

Plate 462: Osmometer working illustration

Osmophillic organism: Organisms that grow in high concentration of sugar and are harmful for concentrated fruit juice, honey, jam, jelly and other sugar solution e.g. *Leuconostus Bacillus cereus, Candida utilis, Aspergillus falvus* etc.

Osmosis: Flow of solution form region of higher concentration to that of region of lower concentration is called osmosis. Or It is spontaneous flow of the solvent through a semi-permeable membrane from a solution of low concentration to one of higher concentration. or The movement of a solvent (water) through a semipermeable membrane (of a living cell) into a solution of higher solute concentration that tends to equalize the concentrations of solute on the two sides of the membrane.

Osmotic dehydration: It is removal of the moisture from the food by the addition of humectants (glycerol, sugar etc) which exert osmotic action on the microorganisms by binding free water and make it unavailable to microbes and resulting in increase in non-refrigerated shelf life. In this dehydration, the water diffuses from dilute solution to concentrated solution (hypertonic solution) through a semi-permeable membrane until concentration equilibrium is reach. In this technique there is simultaneous counter current mass transfer of water from the solution to hypertonic solution. It improves shelf life and quality of the food (Plate 463).

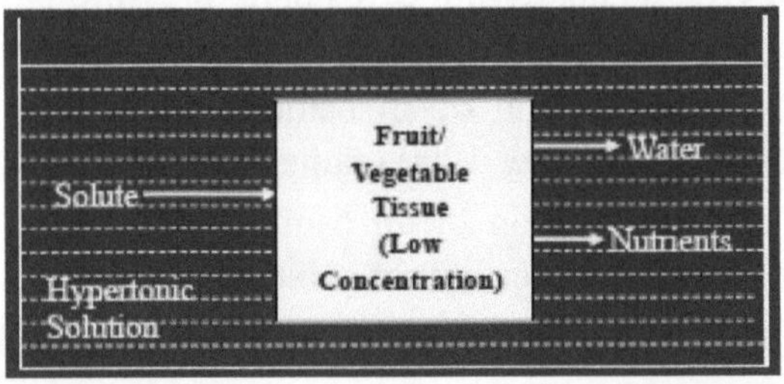

Plate 463: Mass transfer in food during osmotic dehydration

Osmotic pressure: The pressure produced or associated with osmosis and dependent on molar concentration and absolute temperature is called osmotic pressure (Plate 464). It is the maximum pressure develops in a solution separated from a solvent by a membrane permeable or it is the pressure that must be applied to a solution to just prevent osmosis. Or It is a chemical force caused by a concentration gradient. It is a colligative property and the principle behind membrane processing. Or It is the excess pressure applied to a solution to prevent the passage of solvent into it through the semi-permeable membrane.

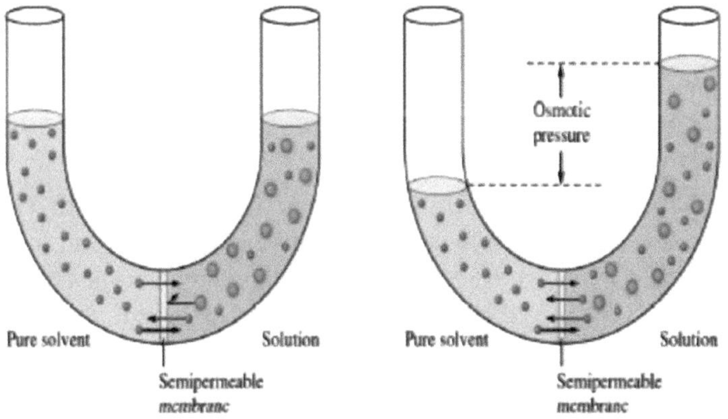

Plate 464: Diagram showing development of osmotic pressure

Osmotic shock: A rapid change in the osmotic pressure by transfer to a medium of different concentration and thus affecting a living system.

Osmovac drying: It is a two stage process involving osmotic drying in combination with vacuum drying. Here high concentration of sugar prevents enzymatic browning.

OTS cans: It is open top sanitary cans comprising one body and two ends made of tin plate.

Otto of Rose: Water distillation of rose flowers by redistillation. It is known as Ruh Gulab.

Overall heat transfer coefficient (U): It is thermal conductivity per unit area. It includes all heat transmission process through the system (W/m^2K).

Overflow method of pasteurization: In this method, the juice is heated to a temperature about 2-5^{0}C higher than the pasteurization temperature and then filled in hot sterilized bottles up to brim, taking care that during filling and sealing the temperature of juice does not fall below the pasteurization temperature. The sealed bottles are pasteurized at a temperature 2.5^{0}C lower than the filling and sealing temperature and then cooled. This method is very suitable for grape juice preservation because it prevents and minimizes the adverse effect of air on the juice quality.

Over-nutrition: Excessive intake of one of more nutrients.

Ovoglobulin: It is an egg protein and act as an excellent foaming agent.

Oxalic Acid: A poisonous strong acid $(COOH)_2$ or $H_2C_2O_4$ that occurs in various plants (as spinach) as oxalates and is used as a bleaching or cleaning agent and as a chemical intermediate.

Oxaloacetic acid: A crystalline acid $C_4H_4O_5$ that is formed by reversible oxidation of malic acid (carbohydrate metabolism via the Krebs cycle) and in reversible transamination reactions (aspartic acid).

Oxalosuccinic acid: A tricarboxylic acid $C_6H_6O_7$ that is formed as an intermediate in the Krebs cycle.

Oxidants: These are constituents (the oxygen of the air) who adversely affect the food e.g. oils are affected by the oxygen of the air and they become rancid. Here oxygen is an oxidant that causes oxidation. The metals like copper and iron are the strong promoters of oxidation.

Oxidase: It is one of enzymes that catalyze oxidations and enable one to react directly with molecular oxygen.

Oxidation: It is the act or process of oxidizing a substance.

Oxidation state: A positive or negative number that represents the effective charge of an atom or element and that indicates the extent or possibility of its oxidation. It is also called oxidation number. The oxidation state of sodium is +1 and of oxygen -2.

Oxidation–reduction reaction or redox: It is a chemical reaction in which one or more electrons are transferred from one atom or molecule to another. Addition of hydrogen or electrons is reduction, and removal of hydrogen or electrons is oxidation (originally applied to combination with oxygen but now including transfer of hydrogen or electrons) (Plate 465). The processes always occur simultaneously i.e. one substance is oxidized by the other, which it reduces. The conditions of the substances before and after are called oxidation states, to which numbers are given and with which calculations can be made. (Valence is a similar but not identical concept.) The chemical equation that describes the electron transfer can be written as two separate half reactions that can in theory be carried out in separate compartments of an electrolytic cell, with electrons flowing through a wire connecting the two. Strong oxidizing agents include fluorine, ozone, and oxygen itself; strong reducing agents include alkali metals such as sodium and lithium.

Oxidize: It is the process to combine with oxygen or to dehydrogenate by the action of oxygen or remove one or more electrons from an atom, ion, or molecule.

Oxidizing agent: It is a substance that oxidizes something by accepting electrons.

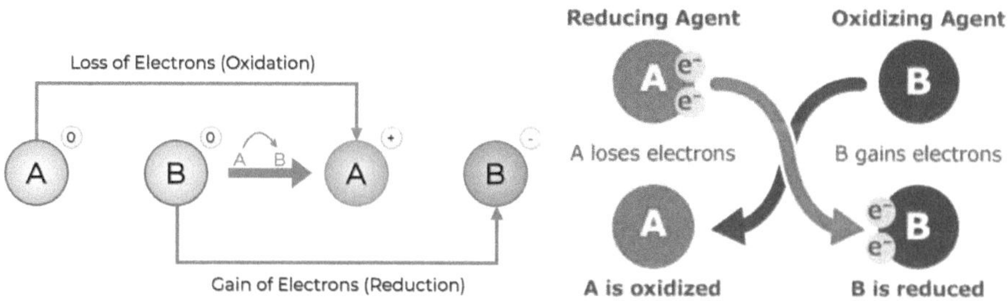

Plate 465: Illustration of oxidation and reduction reaction

Oxido-reductase: These enzymes catalyze oxidation and reduction reaction of the substrate.

Oxygen absorbers: Oxygen absorbers remove the oxygen by physically trapping the oxygen and not by through chemical reaction.

Oxygen isotope analysis: It is the best method to detect the water addition / adulteration in the juices particularly for citrus juice. As citrus juice is the only source of the isocitric acid. If the water is added to the juice from the outside source, then the addition of the water can be determined by measuring the quantity of the minerals in the water which is used in the industry. If the water which is generally used in the industry, contain higher concentration of the minerals as compared to the juice. So if the juice contains minerals higher than the range of the prescribed minerals in the juice it can be said that water is added in the juice. If added water do not contain higher concentration of the minerals, then it becomes difficult to estimate the water addition. So the water addition can be measured by the other methods which involves the measurement of oxygen isotope ratios (R = O^{18}/O^{16} isotope ratios). $S^{18}O$ (ppt) = (R sample – R standard) (1000)/ R standard.

Oxygen scavenging materials: Another approach to extend the shelf life of the foods is through utilizing oxygen absorbing materials. These materials together with modified atmosphere packaging techniques can be effectively used to reduce the oxygen concentration to very low levels to eliminate spoilage by mold and rancidity. For example, a sachet containing mixed iron powder – calcium hydroxide has been commercially used for the scavenging of both oxygen and carbon dioxide in ground coffee to triple the shelf life of the product. E.g. cobalt and antioxidants.

Ozone: Ozone is a powerful oxidizing agent and reacts with ethylene to produce carbon dioxide and water. It is a triatomic very reactive form of oxygen that is a bluish irritating gas of pungent odor, that is a major air pollutant in the lower atmosphere but a beneficial component of the upper atmosphere and that is used for oxidizing, bleaching, disinfecting, and deodorizing.

Ozonization: It is non thermal method of killing micro-organisms using ozone, through the oxidation of their cell membrane (Plate 466).

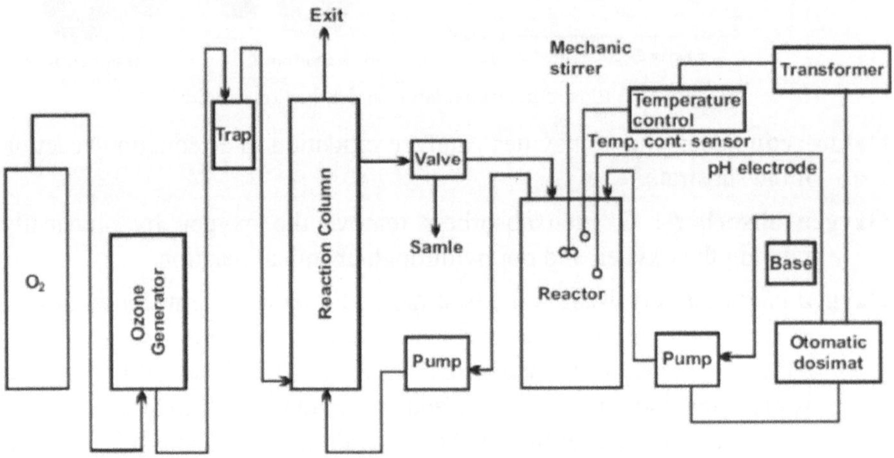

Plate 466: Ozonization system

P

Pack inspection: It is similar to continuous inspection except that here inspector(s) may make in-process checks during different processing operations under contract but not required to be present all the times in the plant during operation.

Package: The package means a pre packed box, tin, barrel, pouch or can etc in which food article is packed.

Package store: A store that sells bottled or canned alcoholic beverages for consumption off the premises.

Packaging: Packaging is an industrial and marketing technique for containing, protecting, identifying and facilitating the sale and distribution of agricultural, industrial and consumer products (Plate 467). Packaging is defined as an enclosure of products, items or package in a wrapped pouch, bags, box, cup, bottle or any container form to perform one or more of the functions as containment, protection and preservation, communication and utility or performance. Technically, packaging is a system or method by which a product from the manufacturing point to the consumer level reaches in a safe and sound condition at an affordable cost. The primary purpose of the packaging is to protect the food product, to keep it in good condition and to preserve the flavour until it reaches the consumer. It is therefore, very essential that a suitable form of the packaging is chosen for a finished product. Packaging is an art, which determines the buying decision of a consumer. or Wrapping / placement of objects in paper / other packaging materials, keeping in box, etc is known as packaging which has two main functions, to assemble the produce / objects in convenient units for handling and to protect them during transporting, marketing and storage which results in extended shelf life of the produce. "Packaging" play a very important role not only in ensuring safety of the product during transportation, storage and distribution and in conveying information about the pack contents to the consumers, but also ensuring desired sensory characteristics and micro-biological soundless till consumed.

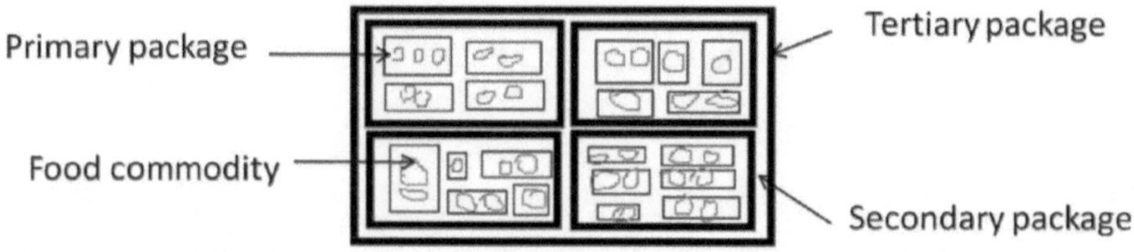

Plate 467: Illustrative diagram for primary, secondary and tertiary packaging materials

Packing: Wrapping or placement of object in paper or other packaging materials is known as packing

Packing house: It is an establishment for slaughtering livestock, processing and packing meat, meat products, and by-products. An establishment used for processing and packing other foodstuffs is also called packing house. It is also called packing plant.

Packman: One who offers merchandise (fresh produce) for sale along the street or from door to door. It is also called pedlar.

Padding materials: These materials are used in the packaging to maintain the placement or immobilization of the packed commodity. Natural padding materials are leaves, straw, grass and coconut husks.

Paddy cleaner: A machine used for cleaning paddy.

Paddy separator: A machine for separation of shelled paddy from those that remains unshelled. This is done after before shelling the paddy.

Paddy Sheller: It is equipment used for removing husk from paddy grain.

PAGE: It stands for Polyacrylamide gel electrophoresis. Polyacrylamide gel electrophoresis (PAGE) is a method of separating DNA fragments/ proteins depending on size, structure, and molecular weight (MW). The gel is prepared by polymerizing acrylamide with the cross-linking agent N,N′-methylenebisacrylamide (bis-acrylamide) (Plate 468).

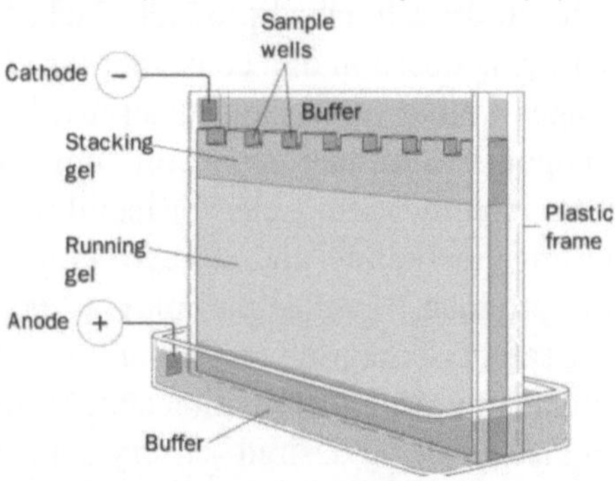

Plate 468: Illustraytion of PAGE

Paired comparison test: In this test, the panelists are given two samples. Out of these two samples, one is standard sample and the other one is experimental sample. Here the panelist is asked to find out the difference between the two samples and to predict whether the experimental sample is similar to that of standard sample or not.

Pak-Gard-Dong: It is a fermented mustard (*Brassica juncea)* leaf product made in Thialand. Thai pickled mustard greens (Pak Gad Dong) are made by pickling Chinese mustard greens in sticky rice water, distilled vinegar, and salt. Once the pickling process is complete, the mustard greens turn sour and salty with a slightly crunchy texture the leafy mustard greens (Plate 469).

Plate 469: Pickle of mustard greens

Palatability: The quality characteristics like colour, flavour and texture of a food product that make an impression on the sensory organs and have significance in determination of the acceptability of the food.

Palletization: The staking of containers / packets of materials for assembling, storing and handling or shipping with a fork lift in a unit load with the help of platform (Plate 470).

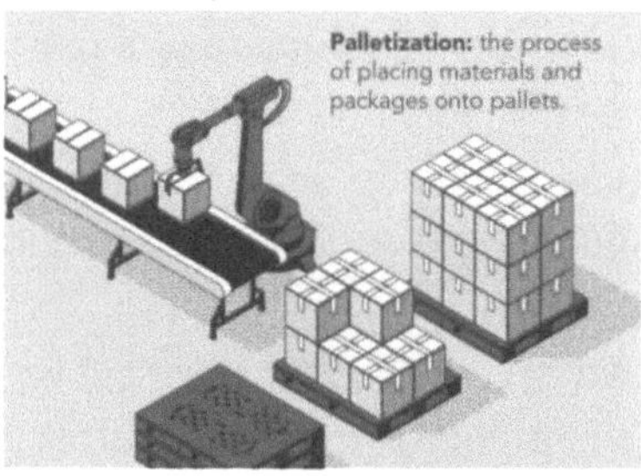

Plate 470: Palletization

Pallets: Devices made up of wood or iron, etc. and used for palletization.

Palm sap wine: It is also called toddy or coyol wine.

Palmitate: It is a salt or ester of palmitic acid.

Palmitic acid: A waxy crystalline saturated fatty acid $C_{16}H_{32}O_2$ occurring free or in the form of esters (glycerides) in most fats and fatty oils and in several essential oils and waxes.

Palmitin: An ester of glycerol and palmitic acid and is a solid ester found in animal fats.

Panelling: Development of flat vertical dents, in large can due practical collapse of the body under high internal vacuum by poor exhausting of cans which does not spoil the products but cans become unmarketable due to its appearance. This can also occur when the tin plate is thin or the cans are pressure cooled at very high pressure.

Papain: Milky latex from the immature papaya fruits is called papain (Plate 471). It is a protease in the juice of unripe papaya that is used as a tenderizer for meat and in medicine for topical debridement of necrotic tissue (removal of lacerated, devitalized, or contaminated tissue by anesthetia). It possesses proteolytic activity.

Plate 471: Papain from papaya

Papaya: A tropical American tree (*Carica papaya* of the family Caricaceae, the papaya family) having an oblong to globose yellow edible fruit with numerous black seeds in a central cavity.

Paper chromatography: Chromatography that uses paper strips or sheets as the adsorbent stationary phase through which a solution flows (Plate 472).

Paraben: It is a food additive similar to benzoic acid but effective at higher pH. Its chemical name is p-hudroxybenzoic acid. It is used for preservation of cheese, maragarines, jams, beverages, pickles and meats.

Parallel flow heat exchanger: Two fluid streams flowing in same direction through the heat exchanger is called Parallel flow heat exchanger (Plate 473).

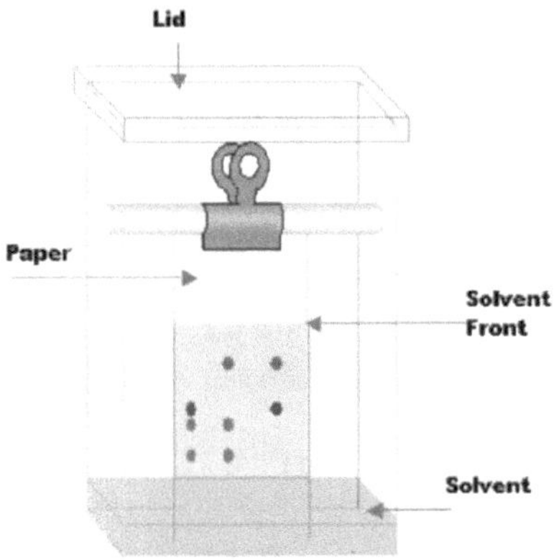

Plate 472: Paper chromatography technique

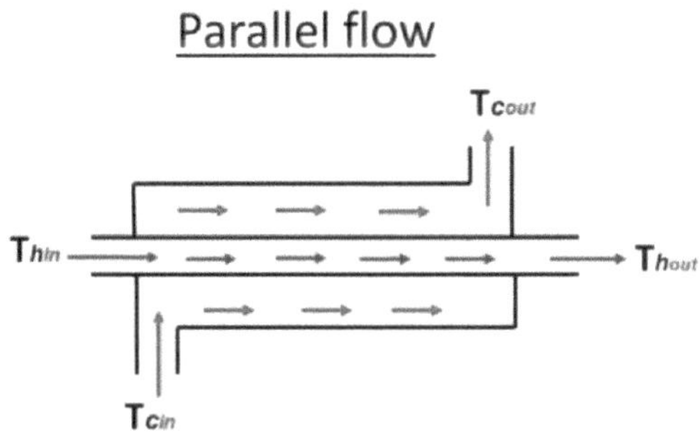

Plate 473: Parallel flow flow heat exchanger

Parasites: The larvae of parasites such as pathogenic flat worms, tapeworms and flukes may infect man by the consumption of flesh of infected pork, beef, and fish. Examples includes *Taenia saginata* (beef tapeworm), *Trichinella spiralis* (nematode in pork) etc.

Parboiled rice: Parboiling is a treatment to which paddy is subjected before milling. It includes soaking the paddy in water, steaming it in hot water and then drying in the sun. It makes husk removal easier, reduces breaking of the seed during milling and improves the keeping quality of the rice. Parboiled rice retains more nutrients while boiling and cooking as food.

Parch: It is the process to toast / shrivel (loss of moisture) food under dry heat or cold.

Parched food: A food deprived of natural moisture or food deficient in moisture content is known as parched food.

Partial pressure: The pressure exerted by a (specified) component in a mixture of gases.

Particle board: A composition board made of very small pieces of wood bonded together.

Particle separation: The process of separating particles from a suspension (Plate 474).

Partition chromatography: The apparatus used for partition chromatography is just like displacement chromatography with one exception. The inert packing is coated with a film of high boiling liquid that is strongly absorbed to the particle surface. This is called stationary phase. Second liquid called the mobile phase, is then passed through the column. The solution selected in this does not dissolve or mix with the stationary phase. Even this solution do not adsorbed to the inert phase as did the stationary phase. Partition chromatography is used in separating hydrophilic homologous when a solution of A and B is placed at the top of a column and the mobile phase S is added, the compounds proceed slowly down the column. The molecules remaining in the mobile phase are carried down to the next particle where the process is repeated again. The inorganic material is used for column packing (Plate 475).

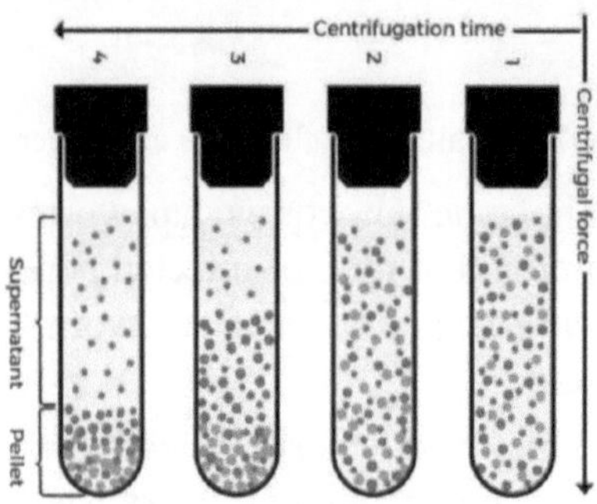

Plate 474: Particle separation by centrifigation

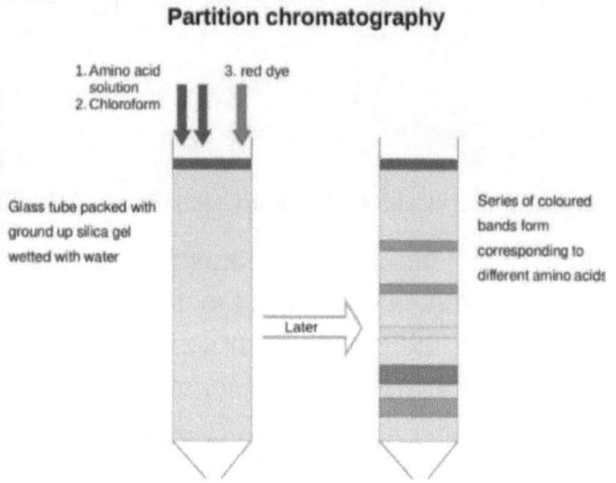

Plate 475: Partition chromatography illustration

Pascal: It is a unit of pressure in the meter kilogram second (mks) system equivalent to one Newton per square meter.

Passivation treatment: Electrolytic tinplate in the final stage of manufacture is treated by chemical or electrochemical means to produce on the tin surface a thin invisible oxidation film which is called Passivation (Plate 476). Passivate mean to make inactive or less reactive or to protect (solid-state device) against contamination by coating or surface treatment.

Pasta: It is a paste in processed form (macaroni) or in the form of fresh dough (ravioli) (Plate 477).

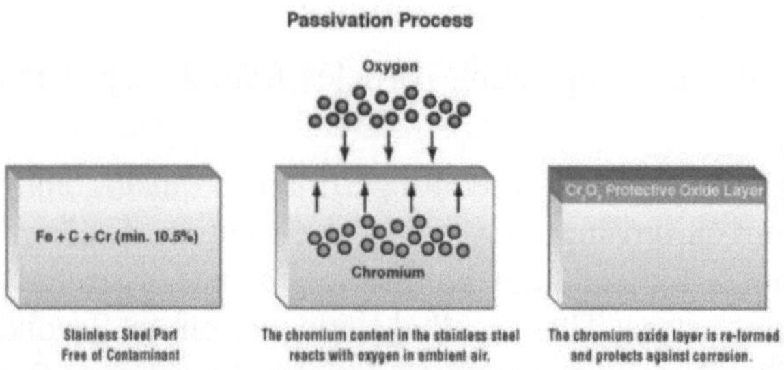

Plate 476: Passivation

Plate 477: Pasta product

Pasteur Effect: Acceleration of sugar utilization in response under condition of low O_2.

Pasteurized milk: The milk is heated at 72°C for atleast 15 seconds to destroy potentially harmful bacteria and cooled rapidly. The nutrient loss is negligible and the pasteurized milk packs or pouches can be kept unopened safely for upto five days in refrigerator.

Pasteurization: Pasteurization is a method / process of food preservation by heating a liquid food or beverage at a controlled temperature (less than 100°C for short time) or low temperature for sufficient time) followed by cooling which will only kill majority (not all of the micro-organisms present), thus preventing /delaying/weaken growth for their survival in the food inside container by sealing it hermetically.

Pasteurizer: It is a type of the pasteurization unit used for pasteurization of the liquid foods. The pasteurization unit may be manually operated on heater /bhatti, or semi-automatically operated by steam and electricity (Plate 478).

Plate 478: Semi-automatic pasteurizer

Pastille troche: A small sweetened and flavored medicated material that is designed to be held in the mouth for slow dissolution and have soothing effect on sore throat. It is also called Lozenge.

Patent: Protected by a trademark or a brand name so as to establish proprietary rights analogous to those conveyed by letters patent or a patent and have exclusive right to make, use, or sell.

Patent flour: Patent flour is high grade wheat flour that consists solely of endosperm.

Patent leather: Patent leather is leather with a hard smooth glossy surface.

Pathogen: It is agent that causes disease.

Pathogenic Gram negative bacteria: The Gram negative pathogenic bacteria typically associated with foods include *Salmonella, Shigella, Escherichia coli, Campylobacter jejuni, Vibrio parahaemolyticus* and *Yersina enterocolitica.* These are usually present in the intestine and faeces of man, animals and birds (Plate 479).

Pathogenic Gram positive bacteria: The Gram positive pathogenic bacteria associated with foods include *Clostridium botulinum, C. perfringens, Bacillus cereaus, Staphylococcus aureus* etc (Plate 479).

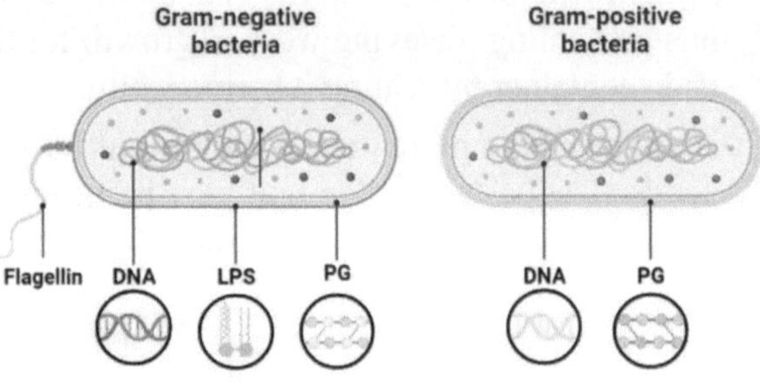

Plate 479: Gram negative vs positive bacteria

Pathogenic organisms: Harmful micro-organisms which cause disease in the body.

Patty shell: It is a shell of puff pastry made to hold a creamed meat, fish, or vegetable filling.

Patty: A small flat cake of chopped food i.e. hamburger or it is a small flat candy of peppermint. A little pie is also called patty.

Patulin: This is a mycotoxin associated with fruit and fruit juice products, produced by several *Pennicillium* spp. It is considered to be carcinogenic and high concentration may cause acute effects such as haemorrhages and Oedema. The presence of patulin in food products is normally associated with the use of mouldy raw materials (Plate 480).

Plate 480: Illustration of Patulin presence in apple juice

PCBs: It stands for polychlorinated biphenyls.

Peaking: Peak or a small ridge formed on the can is known as peaking. It is uneven swelling / deformities of the can near double seam caused by permanent strain during retorting or cooling. It may also be due to thin tin plate and under exhausting.

Peanut oil: It is colorless to yellow fatty nondrying oil that is obtained from peanuts and is used chiefly as a salad oil, in margarine, in soap, and as a vehicle in pharmaceutical preparations and cosmetics.

Pearl essence: A translucent substance that occurs in the silvery scales of various fish (herring) and is used in making artificial pearls, lacquers, and plastics.

Pearl onion: It is a very small pickled onion used as appetizers and as a garnish.

Pearling: It is the process of wet milling where top layers bearing the colour and fiber are removed by abrasion. The polishing of the grains is called pearling.

Pearson square method: It is the method used for estimation of sugar for preparation of the syrup solution. By this method, one can easily calculate the part / portion of syrup required for preparation of a product (Plate 481).

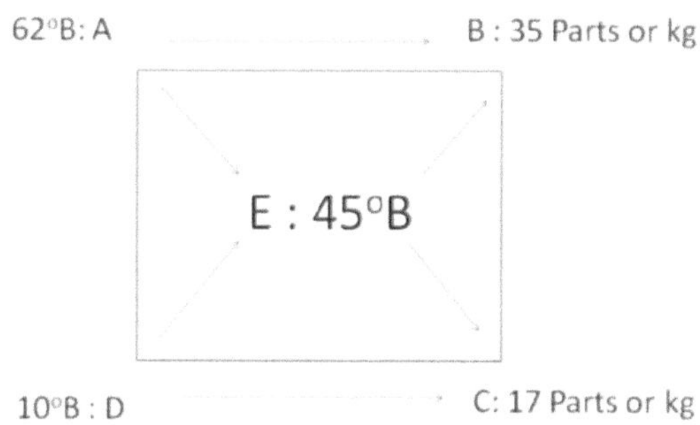

Plate 481: Pearson square method illustration

Pectic acids: It is the polymer of galacturonic acid without any esterification. Substances formed on complete demethyllation (lacking methyl esters) and complete or partial carboxylation of pectin during over ripening or rotting of fruits when pectin is decomposed to pectic acid and methyl alcohol. These are water-insoluble substances formed by hydrolyzing the methyl ester groups of pectins. It is colloidal polyglacturonic acid and mostly free from methyl ester. Pectic acid is simplest in all 50 unit chain and no methylation of C-at 6th position takes place.

Pectic substancs: Collodial carbohydrate derivatives containing a large proportion of anhydrogalacturonic acid units which occur in or are prepared from plants. The carboxyl groups of these acids can be esterifies by methyl groups and partly or completely neutralized by one or more bases.

Pectin: It is one of the water-soluble substances that bind adjacent cell walls in plant tissues and yield a gel which is the basis of fruit jellies. Pectin is a complex carbohydrate derived from pentose in the process of ripening which remains as colloid in the fruit juice. Pectin having > 100 unit chain is most complex, and having > 75% methylation is capable to form jelly with sugar. It is called nature's glue.

Pectin grade: It is gram of pectin required to gel the sugar. 100 pectin grades mean 1 gram pectin / 100 gram sugar.

Pectinate: Pectic acids attached with salt of Ca, Al, Na, and any other salt are known as pectinate.

Pectinesterase: An enzyme that catalyzes the hydrolysis of pectins into pectic acids and methanol is called pectinesterase.

Pectinic acid: It is polymer of galacturonic acid partially esterified with methyl ester groups. Colloidal polygalacturonic acid containing a negligible proposition of methyl ester groups and partly or completely neutralized by one or more bases. Pectinic acid is more complex than pectic acid, 100 unit chain having methylation <75'% of C-at 6th position. It is colloidal polygalactonic acid present in fruits and vegetables.

Peddler /pedlar: One who offers merchandise (fresh produce) for sale along the street or from door to door. It is also called packman.

Peel: It is the skin or rind of a fruit or a thin layer of organic material that is embedded in a film of collodion and stripped from the surface of an object (as a plant fossil) for microscopic study (Plate 482).

Peeler: It is one of the tools that are used to remove the peel. It is long-handled spade-shaped instrument that is used chiefly by bakers for getting something (as bread or pies) into or out of the oven.

Peeling: It is a process to peeled-off piece or strip or to lose an outer layer of food commodity. It is an operation in processing of many fruits and vegetables to remove unwanted or inedible material and to improve the appearance of the food product (Plate 483).

Pekoe: A tea made from young leaves slightly larger than those of orange pekoe is called pekoe.

Pellagra: A disease marked by dermatitis, gastrointestinal disorders, and mental disturbances and associated with a diet deficient in niacin.

Penetrometer / bloom gelometer: The instrument is used to measure the tenderness of foods (Plate 484). This is used to measure the tenderness by determining the distance a cone or other device penetrates the food during a defined period of time and using only gravitational forces.

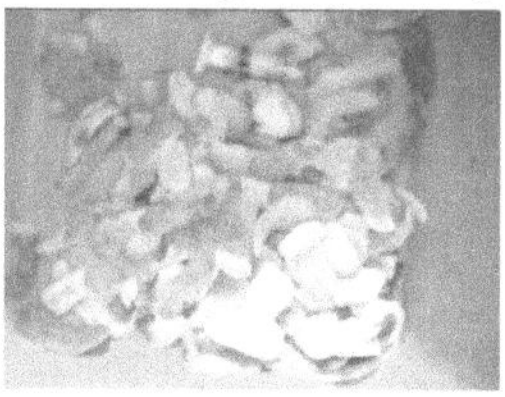

Plate 482: Peel

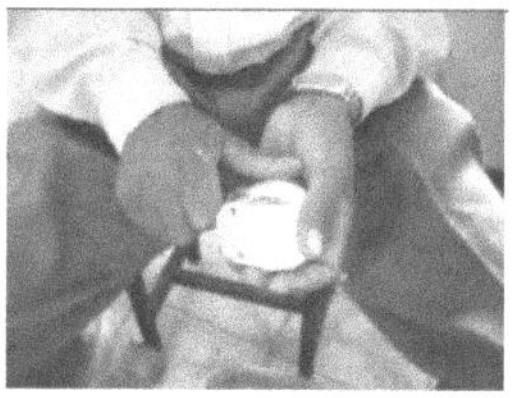

Plate 483: Peeling operation

Plate 484: Penetrometer

Pentosan: It is one of the polysaccharides that yield only pentoses on hydrolysis and occur widely in plants.

Pentose: A monosaccharide $C_5H_{10}O_5$ (ribose) that contains five carbon atoms in the molecule.

Pentose phosphate pathway (PPP): An alternative respiratory pathway that involves five carbon sugar phosphates intermediates for obtaining energy from oxidation of sugars in CO_2 and water (Plate 485).

Pentothenic acid: This vitamin is a type of vitamin B complex. It is unstable as free acid, viscous, yellow oil and available as sodium or calcium, water soluble and slightly sweet is taste. It is widely spread in food and so very rare deficiency symptoms. Deficiency results loss of appetite, indigestion, abdominal pain, mental depression, burning sensation in the feet, insomnia and respiratory infection (Plate 486).

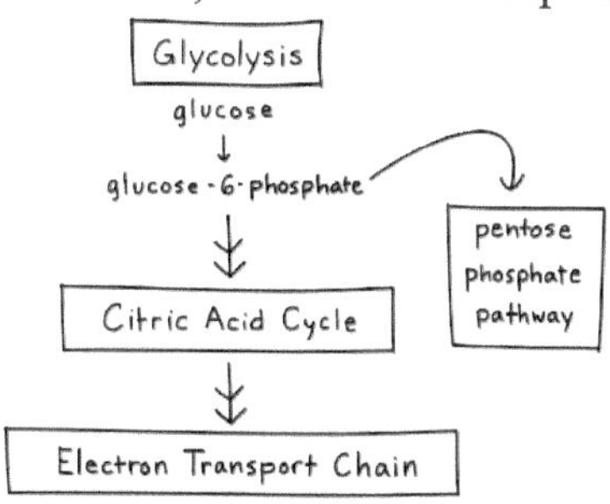

Plate 485: Pentose phosphate pathway

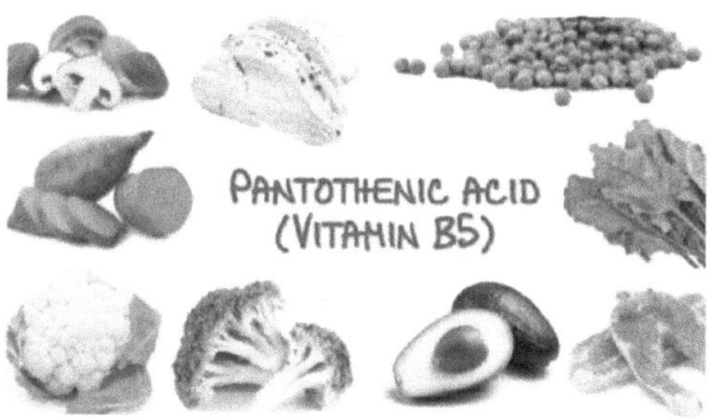

Plate 486: Food rich in pantothenic acid

Pepsin: It is a protease of the stomach that breaks down proteins to polypeptides.

Peptidase: It is an enzyme that hydrolyzes simple peptides or their derivatives.

Peptide: It is one of the amides that are derived from two or more amino acids by combination of the amino group of one acid with the carboxyl group of another and are usually obtained by partial hydrolysis of proteins.

Peptide bond: The chemical bond between carbon and nitrogen in a peptide linkage is called peptide bond.

Peptide linkage: The divalent group CONH that unites the amino acid residues in a peptide.

Percent fat free food: Low / fat free foods are known as percent fat free food.

Percent sag: This test is used to measure the comparative tenderness of the gel. Depth in container minus depth in plate divided by depth in container and multiplied by 100 is called percent sag. This term is used for jelly. The greater the percent sag, tenderer is the jelly. It is also related to jelly grade test.

Percent strength: It is number of parts by weight of solute present per 100 parts by weight of solution.

Percentage humidity: It is also known as degree of saturation. It is the ratio of kg of the water vapour per kg of the dry air and kg of the water vapour per kg of saturated air at the same temperature and pressure. It is expressed in percent unit.

Perfect Markets: A market is said to be perfect, when all the potential sellers and buyers are promptly aware of the prices at which transaction takes place. Any buyer can purchase from any seller. The principle underlying a perfect market expects that there must be a uniform price for any one standardized commodity at a particular time at any one place. Secondly, there should not be any restriction on the movement of a commodity and thirdly, there must be a good number of buyers and sellers.

Perishable foods: The foods that spoil quickly within 1-2 days are called perishable.

Permanent emulsions: Permanent emulsions have amphiphilic molecules i.e. hydrophilic and lipophilic molecules. Permanent emulsions also known as emulsifying agent, emulsifier, and stabilizer.

Permeability: It is the property of a flexible plastic film by virtue of which water vapour and volatile vapours and gases permeate through the film. The rate which a given vapour or gas permeate through a given surface of the film in a given time across a given pressure differential is termed as permeability of the film to the particular vapour / gas.

Peroxide value: The rancidity of the oil due to oxidation of the fatty acid is determined in term of peroxide value. It is also called lea value.

Perry: A wine prepared from pear is called perry.

Pesticides: Pesticides are any chemicals that are applied to control or kill pests and include the following i.e. insecticides, herbicides, fungicides, wood preservations, biocides, bird and animal repellents, food storage protectors, rodenticides, and industrial / domestic hygiene products. These chemicals are used during production to protect crops and improve yield and after harvest they are again used to protect the crops in storage.

PFA: It stands for Prevention of Food Adulteration act (1954) and rule (1955). It protects the consumer against the supply of inferior quality or adulterated foods. It is mandatory law.

pH: It is a quantitative measure of strength of the acidity or alkalinity of a solution, defined as the negative logarithm ($-log^{10}$) of the hydrogen ion concentration $[H^+]$ in moles/litre: $pH = -\log_{10} [H^+]$. A measure of acidity or alkalinity which is expressed as the negative log of the hydrogen ion conc., pH of 7.0 is neutral, less than 7.0 is acidic and more than 7.0 is alkaline. The letters of its name are derived from the absolute value of the power (p) of the hydrogen ion concentration (H). The product of the concentrations in water of H^+ and OH^- (the hydroxide ion) is always about 10^{-14}. The strongest acid solution has about 1 mole/litre of H^+ (and about 10^{-14} of OH^-), for a pH of 1. The strongest basic solution has about 10^{-14} moles/litre of H^+ (and about 1 of OH^-), for a pH of 14. A neutral solution has about 10^{-7} moles/litre of both H^+ and OH^-, for a pH of 7. The pH value, measured by a pH meter, titration, or indicator (e.g., litmus) strips, helps inform chemists of the nature, composition, or extent of reaction of substances, biologists of the composition and environment of organisms or their parts or fluids, physicians of the functioning of bodily systems, and agronomists of the suitability of soils for crops and any treatments needed. The pH is now defined in electrochemical terms. According to Debye-Huckel expression, pH is a function of the concentration of the hydronium ion (H_3O^+).

pH meter: An apparatus for measuring the strength or the amount of acid present in a mixture or solution is called pH meter (Plate 487). It is also called acidimeter.

Plate 487: pH meter

Phar bid system: By this method, one bid is given for all the lots in a particular shop and all the lots are sold at that price.

Pharma food: Food or nutrient that claims medical or health benefits, including the prevention and treatment of disease is called pharma food.

Phase change in potential: It is the transfer of the heat by the potential energy associated with the heat of the phase change such as boiling or freezing.

Phenolase: Phenolase is also known as polyphenoloxidase, tyrosinase, and catecholase. It is an enzyme that causes enzymatic browning of fruit and vegetables. Optimum pH for activity of polyphenolase is pH of 5-7.

Phenols: Phenols are the aromatic compounds having hydroxyl groups and are widely present in the plant kingdom.

Phenolics: These are compounds with benzene rings and various attached groups such as hydroxyl, carboxyl methoxyl, and other non aromatic or ring structures. Phenolics include aromatic amino acid, simple phenols, polyphenols and derivatives like phytoalexins, coumarins, lignin, anthocyanins, flavonals and flavones (Plate 488).

Phenolic meat lacquers: These consist of a synthetic resin dissolved in a solvent consisting of a mixture of alcohol and hydrocarbon.

Phenylalanine: It is an essential amino acid $C_9H_{11}NO_2$ that is converted in the normal body to tyrosine.

Phenylpropanolamine: It is a sympathomimetic drug $C_9H_{13}NO$ used in the form of its hydrochloride as a nasal and bronchial decongestant and as an appetite suppressant.

Phospholipase: These are enzymes that hydrolyze lecithins or phosphatidylethanolamines. It is also called lecithinase.

Phospholipid: A lipid with a phosphate group esterilifed to the third hydroxyl group of glycerol and often other ions are esterified to the phosphate e.g. Choline, glycerol inositol, ethanolamine and acephalin. These are fat like substances consisting of glycerol, 2 fatty acids, a phosphate group and a nitrogenous base e.g. lecithin (Plate 489).

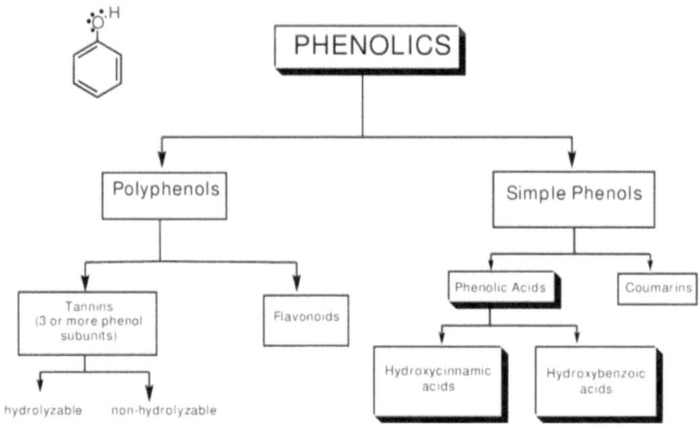

Plate 488: Different Phenolics

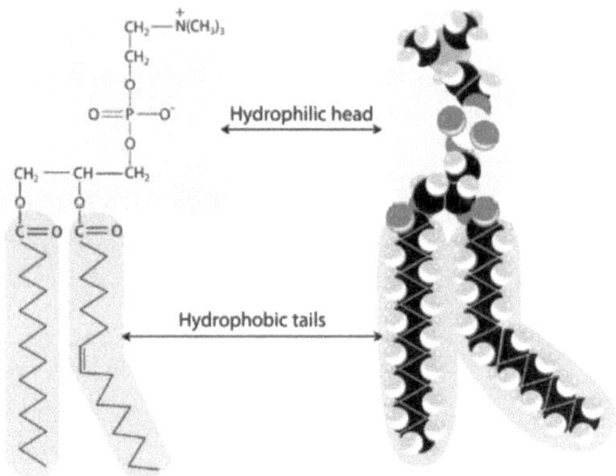

Plate 489: Phospholipid characteristics

Phosvitin: It is an iron carrier in yolk.

Photochemical reaction: A chemical reaction which proceeds at the expense of absorbed light e.g. photosynthesis.

Photodegradable plastics: In photodegradable plastics light sensitive carbonyl base is introduced into plastic.

Photophobia: A disorder or a deficiency disease that cause intolerance to light, especially painful sensitiveness to strong light. It means fear or hatredness of light. It is due to deficiency of riboflavin and characterized by rough eyelid, and sensitivity of the eyes to bright sunlight.

Photophosphorylation: A photochemical reaction (photosynthesis) in which trapped light energy is utilized in the production of high energy phosphate,

ATP by the addition of a phosphate group to ADP. Mitochondria are the site of this reaction. Photophosphorylation is the synthesis of ATP from ADP and phosphate that occurs in a plant using radiant energy absorbed during photosynthesis. It is of two type: cyclic and non-cyclic. When ATP is synthesized during light phase of the photosynthesis. It is utilized in dark phase i.e. Calvin cycle (Plate 490).

Photorespiration: Light dependent O_2 uptake and conversion of photosynthates in CO_2 (distinct from mitochondrial respiration). The C-4 plants photo respire at a much slower rate than do the C-3 plant. Therefore, less energy is wasted in C-4 while half the substrate is wasted in C-3 plants (Plate 491).

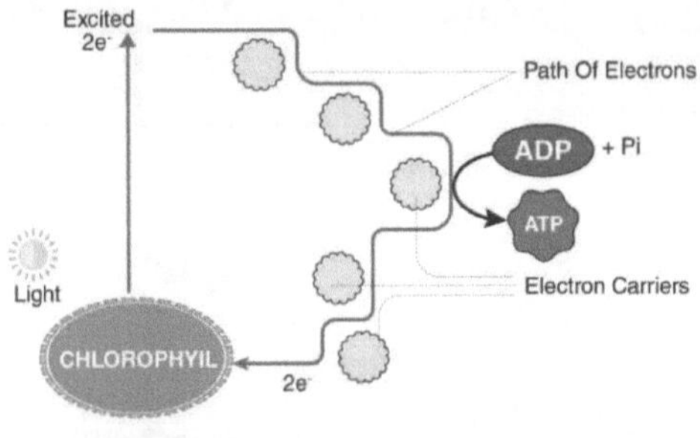

Plate 490: Cyclic Photophosphorylation

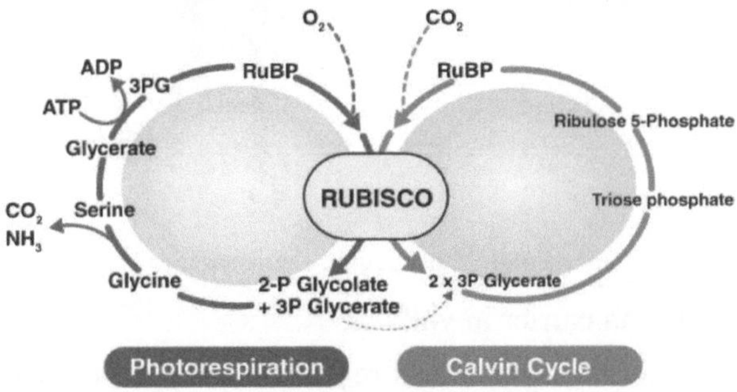

Plate 491: Photorespiration cycle

Photosynthesis: Biochemical reactions leading to synthesis of organic compounds specially carbohydrates from the O_2 of water and carbon of CO_2 in the presence of light as the energy source in chlorophyll containing tissues of plants. Or Photosynthesis is synthesis of chemical compounds with the aid of radiant energy (light) and result in the formation of carbohydrates from carbon dioxide and a source of hydrogen (as water) in the chlorophyll containing tissues of plants exposed to light (Plate 492).

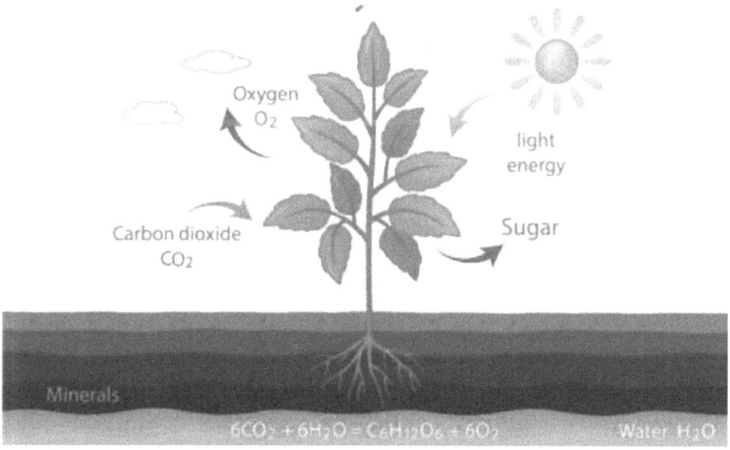

Plate 492: Photosynthesis illustration

Photosynthetic rate: As the gross rate, it is the quantity of CO_2 consumed and O_2 produced in photosynthesis and as the net rate it is the quantity of CO_2 evolved and O_2 consumed in respiration and photorespiration. Net photosynthetic rate represents the gain in dry weight of the plant.

Photosystem: The photosystems act as electron transports that use light as the source of energy in the series of reactions which result in the splitting of water to release oxygen.

Physical hazards: Physical hazards are the most common type of hazard to occur in foods, because of possible presence of foreign material and affect only one or at most a few people. Physical hazards like biological and chemical hazards can enter a food product at any stage in its production. These hazards includes extraneous material such as metal fragments, wooden splinters and stones, which are of foreign origins and not normally found in food, introduced during primary processing, harvesting and cause health risk to the consumers. Physical hazards mainly cause varying degree of injury and rarely death may occur. Physical contaminants include metal piece, insects, thorns, wood, glass, mesh wire, cloth, rust and dirt etc

Physical loss: Physical damage (tissue breakdown) of the perishable due to improper cold storage temperature and relative humidity, undesirable gaseous composition of CA storage and excessive or insufficient heat during processing.

Physico-chemical losses: Postharvest losses of the perishables which is accounted towards the senescence process, transpiration, respiration and sprouting.

Physiological disorders: Disorders in which no primary parasite is involved and they are brought about by abnormal environment conditions like nutrition, weather maturity etc. Physiological disorders refer to the breakdown of tissue that is not caused by either invasion by pathogen or by mechanical damage. They may develop in response to an adverse environment especially temperature or to a nutritional deficiency during growth of organisms.

Physiological disorders in citrus: Physiological disorders in citrus includes rind staining puffiness, Kohansho, granulation (due to formation of granules within vesicle), oleocellosis (oil spotting), peteca (pitting) stylar end breakdown (SEB), watery breakdown, chilling injury, freezing injury, zebra skin, sunburn, wind scar, B and Cu deficiency and ammoniation.

Physiological loss in weight (PLW): Weight loss of fresh fruits and vegetables attributed to many physiological activities like transpiration, respiration etc. which depends on temperature, RH, air circulation and other storage environment.

Physiological losses: Lowering down the quality of the perishables due to normal physiological process after harvest e.g. toughening and sponginess in green beans, sweet corn, carrot, radish etc.

Physiological maturity: Attainment of final stage of biological function by a plant part or the plant as a whole. It is the stage of development when a plant or plant part with continue ontogeny (for the development) even if detached. It refers to the stage in the development of fruit/vegetables when maximum growth and maturation has occurred. It is usually associated with full ripening of the fruit. The physiologically mature stage is followed by senescence.

Physiological salt solution: A cultural media contains not more than 1% sugar and 0.85% NaCl for the growth of the bacteria is called physiological salt solution.

Phytic acid: It is a phosphorous containing compound found in the outer layer of the cereals. Phytic acid is a six-fold dihydrogenphosphate ester of inositol also called inositol hexaphosphate, inositol hexakisphosphate (IP6) or inositol

Phytins: Phytins has a softening action on peas during cooking by acting as a calcium absorbent, consequently preventing the formation of insoluble calcium pectate. Phytins are calcium magnesium salts of phytic acid that occurs in plants as the main phosphorus storage reserve especially in seeds or tubers and is used as a source of inositol (Plate 493 and 494).

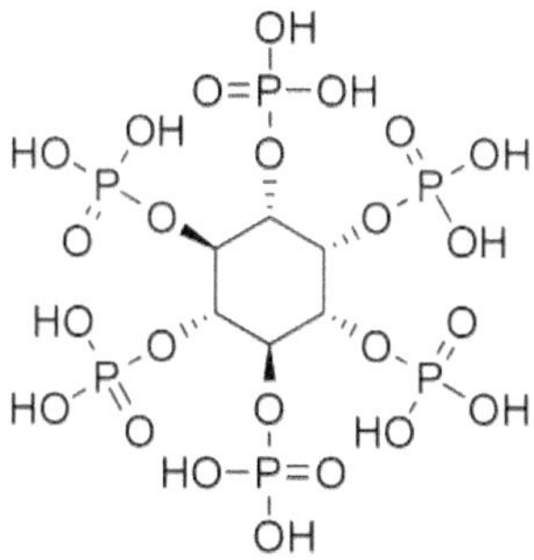

Plate 493: Phytic acid

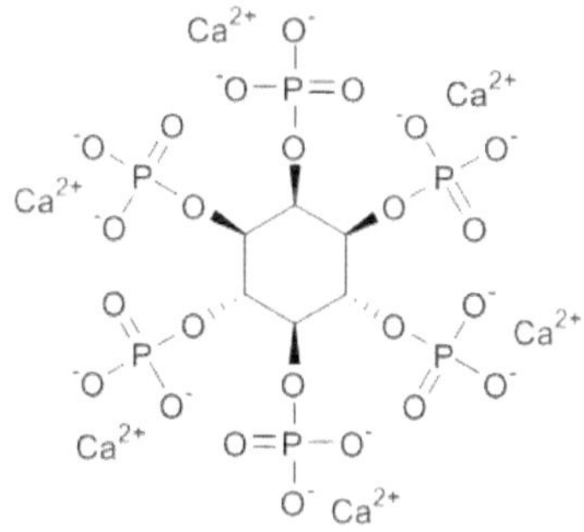

Plate 494: Calcium salts of phytic acid

Phytoalexin: A phenolic substance having antifungal principle, synthesized by plants in response to parasite invasion or infection by certain fungus e.g. pisatin, phaseolin, trifolirhizin, orchinol and isocumarin from pea and bean pods, red clover, orchid tubers and carrot roots respectively. It is one of the antimicrobial chemical substances produced by plants to combat infection by a pathogen (fungus). These are Stress metabolites and are produces due to physiological stresses, microbial invasion or due to chilling injury. These are the compounds which are secreted by the host when it is infected by an organisms and these compound slow down the growth of the pathogens.

Phytochemical: A chemical compound (e.g. beta carotene) occurring naturally in plants. These are important for health to promote various reactions in the body but are not classified as nutrients required for growth and development. These are non-nutrient plant chemicals that contain protective, disease preventing compounds.

Phytosterol: It is one of the sterols derived from plants.

Pick Your Own System: This system is used where the farmer grows the crop and the customers harvest it themselves and pay the farmer.

Pickle: The grain of the corn is called pickle.

Pickles: It is an edible product preserved and flavored in a solution of common salt and vinegar along with spices and oil. Processed fruit / vegetable in common salt, vinegar or edible oil with the addition of spices and condiments, may be unfermented (mostly Indian pickles) and fermented (Western pickle). Pickles are good appetizer and stimulate the flow of gastric juice. Unfermented pickles etc are some of the fermented pickles.

Pickles keep: A solution containing 10% salt, 0.5% turmeric powder, 0.3-0.5% acetic acid to cure the fruits like mango, raw papaya, caronda, amla etc. which gives all sorts of beneficial effects of curing.

Pickling: The preservation of food in common salt or in vinegar is known as pickling. The pickling process is carried out in one of two ways. One method is through anaerobic fermentation in a brine that creates lactic acid. The other method is marinating and storing it in an acid solution, usually vinegar (acetic acid).

Pickling steel: Pickling is a process in which an acid solution is used to remove oxide scale from steel strip, rod, wire, tubing, and discreet shapes. Oxide scale forms on the surface of the steel when steel cools from a molten state. Pickled steel is free from surface carbon steel contamination and embedded iron particles. It typically leaves a dull, matte grey finish.

Pico waved: It is the term used to label foods treated with low level of ionizing radiation.

Pimaricin: It is a preservative of microbial origin and is effective against mould and has application on cheese. It is also used for treatments of juices. It is also called Netamycin. It is a non toxic antibiotic possessing strong antifungal properties. It is propylene obtained from *Streptomyces netalensis*.

Pineapple physiological disorder: The disorders include chilling injury, flesh translucency, bruising, sunburn, malformation.

Pineapple slicer: It is a device used for slicing of the pineapple. The device is equipped with mechanisms to adjust the thickness of slices.

PINK-1: It is a type of teh protein help in normal functioning of the heart and so prevent heart attack. This protein supply required energy for working of cells. Due to its deficiency, the heart cells dies and depends on other cells for getting energy and so put additional pressure on the other cells for getting energy. This condition is called hypertrophy.

Pisatin: Pisatin is antifungal compound. Pisatin is produced in peas if exposed to ethylene level of 0.2 – 20 ppm.

Pizza: Pizza is a dish made typically of flattened bread dough spread with a savory mixture usually including tomatoes and cheese and often other toppings and baked. It is also called pizza pie.

Pizza pie: Pizza is a dish made typically of flattened bread dough spread with a savory mixture usually including tomatoes and cheese and often other toppings and baked. It is also called Pizza.

Plan – Do – Check - Act (PDCA) cycle: A four step process for quality improvement which sometimes referred as Stewart cycle or Deming cycle (Plate 495).

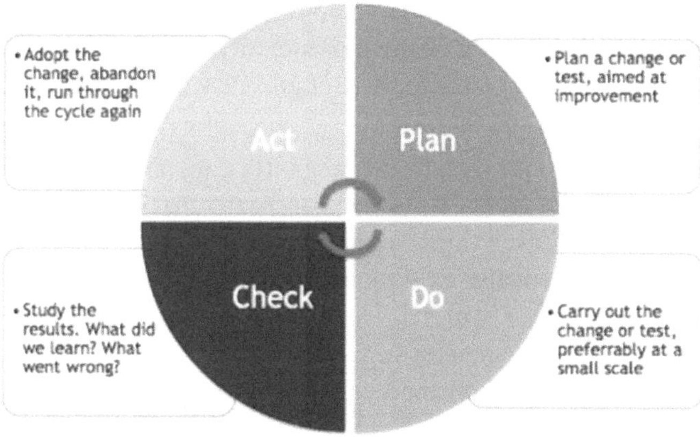

Plate 495: Stewart cycle or Deming cycle

Plant: Plant is the premises, buildings, structure and equipment (including but not being limited to machines, utensils, vehicles and fixtures located in or about the premises) used or employed in the preparation, processing, handling, transporting and storage of fruits and vegetables or the processed products.

Plant layout: It is the disposition of the various facilities and services of the plant within the area of the site selected previously.

Plant Quarantine Order: India introduced the Plant Quarantine (Regulation of Import into India) Order in 2003 to prohibit and regulate the import of agricultural articles.

Plant Sanitation / hygiene: The in-plant sanitation involves quality control and storage of raw products, the proportion of a good water supply, prevention of the contamination of the foods at all stages. The waste material has to be treated to meet the standard limits laid down by the regulatory agencies. The foods for the human consumption should be produced only under good hygienic conditions. No person is interested to eat a product which he knows that it is produced in such a place where there is heavy infestation of rats or other source of infection. So, good plant sanitation reduces losses due to spoilage.

Plantation crops: Perennial crops that are grown in large scale and require some sort of processing before use e.g. cashew nut, tea, coffee etc.

Plasma in food: Plasma is an advanced technique that has attracted scientists and technologists because of its astonishing capacity and potential applications in food industries which include food decontamination, inactivation of food-borne pathogenic microbes, and food packaging.

Plastic: Plastic is a wide group of solid composite materials which are largely organic, usual synthetic resin polymer of natural origin. The plastics possess appreciable mechanical strength. Plastic comprise of two groups i.e. thermoplastic and thermosets.

Plastic water bottle caps: Plastic water bottle caps are colour coded based on the type of water. Blue caps indicate that water is a source of "spring water"; black caps signify that water is alkaline water; green caps signify that water is flavoured water, and white caps mean that water is "processed water." The claim, however, is not true. In the United States, two agencies regulate bottled water: the Food and Drug Administration and Environmental Protection Agency. However, based on Snopes' analysis of FDA and EPA guidelines, neither organization has any rules or regulations regarding color-coded caps. Further, some brands use caps that match the alleged "rules," like the black caps on Essentia Alkaline Water. But many brands break the purported coding system's rules. Fiji Water, which exclusively uses a blue cap, doesn't come from a spring and Crystal Geyser's water *does* come from a spring but generally uses white caps. Pink Caps Sometimes used for water bottles marketed towards breast cancer awareness or charity events. Yellow Caps used for flavoured or vitamin-enhanced water, usually with added nutrients or electrolytes. Red Caps signify sparkling or carbonated water.

Plastic cream: The heavily concentrated cream containing about 80% fat is called plastic cream.

Plasticizers: Certain plasticizers and other plastic additives are toxic and are of concern if they are able to migrate in to food. Migration depends on the constituents present and also on the type of food; for example, fatty foods promote migration more than some other food stuffs.

PLU Codes: Price look-up codes, commonly called PLU codes, PLU numbers. PLUs, produce codes, or produce labels, are a system of numbers that uniquely identify bulk produce sold in grocery stores and supermarkets. The codes have been in use since 1990, and over 1400 have been assigned. A four digit PLU code tells that the item was conventionally raised, meaning pesticides and chemicals were used in its production. The 4-digit PLU codes are assigned randomly within a series of numbers within the 3000 and 4000 series. As use of PLU codes grew, the numbers became standardized around 1990 under the International Federation for Produce Standards (IFPS). All four-digit numbers begin with a 3 or a 4 — which means the product was grown conventionally — and five-digit combinations start with a 9 (then the product's four-digit code), meaning it was organically grown (Plate 496).

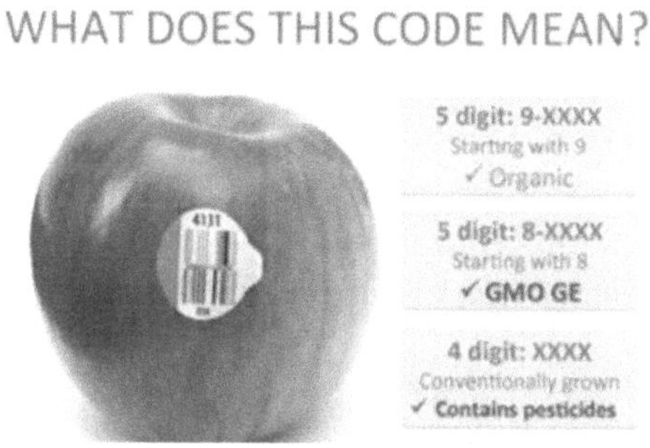

Plate 496: Illustration of PLU code on fruit

Plum brandy: Plum brandy is also slivovitz /sliwowitz.

Plum tomato: It is also known as processing tomato or paste tomato. It is bred for making sauce and packing purpose. It contains fewer seeds compartments and generally has higher solid content which make them suitable for processing. Even very small size tomato is also known as grape tomato or cherry tomato, Roma VF, San Marzano are example of plum tomato varieties.

PME: It stands for Pectin Methylesterase. Some time it is also called as pectin esterase (PE).

Pneumatic Handling: This type of handling is related to use of gas (air or wind) or worked by air pressure or inflated with compressed air.

Poached egg: Half fried egg is called poached egg (Plate 497).

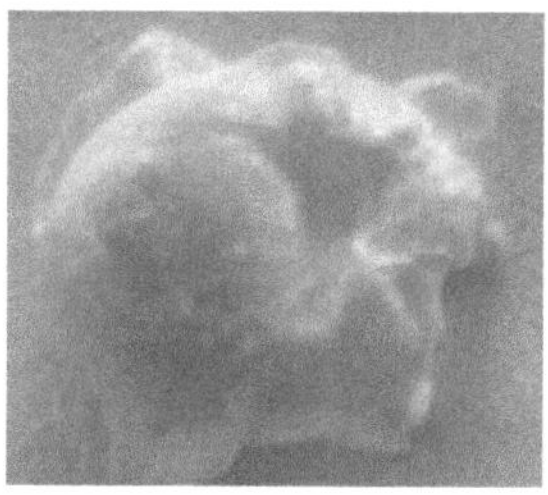

Plate 497: Poached egg

Poise: It is a centimeter-gram-second (cgs) unit of viscosity equal to the viscosity of a fluid that would require a shearing force of one dyne to impart to a $1 cm^2$ area of an arbitrary layer of the fluid at a velocity of one centimeter per second relative to another layer separated from the first by a distance of one centimeter.

Poiseville equation: It is related to viscosity of the drinks.

Poising capacity: The resistance to change a potential of food is called Poising capacity.

Polari scope: Polari scope is used for quantitative analysis of sugar.

Polarization: It is the action of polarizing or state of becoming polarized. It is the action or process of affecting radiation and especially light so that the vibrations of the wave assume a definite form. By polarization process there is increase in the resistance of an electrolytic cell often caused by the deposition of gas on one or both electrodes (Plate 498).

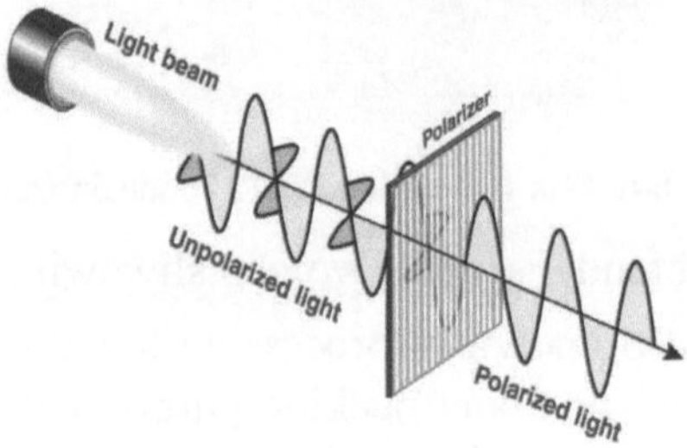

Plate 498: Illustration of polarization

Polyacrylamide: It is a polyamide of acrylic acid.

Polycarbonate films: Polycarbonate is one of the strongest and most rigid thermoplastics. Polycarbonate displays greater than 90% light transmission with minimal haze.

Polychlorinated bi-phenyls (PCB_S): PCB_S are members of a group of organic compounds that have been used in a number of individual applications and the most significant source of PCBs in food stuffs is through absorption from the environment by fish and then accumulates through the food chain and can be found in high level in tissues with high lipid contents.

Polycyclic aromatic hydrocarbons (PAHs): These are compound of benzene rings linked together and are the largest class of known environmental carcinogens. They are found in water, air, soil and food. They originated from coal derived products, charcoal, engine exhaust, petroleum distillates, smoke curing and tobacco smoke.

Polyester films: It is condensation product of ethylene glycol and teraphthalic acid. The films provide excellent barrier properties against oils and odours and are resistant to many chemicals.

Polyhydroses: These are ketones or condensation product of such compounds.

Polyhydroxy: These are the compounds containing more than one hydroxyl group in the molecule.

Polymer: A chemical compound or mixture of compounds formed by polymerization and consisting essentially of repeating structural units.

Polymers are carbohydrates like cellulose, starch, hemicellulose, fibre, lignin, pectin, and lignin. They are not digested as such but they are used to cure many diseases. In most country white flour bread is banned because it does not contain fibre. Chaki flour is preferred as it contains fibres. Starch from plantain, cassava, yam, sweet potato and potato provides the bulk of energy. Lignin, a complex polymer of aromatic compound linked by propyl units, is as a major component of fibre. Starch and cellulose have same composition but starch is formed by α-1,4 linkage, which is easily hydrolyzed by amylase enzyme secreted by man; whereas cellulose is formed by β-1,4 linkages, however, cellulase enzymes are not produced by man.

Polymerization: A chemical reaction in which two or more molecules combine to form larger molecules that contain repeating structural units.

Polymethylpentene (PMP) films: These are crystal clear films having heat resistance, microwavabilityand recyclability.

Polyol: It is a sugar alcohol e.g. sorbitol (hexitol) and xylitol (pentitol).

Polyolefin: It is a polymer of an alkene (polyethylene).

Polyomavirus: It is a family (Polyomaviridae) of double-stranded DNA viruses that produce tumors in specific mammals. It is also called polyoma.

Polypropylene: It is a clear glossy film with good optical properties, high tensile strength and puncture resistance.

Polyphenol oxidase: An enzyme causing oxidation of phenols or polyphenols which maybe the principal reaction in enzymatic browning. Removal of O_2 by blanching or sulphiting of materials or deaeration of juice can help to prevent the activity of this enzyme.

Polysaccharide: A polysaccharide is a carbohydrate that can be decomposed by hydrolysis into two or more molecules of monosaccharides like cellulose, starch, or glycogen and contain many monosaccharide units (Plate 499). These are long linear or branches chains of monosaccharides principally as D-glucose. Polysaccharides are important in plant as storage compound e.g. starch, inulin, mannans and arbinans or structural component like cellulose, hemicellulose. Most abundant cellulose is straight chains molecules of glucose represent one of the most abundant compound present in intact plant. The individual cellulose molecule may contain 1000-10,000 units of glucose. Cellulose is found in primary and secondary cell wall e.g. starch, dextrins.

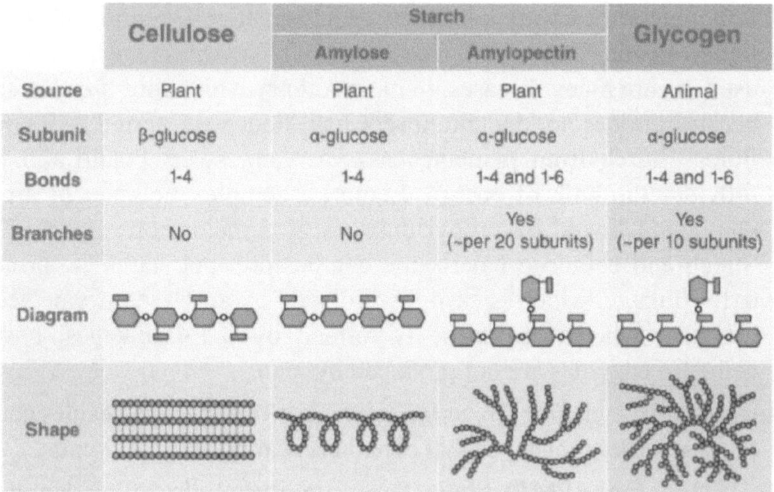

Plate 499: Polysaccharide examples

Polyunsaturated fatty acid (PUFA): A fatty acid containing 2 or more double bonds e.g. linoleic acid, linolenic acid and arachidonic acids (Plate 500).

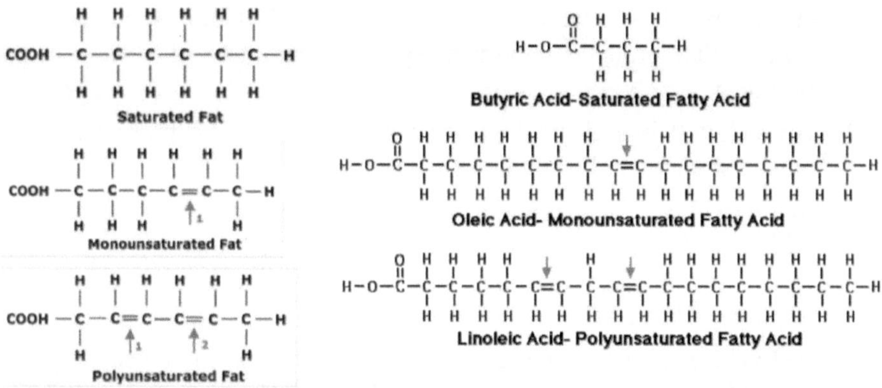

Plate 500: Saturated, Mono-unsaturated and poly-umsaturated fat and fatty acids

Polyvinyl chloride (PVC) films: These films are available in the thickness range of 0.01 to 0.1 mm. It is semi-rigid or high flexible films, stretched or biaxially oriented, high shrink film.

Polyvinylidene chloride: The film made of polyvinylidene chloride has very good moisture, odour and gas barrier properties. It is fat resistant and does not melt in contant with hot fats, making it suitable for "freezer to oven" foods. It is also used for coating films and bottles.

Pomace: It is a residue of crushed apple pulp or other fruit pulp after expressing juice. It is residue of crushed fruit after expressing juice (apple pomace).

Pombe: It is an African beer prepared from millet seeds.

Ponceau Red: It is a red azo dye and used as a food dye. When added to food it is denoted by E125. It usually is the form of disodium salt. It is permitted in fruit peels.

Poonac: Cake that left after extraction of oil from copra

Pop or popping: It is the process that causes to explode or burst open with sharp sound e.g. popcorn.

Popcorn: It is an Indian corn (*Zea mays praecox*) whose kernels on exposure to heat burst open to form a white starchy mass after popping. Popcorn is prepared from mature corn / cob.

Pork: It is meat of swine. Good quality pork is obtained from animal between ages of 3 to 12 months. Pork has more fat than meat.

Port wine: A wine made from grape must and fortified with brandy of good quality.

Post harvest: All the succeeding action or operations are defined as post harvest in case of food / horticultural commodity. Post harvest is the time that begins with separation of commodity from growing of production medium and those actions when the food enters the process of production for final consumption e.g. precooling, grading, sorting, packing transportation, storage and distribution.

Post Harvest diseases of papaya: Post Harvest diseases of papaya fruit are surface rot, stem end rot, internal fruit infection.

Post Harvest losses: All those losses which occur after the healthy commodity has been separated from the stalk / stem or other medium from which it is originated and nourished to growth. Or these are the losses which occur between the period of separation of the healthy commodity from this parent plant until they are consumed.

Post harvest physiology (PHP): This is a science and practice which deals with post harvest biology of perishable horticulture commodity viz., handling, and marketing of plant / plant parts. PHP is a subdivision of plant physiology, horticulture or fruit biochemistry which deals with most critical function and processes in plants /plant matter i.e. fruit after it has been harvest is thoroughly investigated. Or

Post harvest physiology (PHP): PHP is a subdivision of plant physiology, horticulture or biochemistry which aims at understanding the biochemical control of fruit ripening and organ senescence. It deals with all aspect of the ontogeny of the fruit such as maturation, ripening, senescence and storage. Or Physiology of food after harvest because they behave like living organisms and continue to perform the metabolic reactions (Respiration, Transpiration, etc.) and maintain the physiological system which were present when they were attached to the plant and for this metabolic processes, they depends entirely on their own reserves and moisture content. By manipulating the factors affecting maturation, ripening and senescence, we not only improve the storage life of food but also the productivity and quality of product / food.

Post Harvest Technology: Post Harvest Technology is defined as the branch of the science and technology that help to increase the shelf life of harvested produce which includes packaging, use of low and high temperature, chemicals and other suitable methods. or The term post harvest technology means the use of science and technology for the management of food commodities including ornamentals at post harvest level and includes packaging, transportation, storage, processing, and development of improved product, preservation and marketing.

Post processed spoilage: It is the spoilage of canned food after processing which is caused only by thermophiles if products are under processed but it is done by all sorts of microorganisms in case of faulty sealing or leakage.

Postharvest operation: All actions following harvest are known as post harvest operations.

Potable: A liquid that is suitable for drinking especially an alcoholic beverage.

Potable water: Water which is fit for human consumption as per Bureau of Indian Standards.

Potassium alginate: It is a chemical compound that is the potassium salt of alginic acid. It is an extract of seaweed and has empirical formula $KC_6H_7O_6$. It is widely used in food as stabilizer, thickener and emulsifier.

Potassium aluminium sulphate (PAS): It is used for coagulation of the latex of the banana.

Potassium bromated: It is a dough conditioner / bleaching agent and is used in flour and bread.

Potassium sorbate: It is potassium salt $C_6H_7KO_2$ of sorbic acid used as a food preservative.

Potato chip: A thin slice of white potato that has been cooked until crisp and then usually salted or spiced (Plate 501).

Potato pancake: A fried flat cake of grated potato mixed with raw egg and grated onion and spices is called potato pancake.

Potato peeler: It is an apparatus used for peeling of the potato prior to slicing. It may be hand or machine operated potato peeler, machine operated abrasive peeler (Plate 502).

Plate 501: Potato chips

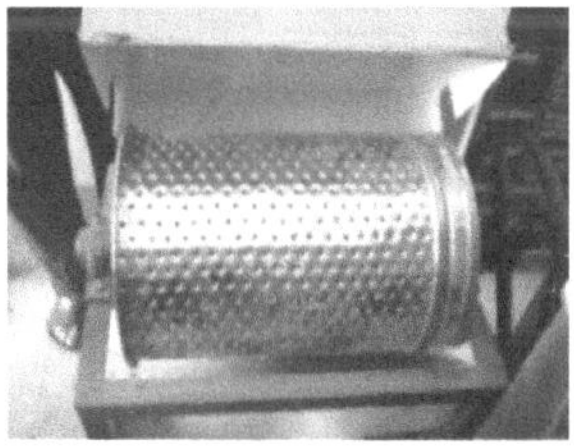

Plate 502: Machine operated stainless steel potato peeler

Potential energy: The energy that a piece of matter has because of its position or nature or because of the arrangement of parts is potential energy.

Pouch pack process: It is also called retort sterilization.

Poultry: This term is usually restricted to domestic fowls or chickens kept for both egg and meat production. It is also applied to ducks, turkeys and geese kept for meat production.

Poultry science: It is the study of principles and practices involving poultry management including breeding, incubating, housing, feeding, treating the diseases, processing and marketing of the produce.

Poussin: The chicken or game bird reared for table purpose, usually 7-8 weeks old weighing about one kilogram live weight.

PP cap: It stands for Pilfer proof cap.

PP cap sealing machine: PP cap sealing machine is used for closing and sealing the cap on the bottles (Plate 503).

Plate 503: PP Cap sealing machine

PPM: It stands for parts per million. PPM means one part of the solute per million part of the solution.

Pre processed spoilage: Spoilage of products (raw materials) in cans before processing which is caused by all types of microorganisms.

Prebiotic: These are chemical or environmental precursors of the origin of life and exist before the origin of life. E.g. a food / drink in which fibre have been added to give health benefits are called prebiotic foods. These are non-digestible food ingredients and have beneficiary effects in the host by selectively stimulating the growth and activity of some of the bacteria in the colon. Prebiotics are not digested in the upper part of gastrointestinal tract and reach the colon, where these are metabolized by bacteria residing in the colon and imparting beneficial effect on the host e.g. *Lactobacillus, Bifidobacteria.*

Precooling: It is an important post harvest operation of vegetable which is accomplished by refrigeration or water immersion with a view to remove field heat, conserve weight and retard ripening and senescence of the harvested vegetables. In short, it is removal of the field heat from the food commodity after harvest. Precooling is prompt cooling after harvest or removal of field heat.

Precursor: It is a compound that can be used by the body to form an essential nutrient.

Preference testing: It is the sensory testing used to determine acceptability of preference between products.

Pre-formed Pouches: The pre-formed pouches / bags are the earliest form of package meant for manual handling, filling, sealing and is widely used by most small scale industries and retail distribution of food grains in consumer packs. The pre-formed bags are available in a wide range of sizes and materials.

Pre-freezing: Pre-freezing is the process of freezing of raw material before sublimation.

Preharvest losses: These losses are caused by insect, fungi, birds, rodents, cultural practices etc during the growth period of the commodity. It is also called production loss.

Pre-packaging: Pre-packaging is defined as packaging the produce in consumer size units either at producing centre before transport or at terminal markets.

Preparation: It means the operations of slaughtering, processing, preserving, packaging and labeling of agricultural products.

Preservation: Method of extending the shelf life of fresh as well as processed food products by means of low temperature, drying, anaerobic condition, chemical irradiation and fermentation.

Preservatives: Substances capable of retarding or arresting the deterioration of food or any substance which is capable of nhibiting, retarding or arresting the growth of micro organisms is known as a preservative.

Preserve: Matured fruit pieces or mature but tender vegetable pieces when cooked in heavy syrup till it becomes soft and transparent and possess soluble solids of at least 68%. Or Preserves are prepared by cooking the fruit as a whole or in large pieces in sugar syrup until the soluble. Solids concentration is reached to about 60% or more (Plate 504).

Plate 504: Aonla preserve

Press liquor: It is the solution expressed from the shred and peel.

Prevention of Food Adulteration Act: One of the early acts to be promulgated in food laws and standards was the Prevention of Food Adulteration Act of 1954, which has been in force since June 1, 1955. The Act prohibits the manufacture, sale and distribution of not only adulterated foods but also foods contaminated with toxicants and misbranded foods. The objective of this act was to ensure that food articles sold to the customers are pure and wholesome. This prevents fraud or deception and encourages fair trade practices. The PFA Act lays down the guidelines for setting up standards for various food items like cereals and cereals products, pulses, ghee etc. All these processed items are expected to conform to these standards. It is mandatory & comes under the Ministry of Health & Family Welfare to operate Central Committee for Food Standards (CCFS). It is applied to domestic and imported products and cover various aspects of food processing and distribution-food colour, preservatives, pesticide residues, packaging and labelling, etc.

Primary colour: Blue, yellow and red colours are known as primary colour. Combinations of two of primary colour give rise to secondary colours.

Primary container: A primary container is one that comes in direct contact with the food, for example, a can or a jar. Primary container must be non-toxic and compatible with the food and causes no colour, flavour or other foreign chemical reactions. Some foods which are provided with efficient primary containers by nature are nuts, orange, eggs, milk powder, dried eggs and fruit concentrates (Plate 505).

Plate 505: Primary container for mango

Primary organism: This is the organism which is grown directly for the purpose in mind.

Primary processing: It involves basic processing of the natural produce like cleaning, grading, washing, dehusking.

Primary settling tank: It is a part of effluent treatment plant. It provides a smooth transition as effluent flows from flocculation tank to aeration tank. The sludge settled out into sludge drying beds built adjacently. The tank is provided with clarifier and scraper.

Primary wholesale markets: These are located in big towns near the centres of production of agriculture commodities, where transactions mostly take place between farmers and traders. These markets are periodically held, either ones or twice a week. Agricultural produce comes from neighboring villages. These markets are also known as local markets and organized by village panchayats and every shopkeeper has to pay certain rent. These markets deal in the sale of fruits, vegetables, food grains, all household requisites etc. e.g. Village market

Priming: It is the process of formation of the "wet steam" in the boiler. These are formed when some droplets of the liquid water are carried along with the steam.

Principle of bomb calorimetry: The heat gained by the cold bodies is equal to the heat lost by the hot bodies.

Principle of conservation of matter: The matter can neither be created nor destroyed during any physical or chemical action. According to this principle the matter can be converted into energy and energy can be converted into matter but the total amount of matter and energy always remains constant.

Principle of freezing: The basic principle of the freezing is the removal of heat by the refrigerant.

Principle of steam distillation: The reduction of pressure lowers the boiling point so that the liquids boil at lower temperature (Plate 506).

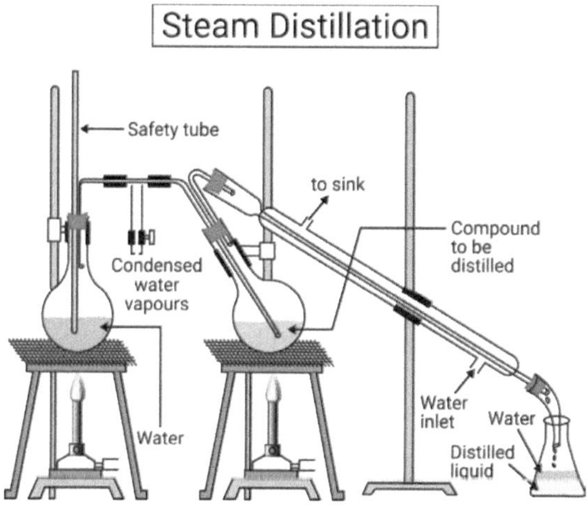

Plate 506: Distillation process illustration

Principle of ultra filtration: It is a sieving process through a membrane having pore size of 5 to 35 mm at a pressure of 500 KPa (5 bar). UF works on the principle of cross flow filtration. It is used for concentration of the proteins (Plate 507).

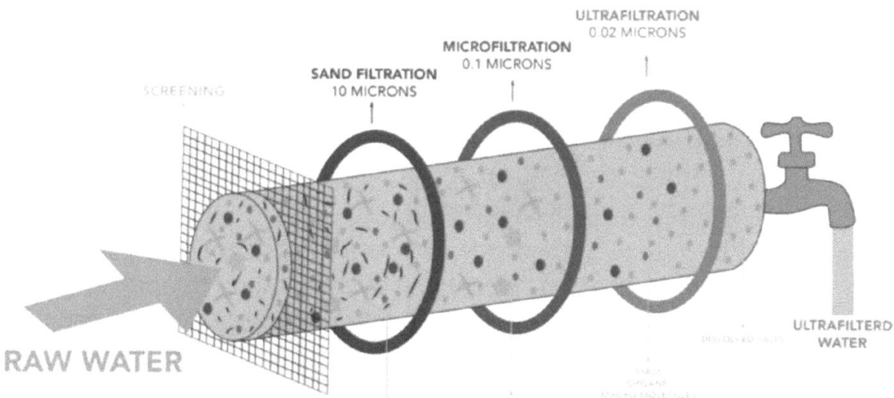

Plate 507: Ultra filtration process illustration

Probability: The ratio of the number of outcomes in an exhaustive set of equally likely outcomes that produce a given event to the total number of possible outcomes or it is the chance that a given event will occur.

Probiotic: A food / preparation (dietary supplement) containing live bacteria (lactobacilli) that are taken orally to restore beneficial bacteria to the body.

Process: It is a series of operation on a physical or chemical change in a material.

Process (thermodynamic): A system has to describe a process whenever it under goes a change in its system. Thermodynamics is the study of the relations between heat, work, temperature, and energy. The laws of thermodynamics describe how the energy in a system changes and whether the system can perform useful work on its surroundings.

Process flow chart: A process flow chart is one in which all incoming and outgoing materials and utilities are shown.

Process layout: Process layout is characterized by keeping similar machines or similar operations at one location. It is also called functional layout.

Process standard: Criteria for the way and method by which products are made.

Processable / processible: A commodity suitable for processing is called processable or processible.

Processed cheese: A cheese made by blending several lots of cheese is called Processed cheese. These are cheese by-products made from a combination of natural cheese and added ingredients, such as stabilizers, emulsifiers and flavour enhancers. It is also called process cheese / cheese food.

Processed flavour: The flavour developed in produce due to heating is called processed flavour.

Processing: Series of operations performed in order to make good, tasty or relishable products from raw materials (fruit and vegetable) usually done by washing, cooking, mixing of different additives, preparation of different products, carbonation etc. or

Processing: The heating of the foods for preserving is known as processing. However in the canning, processing means heating or cooling of the canned foods to inactivate bacteria. Fruits are processed at a temperature of 100°C while non acidic vegetables are processed at 115-121°C.

Processing plant: Processing plant is the premises, buildings, structure and equipment used / employed in the preparation, processing, handling, transporting and storage of any food commodity /fruits and vegetables / the processed products.

Processing water: Processing water is used for post harvest treatment of produce, such as washing, cooling, waxing or transport.

Procurement prices: Procurement price of a commodity refers to the price at which government procures the commodity from producers/

manufactures for maintaining the buffer stock or for the public distribution system. These prices are announced by the govt. of India on the recommendations of the Commission for Agricultural Costs and Prices before the harvest season of the crop. At these announced prices, govt. procures the food grains (wheat, paddy and coarse grains) in the needed quantity either for maintaining the buffer stock or for the distribution through fair price shops. Procurement prices are fixed generally at a level, which is somewhat higher than the level of minimum support prices but lower than the prevailing market prices. The procurement prices are lower in relation to the actual market prices and as such farmers and traders are not willing to sell their stocks voluntarily to the govt. In such circumstances, the govt. procures food grains at the announced procurement prices either by imposing a levy on the farmers, or on the traders or through other methods. Procurement prices are announced before the sowing season. As a result, the procurement price itself becomes the support price at which the govt. purchased all the food grains offered for sale. Procurement prices also become the minimum support prices because the govt. was bound to purchase the food grains offered by the producers for sale.

Producers: Farmers or producers, perform one or more marketing functions. They sell the surplus either in the village or in the local market. Some farmers, especially the large ones, assemble the produce of small farmers, transport it to the nearby market, sell it here and make a profit.

Product brand: It is related to a brand name of the products and has a well known highly regarded or marketable name.

Product layout: Product layout is also called line layout. Various operations on a product are performed in a sequence and the machines are placed along the product flow line.

Product placement: The inclusion of a product in a television program or film as a form of paid advertisement.

Product standard: Specification and criteria for characteristics of products.

Production: Production means the operations undertaken to supply agricultural products in the state in which they occur on the form, including initial packaging and labeling of the product.

Production (pre-harvest) loss: This includes damage caused by insects, fungi, birds, rodents, cultural practices etc. during the growth period of the commodity.

Production control: Systematic planning, coordinating, and directing of all manufacturing activities and influences to insure production of goods well in time, of adequate quality, and at reasonable cost.

Production losses: These losses are caused by insect, fungi, birds, rodents, cultural practices etc during the growth period of the commodity. It is also called preharvest loss.

Productivity: The production rate per unit area or per unit volume is called productivity.

Profiling: It is very detailed word description (usually of flavour) developed by a highly trained panel against which subsequent production is evaluated to maintain quality of production.

Prokaryote / procaryote: It is one of the unicellular microorganisms that lack a distinct nucleus and membrane bound organelles and that are classified as a kingdom (Prokaryotae syn. Monera) or into two domains (Bacteria and Archaea) (Plate 508).

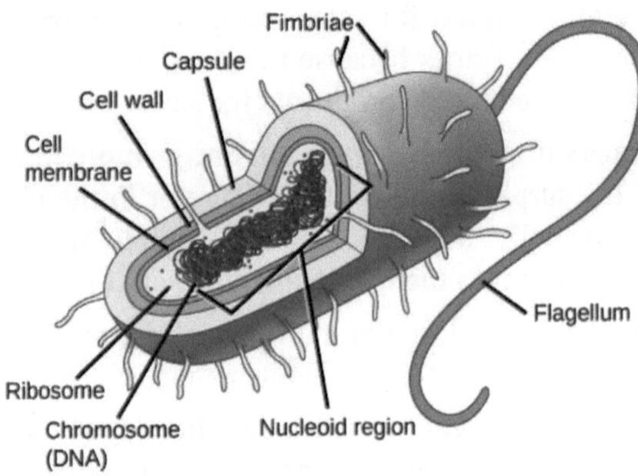

Plate 508: Prokaryotic bacteria

Proline: An amino acid $C_5H_9NO_2$ that can be synthesized by animals from glutamate. Fruits harvested from the interior canopy of the tree have higher praline content in the peel and are more resistant to chilling injury than those harvested from the exterior canopy with lower praline. Early harvested fruits are more chilling sensitive that late harvested fruits.

Promoter: A substance that in very small amounts is able to increase the activity of a catalyst. It is a binding site in a DNA chain at which RNA polymerase binds to initiate transcription of messenger RNA by one or more nearby structural genes. It is similar to enzymes.

Proofing: Shaping and leavening of bread is known as proofing

Propellant: It is a fuel plus oxidizer used by a rocket engine or it is a gas kept under pressure in a bottle / can for expelling the contents when the pressure is released.

Properties of system: The parameters like volume, temperature, pressure and density etc which can be used for identification or describing a system or material are called properties of system.

Propionate: It is a salt or ester of prop ionic acid.

Propionic acid: It is a liquid sharp odoured fatty acid $C_3H_6O_2$ found in milk and distillates of wood, coal, and petroleum and used as a mold inhibitor and flavoring agent.

Propyl gallate: It is isomeric alkyl radicals C_3H_7 derived from propane and often used in combination with gallate (salt or ester of gallic acid).

Propylene glycol: A sweet hygroscopic viscous liquid $C_3H_8O_2$ made of propylene and used as an antifreeze and solvent, in brake fluids, and as a food preservative.

Prosthetic group: It constitutes a nonprotein group of a conjugated protein.

Protease inhibitor: It is a substance that inhibits the action of a protease. It is one of the drugs (indinavir) that inhibit the action of the protease of HIV so that the cleavage of viral proteins into mature functional infectious particles is prevented and is used in combination with other agents in the treatment of HIV infection.

Protease: It is one of the enzymes that hydrolyze proteins and are classified according to the most prominent functional group (serine or cysteine) at the active site. It is also called proteinase.

Protective foods: Fruits and vegetables are known as protective food because they help in prevention of various ailments. E.g. green leafy vegetables, yellow and orange fruits, aonla, harar, behera etc and many more are called as protective foods.

Proteins: Proteins are naturally occurring extremely complex substances that consist of amino-acid residues joined by peptide bonds, contain the elements carbon, hydrogen, nitrogen, oxygen, usually sulfur, and occasionally other elements (phosphorus or iron), and include many essential biological compounds (enzymes, hormones, or antibodies).

Protein efficiency ratio (PER): It is defined as the weight gain of body (g) per unit intake of proteins (g).

Protein equivalent: It is equal to percentage of true protein plus half of the percentage of the non-protein nitrogen.

Protein floc: Bentonite treatment is generally used to remove the portions of must & to prevent after precipitation. It is called "Protein floc". It is usually added prior to the fermentation to clarify the must @ 60-100g/hl.

Protein Hydrolysate: A product of hydrolysis of proteins is called protein hydrolysate.

Protein I: Protein I is called Iron-Protein. It is a dimmer having identical peptides with molecular weight of 64 KG (kilodelton).

Protein II: Protein II is known as Iron-Molybdenum Protein. It is a tetramer consists of 2 beta and 2 alpha subunits with 220 KD. Each protein contains 2 molybdenum, 28-30 Iron and 28-32 S clusters.

Protein kinase C: A protein kinase is an enzyme that catalyzes the phosphorylation of specific serine or threonine amino acid residues or it is an enzyme that catalyzes the transfer of a phosphate group from ATP to one or more amino acids in the side chain of a protein resulting in a conformational change affecting protein functions.

Proteolysis: It is the hydrolysis of proteins or peptides with formation of simpler and soluble products. Proteolysis is the breakdown of proteins into smaller polypeptides or amino acids. Uncatalysed, the hydrolysis of peptide bonds is extremely slow, taking hundreds of years. Proteolysis is typically catalysed by cellular enzymes called proteases, but may also occur by intra-molecular digestion (Plate 509).

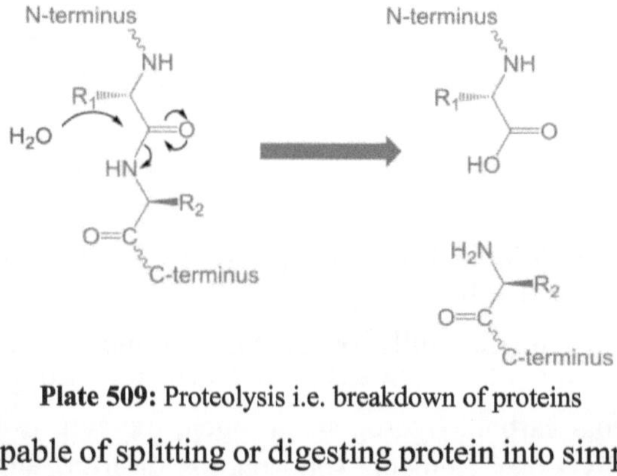

Plate 509: Proteolysis i.e. breakdown of proteins

Proteolytic: Capable of splitting or digesting protein into simpler compounds.

Protopectin: It is water insoluble substance linked with cellulose and hemicelluloses in presence of Al^{+++}. It is a form of the pectin substances found in the unripe fruit and some vegetables. It is long methylated polymer of galacturonic acid. It is insoluble precursor of pectin.

Protozoa: Protozoa such as *Taxoplasma gondii, Giardia intestinalis, Cyclospora* and *Cryptosporium parvum* produces larvae which subsequently infect man on ingestion. Infected meat and raw milk serve as a source for Taxoplasma whereas raw milk serves as a source of *Giardia*, *Cyclospora* and *Cryptosporidium*.

Provolone cheese: A firm pliant (bend freely or repeatedly without breaking or plastic) often smoked hard cheese of Italian origin (Plate 510).

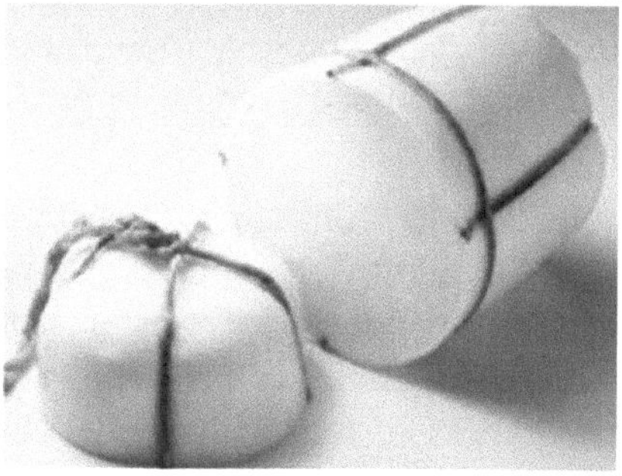

Plate 510: Provolone cheese

Proximate analysis: It is used to give information pertaining to nutritional and biochemical component (Plate 511).

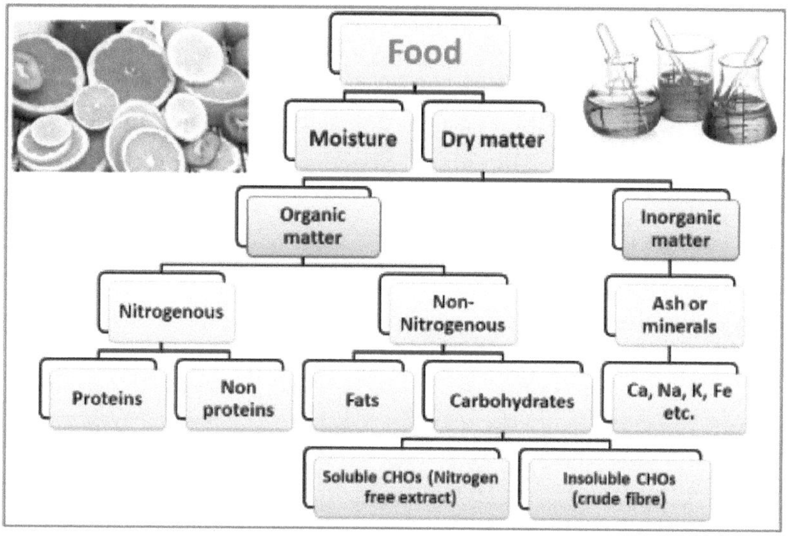

Plate 511: Proximate analysis

Pseudo-yeast: It is like true yeast but do not form spores.

Pseudoplasic fluids: These are also called shear thinning fluids. At low shear rates (dv/dy) the shear thinning fluids is more viscous than the Newtonian fluid and at high shear rates it is less viscous. These are the fluids those have apparent viscosity decreases with an increase in shear rate. In other words, the concentrates moves more rapidly through a pipe, its apparent viscosity decreases.

Psig: It stands for pound per square inch gauge.

Psychological colour: The colour of the sample evaluated by a person by his perceptions is called psychological colour.

Psychological loss: Loss of the commodity or food (fruit and vegetables etc) at the consumer level due to human aversion or refusal to eat for religious or personal reason.

Psychrometer: A hygrometer consisting essentially of two similar thermometers with the bulb of one being kept wet so that the cooling that results from evaporation makes it register a lower temperature than the dry one and with the difference between the readings constituting a measure of the dryness of the atmosphere. **Or** Psychrometer is an instrument used to measure the amount of water in air / humidity of the atmosphere.

Psychrometric Chart: A *psychrometric chart* presents physical and thermal properties of moist air in a graphical form (Plate 512).

Psychrometric properties: The properties of the moist air are called psychrometric properties and the subject which deals with the behaviors of moist air is called as psychrometry. These includes Dry-bulb temperature (DBT), Wet-bulb temperature (WBT), Dew point temperature, Humidity, Specific enthalpy, Specific volume, Psychrometric ratio, Humid heat.

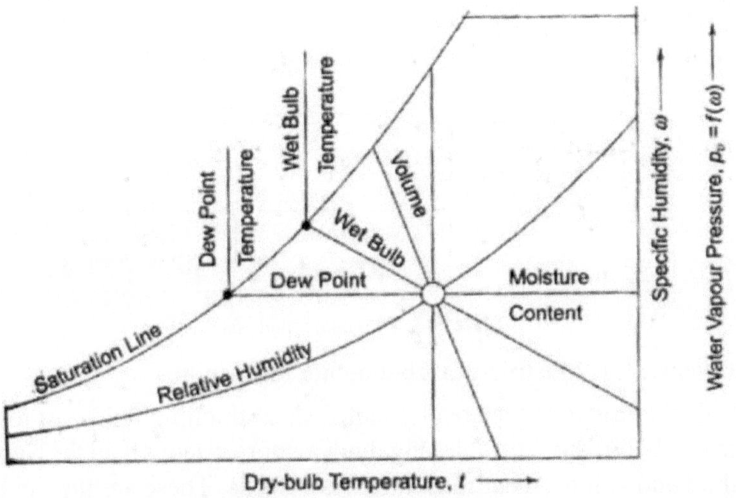

Plate 512: Psychrometric chart

Psychrometric ratio: It is the ratio of heat and mass transfer coefficient.

Psychrometric relation: The relationship between moisture content and temperature of the air during drying process in known as psychrometric relation.

Psychrophile: The organisms thrive at a relatively low temperature. Psychrophile includes majority of gram negative bacteria like *Pseudomonas, Achromobacter, Flavobacterium* and only a few are gram positive e.g. *Micrococci.*

Psychrotrophs: These are the microbes which fairly grow well at refrigeration temperature.

Ptyalin: It is an amylase found in the saliva of many animals that converts starch into sugar.

Pucca Arhatias: He is the real purchaser in the wholesale market on his own behalf of acting for some businessmen, firms in consuming markets. Big mills (rice, oils, cotton etc.) play role as their agent and order him to purchase certain quantity within a given range of price. When pucca arhatia trades on his own, he dispose of his produce brought by him through dealers in different parts of country.

Puff pastry: Pastry dough contains many alternating layers of butter and dough or the light flaky pastry made from butter and dough is called Puff pastry. It is also called puff paste.

Puffer fish: It belongs to family Tetraodontidae which is chiefly tropical scaleless marine bony fishes. It distends them to a globular form and most of which are highly poisonous. It is also called blowfish / globefish / swellfish.

Puffiness: Separation of the peel from the pulp during storage is called puffiness.

Pulp wash: The juice solids obtained by washing and refinishing the pulp several time is known as pulp wash.

Pulper: A machine that chops up whole fruit, vegetable and screen out the large particles and seeds to form a pulp that can be used directly or fed into the finisher to obtained pulp / juice (Plate 513).

Pulque: It is a wine prepared by spontaneous fermentation of agave juice. When fermentation starts, it is ready for consumption even after 24 hours. Shelf life of the pulque is 1-2 days. It contains 6% alcohol.

Pulse electric field processing: The processing of the food with the application of the high electric field strength (10-80kV per cm) with short pulses

(1-20 μs) to samples placed between the electrodes (Palte 514). This operation takes place at low temperature of less than 50°C. It is generally used for pasteurisation of the juices. The treatment with PEF also used to increase the yield of the juice during extraction.

Pulveriser: It is an apparatus used for grinding of the food ingredients in powdered form. The apparatus may be from small scale capacity to large scale capacity (Plate 515).

Pummelling: It is a process to strike dough repeatedly with the fists.

Plate 513: Pulper

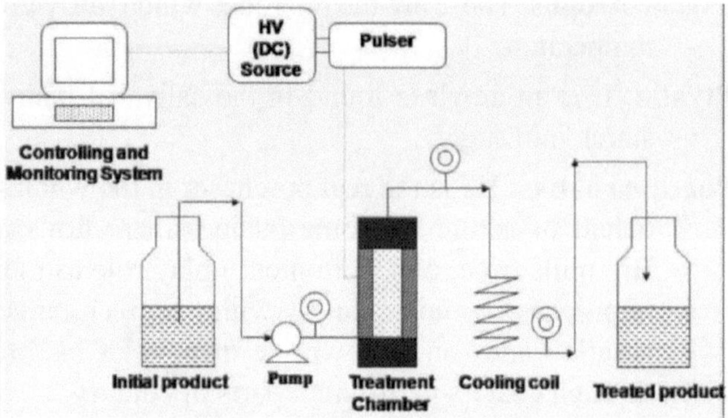

Plate 514: Continuous Pulse electric field system

Plate 515: Pulverizer

Punch: It is an alcoholic or non alcoholic beverage variable in comparison but usually based on wine or some sort of distilled liquor, fruit, soda or fruit juices with sugar, lemon and other ingredients added to form a flavoured combination.

Puree: Concentrated fruit and vegetable pulp without seed and skin, with or without salt, may be sweetened by the addition of sugar and preserved by adding chemical preservatives and the concentration of salt free solid in the finished product varies between 8.37 to 12.0 % (9% optimum).

Purification: The act or an instance of purifying a food product especially drinks.

Putrification: Decomposition of proteinaceous materials particularly sulphur containing proteins leading to the development of foul smell (Putrid odour) which contain mainly H_2S and CO_2.

Putty knife: The implement with a broad flat metal blade used for scraping putty is called putty knife.

Putty: It a dough like material of fruit paste having a solid consistency.

PVC shrink sleeves: It is used for shrink-wrapping type of the packaging since it has better tear properties and is economical.

Pycnometer: It is a device used for measurement of the relative density (Plate 516). It is a standard vessel often provided with a thermometer for measuring and comparing the densities of liquids or solids. It determines the particle density which is obtained by dividing the sample weight by the particle volume.

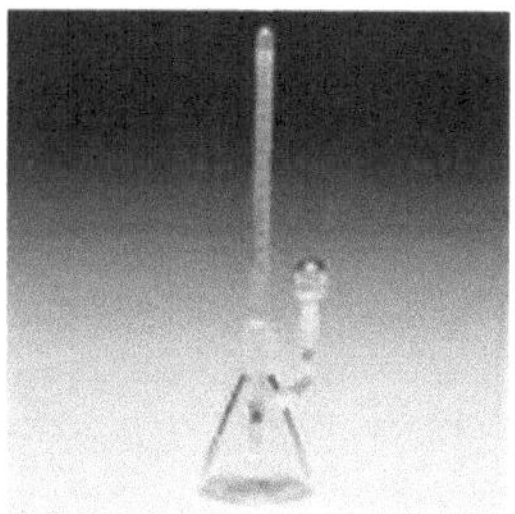

Plate 516: Pycnometer

Pyrenoid: It is centre of starch formation on certain chloroplasts of algae.

Pyrex: It is the glass made up of SiO_2 (85%), B_2O_3 (12.9%), Na_2O (3.8%), K_2O (0.4% and Al_2O_3 (2.2%). It is called Boro-silicate glass.

Pyridine: It is a nitrogenous base of many organic compound e.g. nicotine.

Pyridoxal: It is a crystalline aldehyde $C_8H_9NO_3$ of the vitamin B_6 group that occurs as a phosphate and is active as a coenzyme.

Pyridoxamine: It is a crystalline amine $C_8H_{12}N_2O_2$ of the vitamin B_6 group that occurs as a phosphate and is active as a coenzyme.

Pyridoxine: It is a crystalline phenolic alcohol $C_8H_{11}NO_3$ of the vitamin B_6 group found in cereals and convertible in the organism into pyridoxal and pyridoxamine.

Pyrolysis: Chemical change brought about by the action of heat is called pyrolysis. It is thermal degradation of volatile and non-volatile products prior to combustion.

Pyrometer: It is apparatus used for measuring the high temperature.

Pyruvate kinase: It catalyses the 3rd irreversible step in glycolysis.

Pyruvic acid: It is a 3 carbon acid ($C_3H_4O_3$) that in carbohydrate metabolism and is an important intermediate product formed during glycolysis (Plate 517).

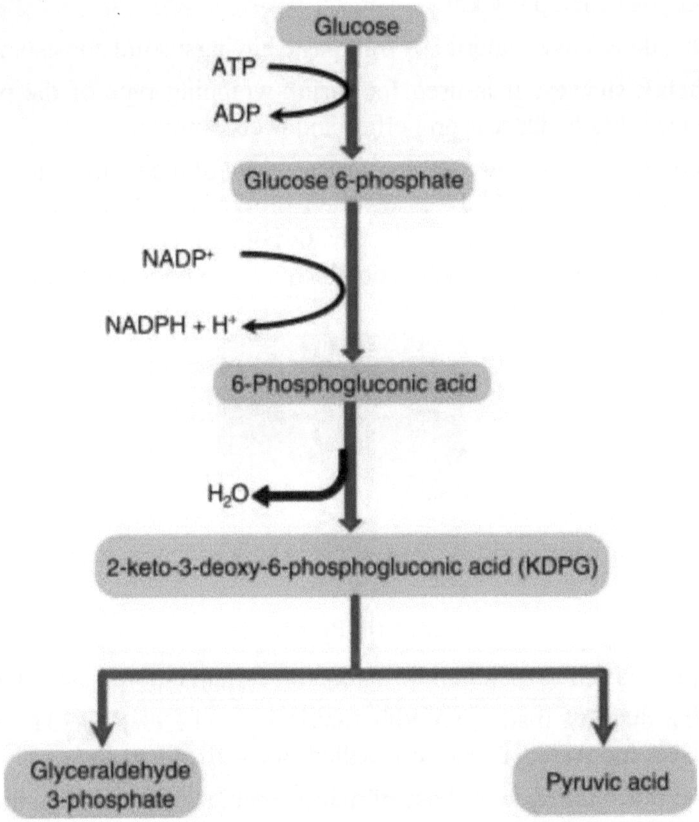

Plate 517: Formation of Pyruvic acid metabolic product during glycolysis

Q

Q_{10} value: The number of times a reaction rate change with a 10°C change in temperature is called Q_{10} value. If reaction rate doubles with 10°C change in temperature, $Q_{10} = 2$. It is also called temperature quotient.

Q-enzyme: It is a branching enzyme responsible for formation of 1-6 bonds in amylopectin.

QHV process: It stands for Quality Has Value Process.

QO_2: It is a unit for measurement of respiration. Or It is the oxygen uptake in microlitres per milligram on dry weight per hour.

Quailing: In this practice the oranges are kept in the baskets in which they are harvested to allow the fruit to loose little moisture. It is done for a couple of hours before transport to packhouse. This practice would allow the fruit to lose a little moisture which is called quailing. This operation is done for oranges to prevent Oleocellosis.

Qualitative measurement: It is a subjective / objective measurement of the quality of the product for presence of certain compound. It gives the idea of the presence or absence of the particular compound. It deals with the nature of the substance and not its amount or quantity.

Quality: Quality is a measure of the degree of excellence or degree of acceptability by the consumer. The quality may also be defined in term of end use and may vary depending upon consumer's perception and need. Or quality is the totality of features and characteristics of a product that bear on its ability to satisfy stated or implied needs. It is a term defined by the consumer, buyer, grader, or any other client based on subjective and objective measurements of the food product.

Quality control: It is generally defined as the regulation by law of food manufacture, distribution and sale in order to prevent health hazards and fraud to the consumers. Thus, it becomes a criminal offence to sell adulterated, filthy or contaminated foods. Quality control is the scientific control of production.

Quality management system: Quality management system is based on four important pillars for successful quality control. These quality management systems are management responsibility (defined responsibility of the key authority from top management), resource management (trained, competent and qualified personnel's having all types of information about process control), product/service realization (implementation of standards and to ensure the process operate under controlled condition), and measurement, analysis & improvement (essential to measure and monitor system performance, customer satisfaction, process performance, characteristics of products and services). This management system is also important for ISO certification in an organization / industry.

Quantitative Descriptive Analysis: It is used for thorough description of the product and for quantification of their intensity.

Quarantine: It is a measure during which a ship arriving in port and suspected of carrying contagious disease is held in isolation from the shore or it is a restraint upon the activities or communication of persons or the transport of goods designed to prevent the spread of disease or pests or it is a period of 40 days.

Quercetin: It is a yellow crystalline pigment $C_{15}H_{10}O_7$ occurring in the form of glycosides in various plants. It is found in citrus. It is used as anti-oxidant and absorber of the ultraviolet rays. It is used in vegetable oils.

Queueing: A sequence of messages or jobs held in temporary storage awaiting transmission / processing **or** a data structure that consists of a list of records such that records are added at one end and removed from the other (Plate 518).

Quiche: It is an unsweetened custard pie usually having a savory filling by spinach, mushrooms, or ham (Plate 519).

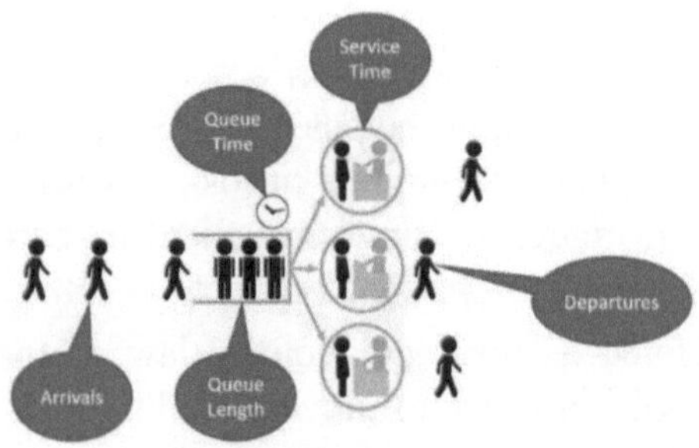

Plate 518: Queueing

Plate 519: Quiche

Quick freeze: It is a process to freeze food for preservation so rapidly that ice crystals formed are too small to rupture the cells and the natural juices and flavor are preserved.

Quick freezing: Freezing of products in a relatively short time (30 minutes or less) maybe done by direct immersion of the food (packed or unpacked) in a refrigerant (Liquid N_2 or liquid CO_2) by indirect contact with the refrigerant (Air blast freezing).

Quick lime: It is unslaked lime or Calcium oxide (CaO).

Quick process: Quick process is for vinegar production and is similar to generator process. An aqueous solution of ethyl alcohol is mixed with oxidized in the presence of bacteria. The wooden vat is fitted with two wooden slits with holes. Between these plates is a filling of beech wood, moistened with an old vinegar solution that is a major source of Mycoderma aceti. A 10% aqueous solution of ethyl alcohol is gradually reduced from the top of the vat and the air is transferred in a controlled amount through holes near and below the vat. Ethyl alcohol is added to acetic acid. This process is called the rapid vinegar process because vinegar forms very quickly. This process is also called as packed generator process or trickling generator process or fringe generator method. This method was developed in Germany for rapid microbial oxidation of ethanol to acetic acid. The quick vinegar process makes the use of fringe generators (Plate 520).

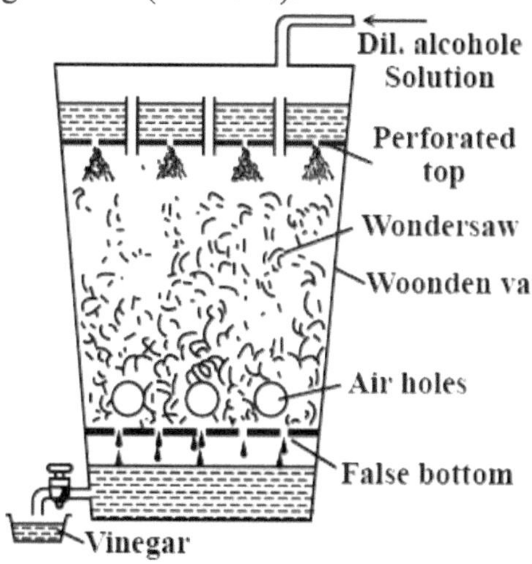

Plate 520: Quick vinegar process illustration

R

R - Enamel can: The acid resistant, gold coloured enamel treated can are called R enamel cans. It is used for canning of acidic food which have soluble colouring pigments.

R value: R value is a measure of resistance to the flow of heat through a given thickness of a material (insulation) with higher numbers indicating better insulating properties. R-Value is a measure of insulation's ability to resist heat traveling through it. The higher the R-Value the better the thermal performance of the insulation.

Rabbitto and putridity: The surface taint of the milk caused by Pseudomonas putrefaciens is called rabbitto and putridity.

Racking: An important operation for the wine production in which wine is siphoned off and is transferred to a fresh vat with the help of a false bottom. Racking is repeated decanting to remove the slowly setting cloud of the freshly fermented young wine. It is also called clarification. It is synonym to siphoning.

Rad: It is the unit of radiation dose being equivalent to the absorption of 100 erg/g of irradiated material.

Radappertization: It means complete sterilization of foods by using irradiation. It is equivalent to commercial sterility. High dose of the radiation (25-45 or 30 - 40 kGy) is required for killing all microbes.

Radiation: Emission and propagation of energy through the medium of space or material. Radiation may be ionizing or non-ionizing. Ionizing radiation are of two types, electromagnetic (X-rays, γ rays) and particulate (α, β, proton and neutrons) radiation (Plate 521, 522 and 523). Non-ionizing radiation includes UV rays. It is a cold process and is additive free.

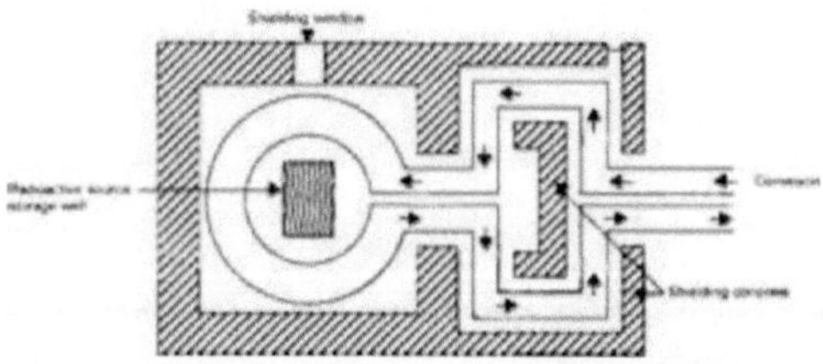

Plate 521: Typical diagram of irradiation plant

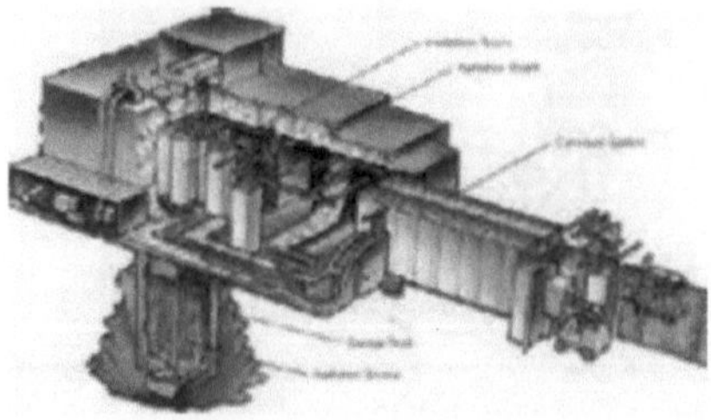

Plate 195: Irradiation plant

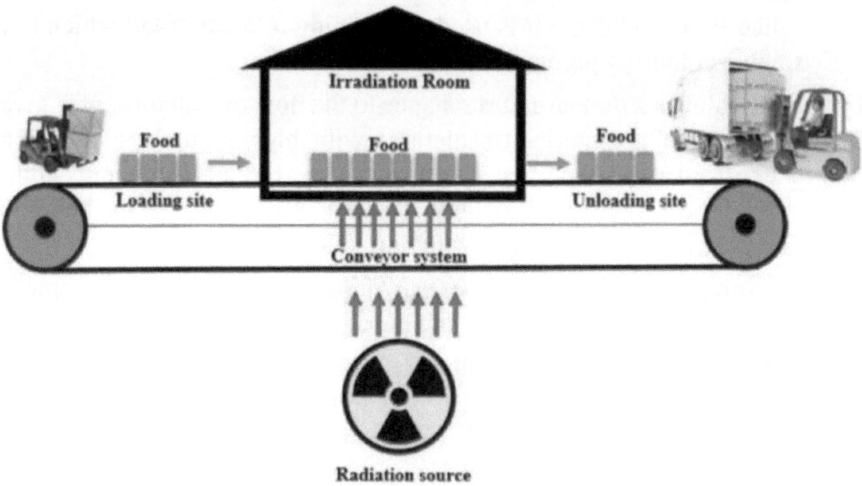

Plate 523: Process of food irradiation

Radiative properties: When radiation strikes a surface, a portion of it is reflected and rest enters the surface. The portion of the radiation that enters the surface, from that portion, some of the radiation is absorbed and remaining radiation is transmitted through it. The three radiative properties viz. absorptivity, reflectivity and transmissivity all have values between 0 and 1.

Radicidation: It is low level of the radiation dose (2.5-10.0 kGy) to destroy viable non-spore forming micro-organisms e.g. Salmonella.

Radiobiology: It is the scientific principles, mechanisms and effects of interaction of ionizing radiation with living matter.

Radiocytology: It is the branch of radiobiology that deals with effect of radiation on living cells.

Radurization: It means pasteurization by using irradiation. It have low level of radiation dose (0.75 – 2.5 kGy) and is sufficient to kill pathogenic micro-organisms.

Raffinose: It is a white, crystalline trisaccharide slightly sweet sugar $C_{18}H_{32}O_{16}$ obtained commercially from cottonseed meal and present in many plant products like sugar beets, molasses. Raffinose on complete hydrolysis yields fructose, glucose and galactose (Plate 524). It is also called gossypose or melitose. Raffinose is a trisaccharide composed of alpha-D-galactopyranose, alpha-D-glucopyranose and beta-D-fructofuranose joined in sequence by 1->6 and 1<->2 glycosidic linkages, respectively.

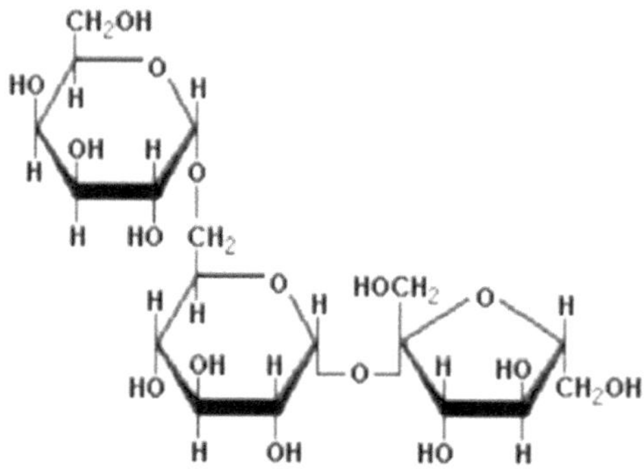

Plate 524: Raffinose structure

Raisin: It is dried seedless grape. Good raisin grape varieties should possess soft texture, marked pleasing flavor and after drying they should have more than 17% moisture e.g. Pusa seedless, Thompson seedless, Sultana etc.

Rancidity: The chemical deterioration of the fats caused by the uptake of oxygen or water (hydrolysis). The development of off-flavour due to oxidation of fats is called rancidity. The unpleasant taste and odour is due to the formation of peroxides (Plate 525).

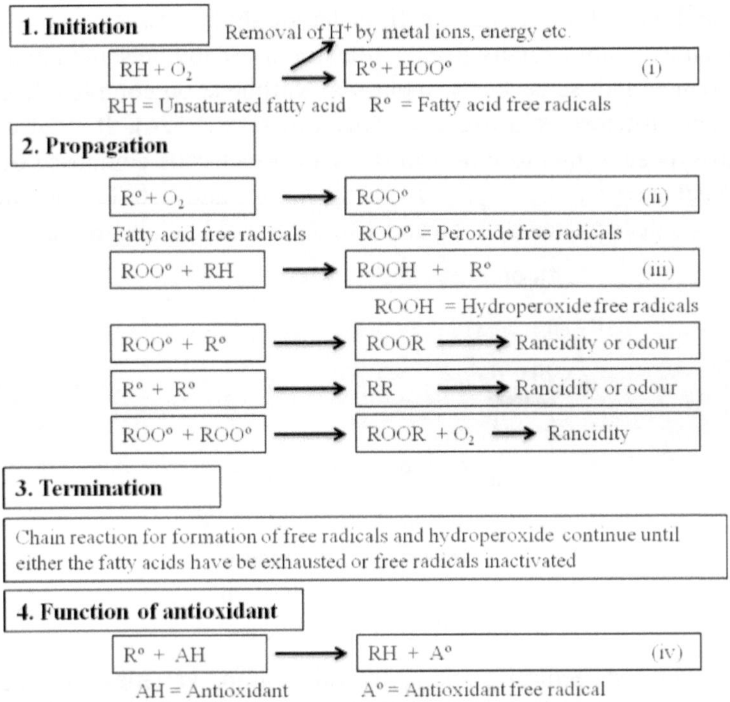

Plate 525: Mechanism of rancidity and its prevention by anti-oxidant

Raoult's law: At a constant temperature, the partial vapour pressure of volatile component of a solution is equal to the mole fraction of that component in solution multiplied by the vapour pressure of pure component. It is vapour pressure of the solution relative to that of the pure solvent is equal to the mole fraction of the solvent.

Rating tests: The rating tests are difference tests with a quantitative aspect through direction and degree of judgements. A suitable scale or score is used for giving quantitative judgements of the food depending upon the sensory evaluation. This test is also known as ranking test. In a ranking test, the panelists are asked to rank a series of samples in the increasing or decreasing order for specific characteristics like colour, flavour, odour / aroma, texture and overall acceptability. For giving quantitative judgements various methods are used.

Ravioli: The pasta in the form of little cases of dough containing a savory filling of meat or cheese is called ravioli.

Raw foods: The foods in the earliest or primary stage after harvesting of slaughter and not have been subjected to any treatment apart from cleaning and size grading.

RDA: It stands for Recommended Daily / Dietary Allowance. The RDA as per ICMR is 120 g fruits per day/head, 280 g vegetables / day / head, 400 g cereals per day/head and 85 g pulses per day /head. While in India per capita consumption is 40, 120, 436, and 38 g per day per head, respectively.

RDI: Recommended dietary intake for nutrients.

Ready to serve (RTS): It is a type of the fruit beverage which contains at least 10% fruit juice and 10% TSS besides about 0.3% acid.

Reaper: It is a mechanical harvester for harvesting of the crops.

Reconditioning: The exposure of the cold stored potatoes or onions to a temperature of 22±2°C for 2-3 weeks prior to processing is known as reconditioning. It helps conversion of sugars formed during cold storage into starch and thus prevents non-enzymatic browning. For prolonged winter storage the potatoes are kept in the darkness at low temperature (4-5°C) and high RH (80-90%). However, this favours a slow breakdown of the potato starch to sugar. For the manufacture of potato chips, high sugar content would result in excessive browning upon frying, which can be avoided by a reconditioning treatment. Reconditioning helps in using up the sugars and their conversion back to starch.

Reconstitution ratio: The ratio of dried material to rehydrated material.

Recrystallization: Recrystallization is a physical change in which many small ice crystals combine to form a smaller number of large crystals.

Recycling: It is a process in which part of the output stream is returned to join the feed streams.

Red algae: A division (Rhodophyta) of chiefly marine algae that have predominantly red pigmentation. The extracts (dried polysaccharide) of red algae are used in the manufacture of agar.

Red hot: A small red candy strongly flavored with cinnamon is called red hot.

Red meat: The meat of beef and mutton is called red meat. It is distinct from white meat of chicken and other game-birds (cold blooded).

Red sauce: Tomato sauce is known as red sauce.

Red wine: Red wine is produced by extraction of juice from coloured grapes. Maceration of the skin is done for red wine. Red wine involves maceration, carbonic maceration, and thermovinification.

Reducing: It is a process to change (an element or ion) from a higher to a lower oxidation state or to add one or more electrons to an atom / molecule.

Reducing agent: A substance that reduces a chemical compound usually by donating electrons.

Reducing sugars: Sugars possessing a free aldehyde or ketone group e.g. maltose, glucose etc. Reducing sugars in many cases are responsible for undesirable browning reactions used for processing purpose e.g. dehydration and frying. These sugars reacts with amino acids of proteins mostly lysine and causes non-enzymatic browning. Those sugars which act as reducing agents are called reducing sugars. They contain an aldehyde (- CHO) are or a ketonic C = O all monosccharides and disaccharides (except sucrose) are reducing sugars, e.g., glucose, fructose, lactose etc. They reduce Fehling solution and Tollen's reagent (Plate 526).

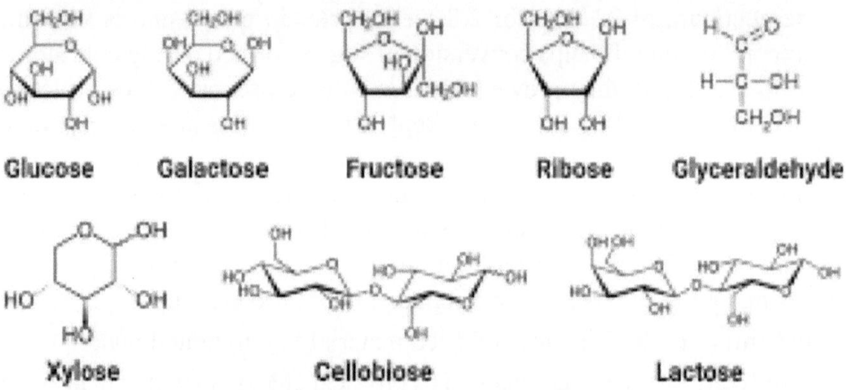

Plate 526: Examples of Reducing sugars

Redurization: Combination of pasteurization and ionising radiation which greatly extends the shelf life of the perishables.

Reference man: It is a man having an age of 25 years and weight 55 kg.

Reference woman: It is a woman having an age of 25 years and weight 45 kg.

Reflux: The flow back or return of the vapours formed by heating and condensation. The vapors formed condense and return to be heated again.

Refluxing: It is a process that causes flow back or return of the vapours formed by heating and condensation. The vapors formed condense and return to be heated again.

Refraction: It is deflection / bending of the light rays / energy wave from a straight path in passing obliquely from one medium (air) into another (glass) in which its velocity is different (Plate 527).

Refractive index: It is the ratio of the speed (velocity) of radiation (as light) in one medium (as air, glass, or a vacuum) to that in another medium. It is the ratio of the sine of the angle between the incident ray and the line perpendicular to the surface of the prism to the sine of the angle between the reflected ray and line perpendicular to the surface of the prism. (sin θ incident ray /sin θ reflected ray).

Refractometer: It is an instrument for measuring refractive indices of food commodities for determination of total soluble sugars. Or Refractometer is an instrument used to determine the refractive index of liquids. It is employed as a production line tool. Total soluble solids are estimated by this instrument. It requires 2-3 ml of the sample for estimation (Plate 528).

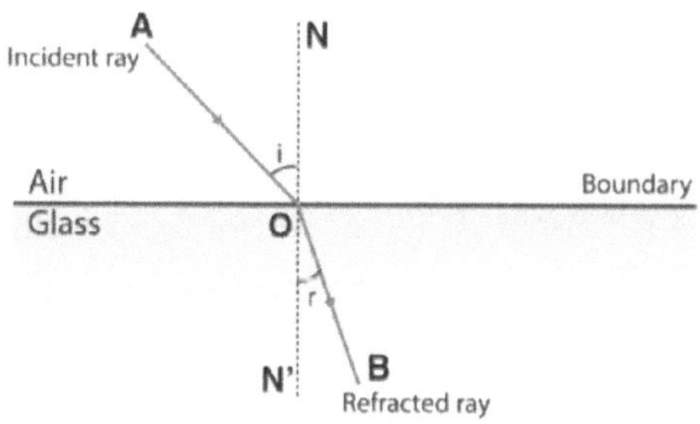

Plate 527: Refraction of light

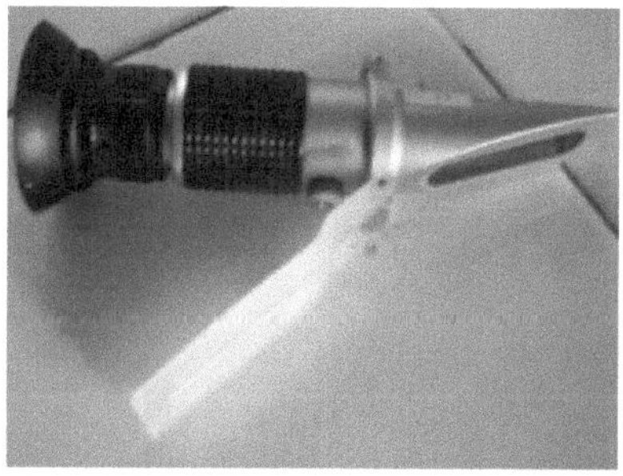

Plate 528: Hand Refractometer

Refried beans: Beans cooked with seasonings, fried, then mashed and fried again are known as refried beans.

Refrigerant: A gas that refrigerates the products placed in a refrigerator by absorbing latent heat from the surrounding is called refrigerant. Temperature for evaporation of the gas in liquid form is below the freezing point of water and the latent heat of evaporation is also relatively large. The first refrigerant used in a practical system is ammonia gas. Freon's are now used which are non toxic, non combustible, non explosive and leakage of refrigerator usually causes no damage to food. CO_2 and SO_2 may also be used as refrigerants.

Refrigerate: To freeze or chill a food for preservation.

Refrigerated storage: Storage at a temperature below those normally available throughout the year. Refrigerated storage is generally done between about 0 to 8^{0}C.

Refrigeration / refrigerate: It is a process to make or keep cold / freeze / chill (as food) for preservation. Or Refrigeration is the art of producing cold which is done by removal of heat from the insulated chamber by means of refrigerants. It is the process of lowering the temperature of substance to that of its surrounding atmosphere by using refrigerant.

Refrigerator: It is an apparatus / room or appliance that is used for keeping food or other items in cool / refrigerated conditions. It is an insulated cool camber in which edibles, medicines etc are kept to preserve their freshness. It has major parts like, receiver, compressor, evaporator and condenser. The motor compressor compresses a suitable gas (Refrigerant) which liquefies and expands in a closed system and in this process it absorbs the latent heat from the surrounding resulting in cooling of the chamber (Plate 529).

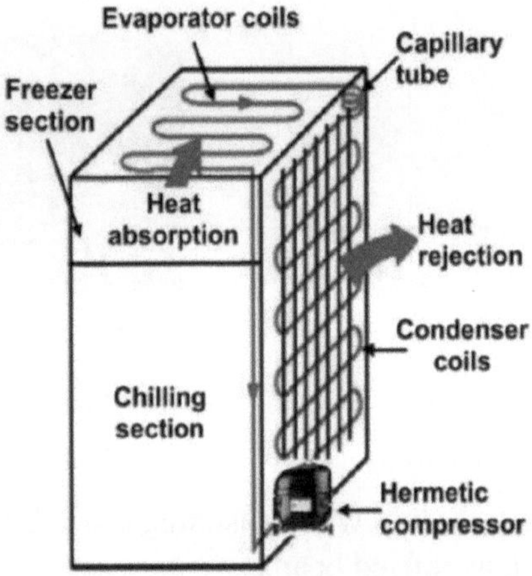

Plate 529: Component of refrigerator

Regenerator: The heat exchanger in which the two fluids streams flow alternatively through matrix and exchange heat energy is called regenerator.

Regulated markets: Here business is done as per the rules and regulated by statutory market organization. Market charges are standardized, fixed and regulated by Agricultural Produce Market Committee (APMC).

Rehydration coefficient: It is the ratio of Drained weight of dehydrated sample x {100 - Moisture content of the sample by drying) to the {Weight of the dried sample taken for rehydration - Amount of moisture present in the dried sample for rehydration} X 100.

Rehydration Ratio: It is the ratio of weight of rehydrated material/ weight of dehydrated material.

Relative Coefficient of Performance (COP): It is the ratio of actual COP to that of theoretical COP.

Relative density: It is the ratio of weight of the given volume of the substance to the weight of the same volume of the water at a fixed temperature.

Relative humidity (RH): It is the ratio of water vapour pressure in air to saturation vapour pressure at the same temperature. Or It is the ratio of absolute air humidity w.r.t. the absolute air humidity at saturation point at same dry-bulb temperature (Plate 530).

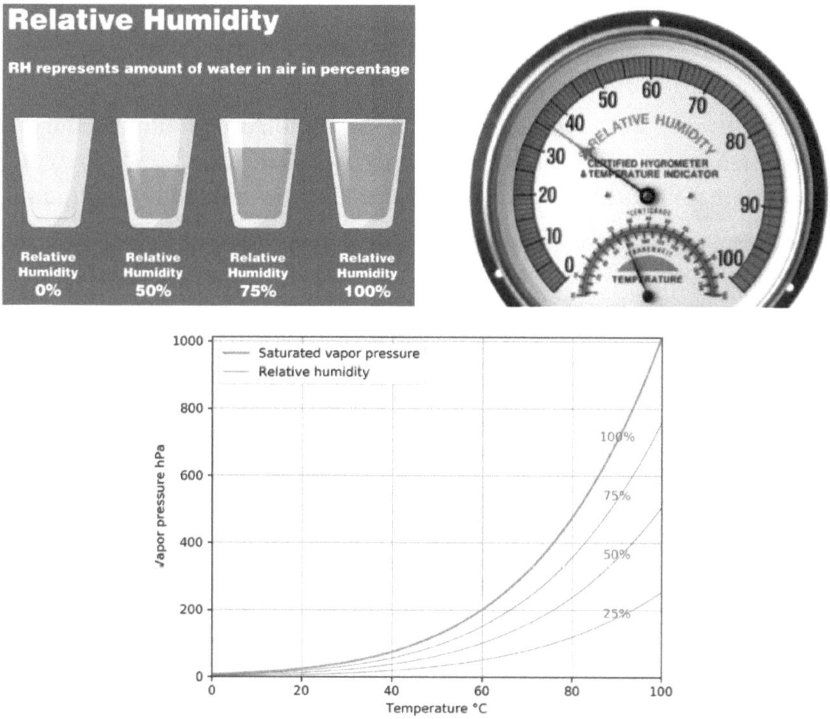

Plate 530: Relative humidity illustration

Rendering: Process of removal of fats from animal cells / tissue / meat is known as rendering.

Rennet: The crude chymosin or rennin enzyme is known as Rennet. Cheese made with rennet is a much better source of calcium than cheese acid precipitate. It is the first genetically engineered enzyme to be use in food is calf chamois (Rennet) used in manufacture of cheese. It is by use of *E. coli.*

Rennin: It is a gastric enzyme that coagulates milk and is used in making cheese. It is obtained from the mucous membrane of the stomach of a calf.

Resin: It is translucent amber colored to almost black brittle friable resin that is obtained from the oleoresin or deadwood of pine trees or from tall oil and used in making varnish. It is also called resin. **or** These are solid or semisolid amorphous fusible flammable natural organic substances that are transparent or translucent and yellowish to brown, are formed in plant secretions. These are soluble in organic solvents (ether) but not in water. These are electrical nonconductors, and are used chiefly in varnishes, printing inks, plastics etc.

Resinating: It is the process to impart flavor in some food articles with the addition of resin.

Respiration: The enzymatic oxidation of sugar to carbon dioxide and water accompanied by a release of energy is called respiration. Changes in proteins, lipids and organic acids also take place through respiration.

Respiratory climacteric: The rapid increase in the rate of respiration and the generation of ethylene by the fruit at a given point in its development at the onset of climacteric ripening (Plate 531).

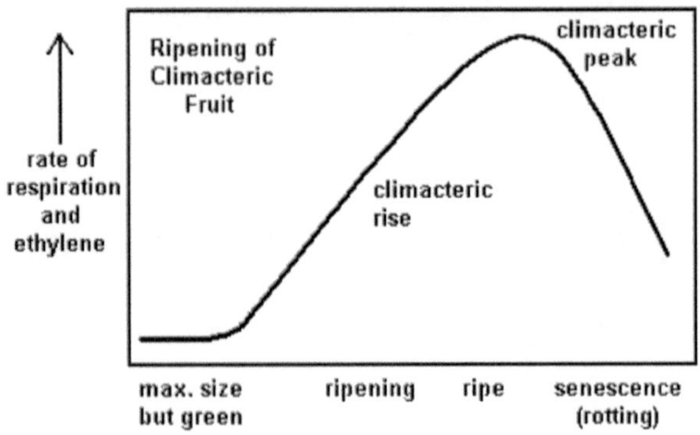

Plate 531: Respiratory climacteric illustration

Respiratory Quotient (RQ): It is the ratio of carbon dioxide (ml) produce to oxygen (ml) consumed. It is useful in knowing the nature of the respiratory substrate used. RQ is 1 for sugars and less than 1 for proteins and fats.

Resveratrol: It is a trihydroxy stilbene derivative $C_{10}H_{12}O_3$ that is found in some plants, fruits, seeds, and grape-derived products (red wine) and has been linked to a reduced risk of coronary disease and cancer. It is a powerful phytonutrient found in the grape skin and has very good anti-oxidant ability.

Retail markets: Here commodities are brought by and sold to the consumers as per their requirement. These markets are spread all over the city or town subject to municipal control. They generally deal in all types of produce and serve the needs of the city people as well as of the surrounding villages. Particular type of market is located in particular locality. Cloth market is one locality and grain, vegetable are in different localities. There is direct selling to consumer.

Retailers: He is the last link in chain of middleman, who sells directly to consumer (Plate 532). He takes little of goods, sells and sets up business usually amidst the consumer's groups. He buys his requirement usually from the wholesalers. Retailers in producing areas may have direct contact with producers and buys goods from them for resale.

Plate 532: Retailer selling fruit and vegetable in retail market

Retailing: The activities involved in the selling of goods to ultimate consumers for personal or household consumption is known as retailing.

Retinaldehyde: Retinaldehyde or retinal is a yellowish to orange aldehyde $C_{20}H_{28}O$ derived from vitamin A that in combination with proteins forms the visual pigments of the retinal rods and cones.

Retinol equivalent (RE): It is a unit used to express vitamin A activity. One RE is equal to 3.33 IU.

Retinol: It is also called Vitamin A. It is fat soluble alcohol, most abundant in fatty fish and especially in fish liver oils. It is not found in plants, but many vegetables and fruits contain β-carotene, which is readily converted in the body to vitamin A. It functions directly in vision, especially in night vision. A derivative, retinaldehyde, is a component of the visual pigments, including rhodopsin, in the retina. Humans require vitamin A in very small amounts. Unlike carotenes, it is toxic in large amounts and is readily destroyed by exposure to heat, light, or air.

Retortable Pouches: The retort pouch process is a technique for the long term preservation of food stuff. Flexible materials can be combined to withstand even the adverse conditions of retorting encountered with low acid foods. This is analogous to canning with the can being replaced by a heat resistant, flexible pouch. This had led to retort pouch being referred as a "flexible can". The shelf life of a retort pouch will depends to some extend on the nature of the product, but will not be less than two years. Such "flexible can" have become standard container for some applications such as providing food to soldier in the field. Retortable pouches are constructed of a three ply laminate; outer layer of polyester film for high temperature resistance, strength and printability, middle layer of aluminium foil for barrier properties and an inner layer of polypropylene film that provide heat seal integrity (Plate 533). Retortable trays are constructed from multilayers of polymers, one of which is ethylene-vinyl alcohol to provide an oxygen barrier. These trays are often sealed with a polymer foil laminate film. Flexible retortable pouches offer good shelf life of the product does not require refrigeration or freezing for storage, heat penetration is fast, easy to open and dispose, require less storage space and light in weight resulting in reduction of transportation and energy cost.

Retrogradation: It refers to the association and crystallization of starch in water causing effects, such as precipitation, gelation and changes in consistency. It describes the charges that occur as a result of the cooling and storage of the starch products.

Reverse osmosis: It is reverse of osmosis. It resists micromolecules and allows passage of H_2O only. It is used for concentration of the fruit juices. It allows passage of water only and restricts everything except water. Or Reverse osmosis is the movement of freshwater through a semi-permeable membrane when pressure is applied to a solution (seawater) on one side of it (Plate 534).

Plate 533: Retortable Pouches

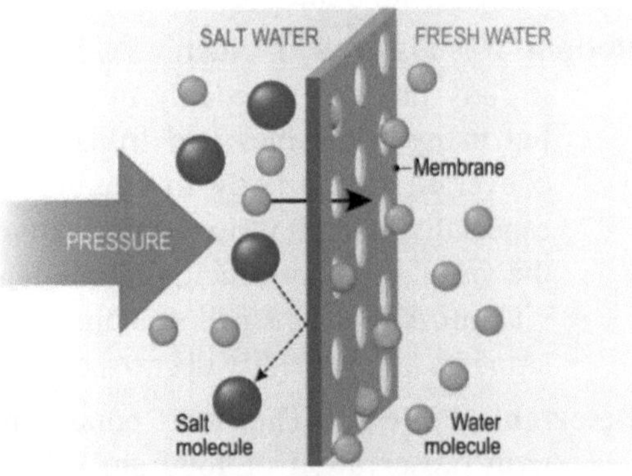

Plate 534: Reverse osmosis

Reversible process: A process is called reversible if at the end of the process, the initial states of all system taking part in the process can be restore, without any outstanding changes in the state of other systems. Otherwise the process is called irreversible.

Reversion: The development of off-flavour in soybean is called reversion.

Reynolds number: A number which is characteristic of the flow of a fluid in a pipe is called Reynolds number. It is a dimensionless quantity that can be used as an index of laminar or turbulent flow. Re = DVρ/μ, whetre D = tube diameter, V = average velocity, ρ = density of the fluid, and μ = viscosity. For value of RE below 2100, the flow is laminar and the pressure drop per unit length of the pipe can be determined by Poiseville equation. For fluid RE above 2100, the flow is turbulent (Plate 535).

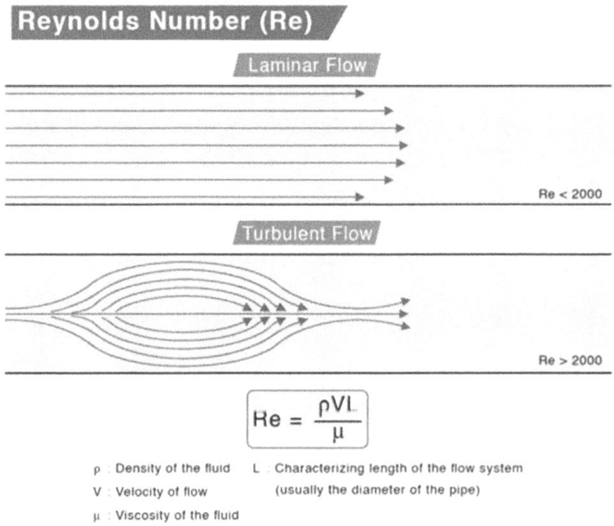

Plate 535: Reynolds number illustration

Rf value: It is defined as the ratio of distance the solute moves to the distance of the solvent moves. This is a parameter that measures how far a material moves with respect to how far the solvent moves at the same time (Plate 536).

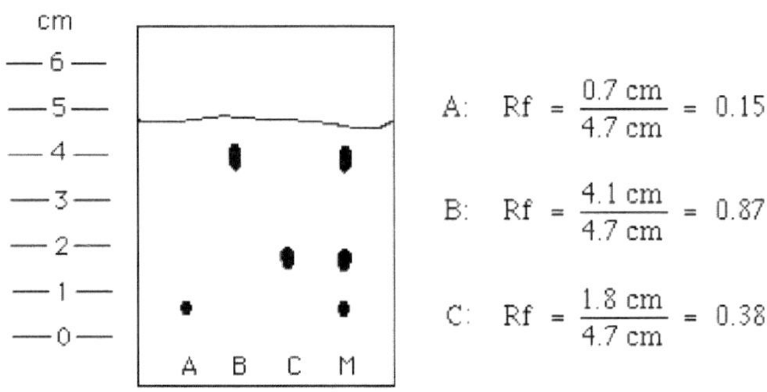

Plate 536: Rf value calculation

Rhamnose: A crystalline sugar $C_6H_{12}O_5$ that occurs in the form of a glycoside in many plants and is obtained in the common dextrorotatory L form.

Rheology: Rheo is a Greek word meaning "to flow". Thus, a science dealing with the deformation and flow of matter is called rheology. or Rheology is the science of flow and deformation of materials i.e. solids and liquid (Newtonian and non Newtonian) under pressure which include three aspects of viscosity, plasticity and elasticity. It is a measurement of texture.

Rheometer: An instrument used for measuring flow of viscous substances (Plate 537).

Rheopectic fluids: The dynamic viscosity increases with the time for which shearing forces are applied e.g. gypsum suspension in water.

Rheostat: A resistor for regulating a current by means of variable resistances is called rheostat (Plate 538).

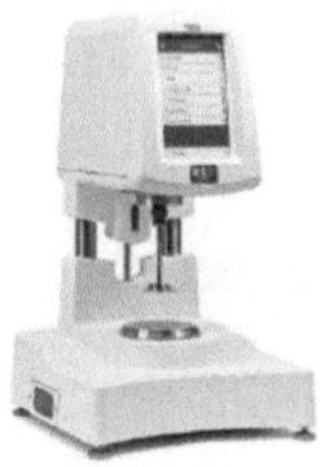

Plate 537: Rheometer

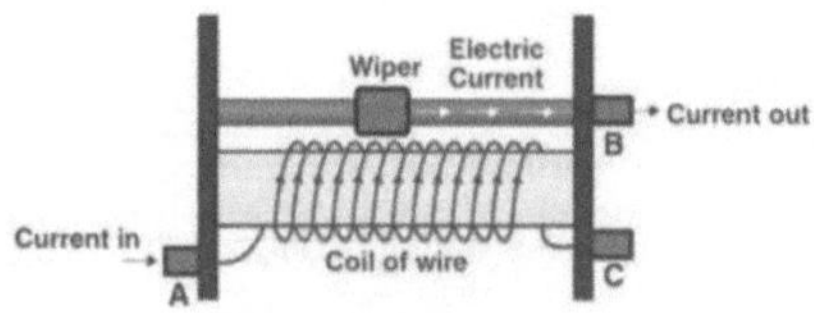

Plate 538: Rheostat

Riboflavin: A yellow crystalline compound $C_{17}H_{20}N_4O_6$ of the vitamin B complex that occurs both free (in milk) and combined (in liver) as a component of coenzymes (FMN) which are essential to normal metabolism. It is also called vitamin B_2. Riboflavin (yellow green pigment) is quite resistant to heat but very sensitive to light.

Rice: It is an herbaceous plant and is a rich source of starch but poor in proteins, fats and minerals.

Rice glazing: The polishing of the rice grain with talc or glucose is called rice glazing.

Rickets: It is a deficiency disease caused by the absence of vitamin D or calcium or both which affects the skeletal system.

Ricotta cheese: A white unripened whey cheese of Italy that resembles cottage cheese or a cheese made in the U.S. from whole or skim milk.

Riddling: Riddling is process used for making sparkling wine. In this process; bottle neck turns upward and down ward daily (clockwise and anticlockwise) frequently and inclined in such a way that the yeast sediments slide towards the bottle neck and settle on the stopper. This process is called remuage or riddling (Plate 539). The riddling helps to make the wine clear. It is carried out at 10-18°C for 4 months. After riddling disgorgement process is done. Volume of wine lost during disgorgement is then replaced by addition of dosage syrup, which is acidified with citric acid and lightly sulfited.

Plate 539: Process of remuage or riddling

Ridgelimeter: It is a device used for the estimation of percent sag of the jelly.

Rigor mortis: It is stiffening / rigidity of muscles fibres after death or slaughtering of animals.

Rin Gene: Non ripening mutants in tomatoes are Rin Genes. It does not produce red pigment lycopene and so climacteric fruit behave like non climacteric fruits.

Ripe gas: An explosion proof ethylene mixture containing six percent ethylene in inert gas (CO_2 w/w) which is used to speed up the ripening process.

Ripening: It is the composite of the processes that occur from the latter stage of growth and development through the early stage of senescence and result in characteristics, aesthetic and/ or good quality as evidenced by changes in composition, colour, and texture or other sensory attributes. Or Ripening is process that involve sequential changes in sensory factors of colour, texture and taste which render the fruits acceptable to eat, and marks the completion of development and commencement of senescence. This is a critical period of transition from the stages of cell disorganization and death. Or Ripening is fermentation of products like cream and cheese with the help of desirable starter and culture.

Ripening enzymes: The enzymes which result change of pectic substances with commencement of ripening of fruit and vegetable are called ripening enzymes. Specific ripening enzymes are pectin methyl esterases (PME) (de-esterify pectin), polygalacturonases (PG) (break down the pectin chains) and cellulases (break down cellulose).

Risk analysis: A process consisting of three components: risk assessment, risk management and risk communication.

Risk assessment: A scientifically based process consisting of hazard identification, hazard characterization, exposure assessment and risk characterization.

Risk profile: Risk profiling is the process of describing a food safety problem and its context in order to identify those elements of the hazard or risk which are relevant to various risk management decisions.

Roast: To cook / dry and parch a food by exposing to dry heat (oven or before a fire) or by surrounding with hot embers, sand, or stones. To dry and parch coffee beans by exposure to heat is called roasting.

Roaster: A bird fit for roasting usually a young chicken larger than a broiler. An individual that roasts or a device for roasting is also called roaster.

Roasting ear: It is an ear of young corn roasted or suitable for roasting or boiling or steaming.

Rochelle salt: A crystalline salt $KNaC_4H_4O_6 \cdot 4H_2O$ that is a mild purgative is derived from the winery by product is called Rochelle salt. It is also called argol.

Roentgen: It is the quantity of gamma or x-rays which produces one electrostatic unit of electric charge of either sign in a cubic cm of air and under the standard conditions.

Roentgen equivalent physical (rep): It is the amount of ionizing energy which produces an amount of ionization equivalent to a roentgen. REP is a measure of absorbed energy that is effective within the food.

Rolling test: The package, standing on its base, is slowly pushed to rotate about one edge of the basic and allowed to fall over on the side face. After this impact, the package is lifted, balanced on its edge and then allowed to fall over on the top. The procedures repeated until the package once again stands on its base. This test helps to evaluate the overall strength of container and to design a suitable container.

Romano: A hard sharp cheese of Italian origin that is often served grated.

Room temperature: The normal temperature inside the house is called room temperature. It is generally considered up to 20^0C.

Rope: It is a bacterial disease of the bread.

Roquefort cheese: It is a pungent French blue semi-soft cheese made from sheep's milk.

Rosing machine: It is a machine used for extraction of juice from the citrus fruits. It consists of a stand shaft, 2 stainless steel roses and 2 numbers of juice extractor (Plate 540).

Roster bid system: Bidding starts from a point in the market at a notified time about which the prospective buyers are given information in advance. The bidding party after the auction of the produce at one shop moves to the next in a clock-wise or anti-clockwise direction till the auction of the produce at all shops is over, or the scheduled auction time expires. On the following day, the auction starts from the next point, and so on. The auction is supervised by the auction clerk or the person nominated by the market committee.

Roughage: It is the mostly indigestible material in food that stimulates the intestine to peristalsis. It is also called Fiber.

Roughness: Degree of abrasiveness of product's surface perceived by the tongue.

Rum: It is an alcoholic beverage distilled from a fermented sugar cane product (molasses) (Plate 541). It is popular in Jamaica.

Plate 540. Rosing machine

Plate 541: Rum bottles

Rotary machine: It is a [illegible] for extraction of juice from the [illegible]. It consists of [illegible] rollers [illegible] (Plate [illegible]).

Rotary bid system: [illegible] a point or the [illegible] about which the [illegible]. The bidding party after the auction [illegible] the produce [illegible] to the next [illegible] in clock-wise [illegible] all the [illegible] done. [illegible] the [illegible].

Roughage: It is the mostly indigestible material in feed that stimulates the intestine to peristalsis. It is also called fibre.

[illegible]

[illegible]

[illegible] rolling machine

Plate: [illegible]

S

Saccharification: The process of breaking a complex carbohydrate (starch or cellulose) into its monosaccharide components is called saccharification.

Saccharimeter: A device (polarimeter) used for measuring the amount of sugar in a solution is called saccharimeter. It is also known as saccharometer. It is a hydrometer with a special scale.

Saccharin: A crystalline compound $C_7H_5NO_3S$ that is unrelated to the carbohydrates, is several hundred times sweeter than sucrose, and is used as a calorie free sweetener.

Saccharomyces: It is one of the genus (Saccharomyces) of the family Saccharomycetaceae and is usually unicellular yeasts (brewer's yeast) that are distinguished by their sparse or absent mycelium and reproduced asexually by budding.

Safe food: Food that provide maximum nutrition and quality while revealing a minimal hazard to public health and they assume any risk present to be minimal.

Safety: Safety is a component of quality. Many experts believe that food safety is the most important quality component since a lack of safety can result in serious injury and even death of the consumer.

Safety range of ERH: In ERH, it is point or a portion of curve between danger point (D) and critical point (C).

Safety test: This test is used to check the safety of the food prior to its consumption. Food for safety can be checked by appearance (like can swell, breathing of can etc.), taste and smell (flat sour of canned product) etc. This type of test can be applied at industry level during production or storage of the product or at consumer level prior to consumption of the food. This test is applied for acceptance or rejection of the food product.

Safrole: Poisonous oily cyclic ether $C_{10}H_{10}O_2$ that is the principal component of sassafras oil and is used chiefly in perfumery. It is prohibited food additive.

Sage cheese: It is a cheese similar to mild cheddar flecked with green and flavored with sage.

Sago palm: It is a plant that yields sago. It is one of the lofty pinnate-leaved Malaysian palms (genus *Metroxylon*).

Sake: It is a Japanese alcoholic beverage of fermented rice often served hot. It is a type of rice beer. It contain 18 to 24% alcohol content.

Salad: The raw greens (lettuce) often combined with other vegetables and toppings and served with dressing or small pieces of food (pasta, meat, fruit, or vegetables) usually mixed with a dressing (mayonnaise) or a green vegetable / herb grown for salad is called salad.

Salad oil: An edible vegetable oil (olive oil) suitable for use in salad dressings is called salad dressing. These oils remains liquids when kept in a refrigerator at 5°C (41°F) and are used for salad dressing, for making mayonnaise etc.

Salami: A highly seasoned sausage of pork and beef either dried or fresh is called salami.

Salmonella spp: These are common contaminants of raw meat, poultry, eggs and dairy products. Dried eggs and dairy products are most sensitive.

Salometer: An instrument used for measuring the salt concentration of any substance (usually liquid) in terms as a degreed salometer (5% salt = 20 salometer reading). It is a special hydrometer. It is also called salimeter or salinometer. Used in pickling and meat packing industry.

Salsa: It is a spicy sauce / chutney of tomatoes, onions, and hot peppers. It is popular in Spanish.

Salts: A crystalline compound NaCl that consists of sodium chloride, is abundant in nature, and is used to season or preserve food. It is also called common salt. The other salts may be a mineral or saline mixture (as Epsom salts), smelling salts etc.

Salt spray test: The purpose of the salt spray test is to evaluate the resistance of the package to corrosion by salt spray and the package is exposed for 50 hours, to a wet dense fog generated by automation of 20% water solution of NaCl. The solution shall be maintained at a pH of 5-7.2 and temperature of fog at 95°F.

Salt stock: Preserved material (mostly vegetables and green fruits) in 15-20% slat solution which is not considered as a consumer commodity and when desired, salt stock is freshened and prepared into consumer types by leaching the salt from the cured materials with warm water. Alum may be added to firm the tissue and turmeric powder may be added to improve the colour.

Sample unit: Sample unit is a container and/or its entire contents or a portion of the content of a container or other unit of commodity or a composite mixture of a product to be used for inspection.

Sample(s): Sample(s) is/are any number of sample units to be used for inspection. Samples are representatives of a population.

Sampling: Sampling is the act of selecting samples of the processed products for the purpose of inspection under the regulation.

Sanitation: Sanitation is the application of chemical agents followed by washing which involves the reduction of microorganisms of public health concern, as well as other undesirable microorganisms, without adversely affecting the quality of the product or its safety to the consumer. Sanitation is broadly defined by "all precautions and measures, which are necessary in the production, processing, storage and distribution, in order to assure an unobjectionable, sound and palatable product which is fit for human consumption". Sanitation is a multistep process that involves cleaning and sanitizing as two very important and separate steps. Cleaning and sanitizing can apply to various aspects of a food plant such as equipment, packages, walls, and floors. Cleaning removes visible soil and organic matter and more than 90% of harmful microorganisms. A sanitizer is applied after a surface has been properly cleaned. The sanitizer further reduces the level of microorganisms to a safe and acceptable level. Common sanitizers in the food industry include chlorine, iodophors, and quaternary ammonium compounds.

Sanspan: It is a foaming agent obtained from *Asparagus adscendeus*.

Saponification: It is alkaline hydrolysis of oil and fats. It is a process to hydrolyze fats with alkali to form a soap and glycerol.

Saponins: It is one of toxic glucosides that occur in plants (soapwort or soapbark) and are characterized by the property of producing a soapy lather. It is a hygroscopic amorphous saponin mixture used as a foaming and emulsifying agent and detergent. Present in soyabean cause nausea and vomiting.

Sapsago: It is a hard green cheese made from skim milk and sweet clover of *Melilotus* species.

Saturated air: Air in which water vapour is in equilibrium with liquid water at a given temperature and pressure. When air holds as much water vapor as it can for a given temperature (100% relative humidity), it is said to be saturated. If saturated air is warmed, it can hold more water (relative humidity drops), which is why warm air is used to dry objects as it absorbs moisture (Plate 542).

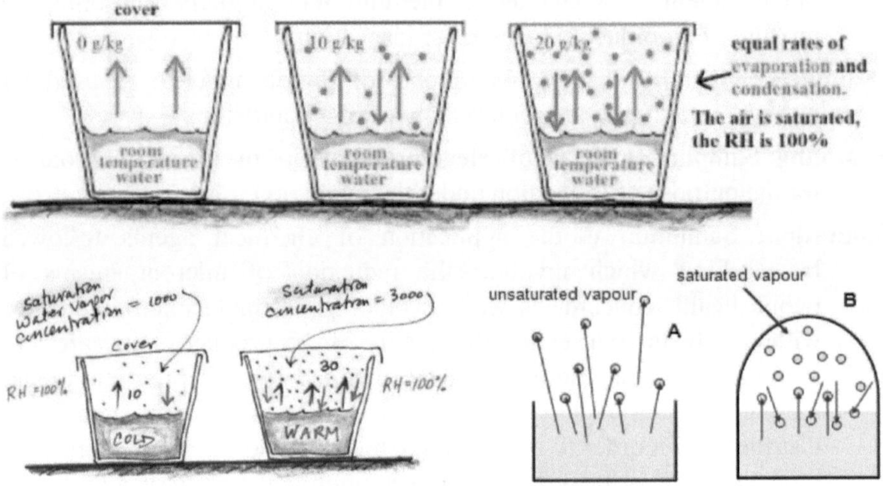

Plate 542: Saturated air illustration

Saturated Fat free food: Food having less than 0.5 gm of the saturated fat per serving and not more than 0.5 gm of trans fatty acids per serving or per 100 gm.

Saturated Fats: Saturated fats are solid at RT. Saturated fats are obtained from most land animals (Cattle, hogs, and poultry). These have -CH_2-CH_2 bonds in their fatty acid structure.

Saturated fatty acids: It is fatty acids in which the carbon bonds are fully saturated.

Saturated humidity: It is absolute humidity of air when it is saturated with water vapour at given temperature. The saturated humidity increases with temperature.

Saturated solution: The solution containing as much solute in a solution as is possible to dissolve at a given temperature is called saturated solution.

Sauce: Concentrate pulp of fruit and vegetable without seed and skin and sugar salt and other spices are added to the extent that flavour of one may not dominate the other and the finished product should have at least 15% total soluble solid, preserved with chemical preservatives. It should have 15% TSS, 1.2% acidity and 750 ppm sodium benzoate.

Sauerkraut: Clean, sound product of characteristic flavor obtained by full fermentation chiefly lection of properly prepared and shredded cabbage in the presence of 2-3% salt and upon completion of the fermentation it contains not less the 1.5% of acid, expressed as lactic acid. It is highly popular in some parts of Europe and USA (Plate 543).

Sausage: A highly seasoned minced meat (pork) usually stuffed in casings of prepared animal intestine (Plate 544). Or It is a seasoned minced meat etc. in a cylindrical edible skin.

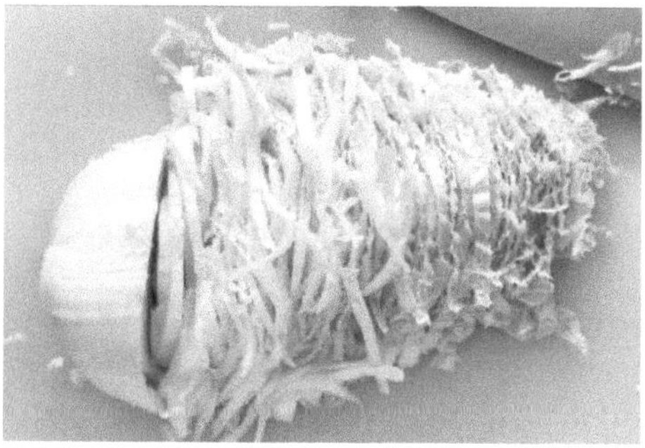

Plate 543: Cabbage for Sauerkraut preparation

Plate 544: Sausage

Sausage meat: It is chopped or minced meat mixed with fat, cereals etc seasoned and stuffed into a tube of gut. Or Sausage(s) is/ are comminuted ground seasoned meats, stuffed/unstuffed into casing; may be smoked, cured, fermented and heated. Hamburger (fresh or uncooked), frankfurter, wiener (cooked and smoked), meat loaf (cooked meat specialities), salami (fermented), pepperoni, chorizo (semi-dry) are the example of different type of sausages.

Sausage roll: It is sausage meat in a pastry roll.

Saute: The frying of potatoes / food quickly in a little fat is called saute. This process is called sauteing. The cooking in this process is done by conduction.

Sauternes: Sweet white wine from Sauternes in the Bordeaux region of France.

Savory / savoury: A dish of stimulating flavor served usually at the end of dinner but sometimes as an appetizer.

Scalding: It is similar to blanching. It is hot enough to scald in water. It is producing the feeling of being burned.

Scavenger: A chemically active substance acting to make innocuous (harmless) or remove an undesirable substance is called a scavenger.

Schrodinger equation: It is an equation that describes the wave nature of elementary particles and is fundamental to the description of the properties of all matter. The Schrödinger equation is a *linear partial differential equation that governs the wave function of a* quantum-mechanical system (Plate 545).

$$\frac{-\hbar^2}{2m}\nabla^2\Psi(\mathrm{r}) + V(r)\Psi(\mathrm{r}) = E\Psi(\mathrm{r})$$

$$\begin{matrix}\textit{Kinetic} \\ \textit{Energy}\end{matrix} + \begin{matrix}\textit{Potential} \\ \textit{Energy}\end{matrix} = \begin{matrix}\textit{Total} \\ \textit{Energy}\end{matrix}$$

Plate 545: Schrodinger equation

Scotch whiskey: It is prepared from barley malt which is dried over a peat and coke fire giving the grain its smoky flavour which passes into mash and upon fermentation and distillation into the whiskey. It may be distilled in pot still or patent still.

Scott method: It is a method used for measurement of oil in citrus. It is easier, fast and more accurate method. It uses bromide bromated salt solution to titrate the acidified alcoholic distillate from the juice sample and express on percent basis.

SCP: It stands for Single cell proteins. SCP is the dried microbial biomass grown for its protein content which in term may be of bacterial, yeast, algal or fungal origin. Since, majority of the microorganisms used for SCP production are unicellular, the protein content from them is called SCP. In addition to their high protein content the microbial cells also contain some other important nutrients e.g. vitamins, fats, carbohydrates, minerals etc.

Scurvy: A disease caused by a lack of vitamin C and characterized by spongy gums, loosening of the teeth, and a bleeding into the skin and mucous membranes.

Sea food: The edible from sea which are used as human food e.g. lobster and prawn etc.

Seaboard markets: These are located near seashore and are mainly meant for import and export of goods.

Sealing: It is process by which the filled cans are sealed air tight after exhausting by using can sealer / double seaming machine.

Search foods: These are the foods where the consumers can determine the quality of food before they buy.

Second law of thermodynamics: According to Kelvin-Plant, it is impossible to construct an engine in a cyclic process, whose sole purpose is to convert heat energy from a single thermal reservoir into an equivalent amount of work. It means that there is no degradation of energy in the process of producing mechanical work from heat applied. The second law of thermodynamics is a physical law based on universal empirical observation concerning heat and energy interconversions. A simple statement of the law is that heat always flows spontaneously from hotter to colder regions of matter (or 'downhill' in terms of the temperature gradient). Another statement is: "Not all heat can be converted into work in a cyclic process (Plate 546).

Second order transition: It is the physical state of the material which changes without release or absorption of the latent heat. E.g. Amorphous state of glass state (Plate 547)

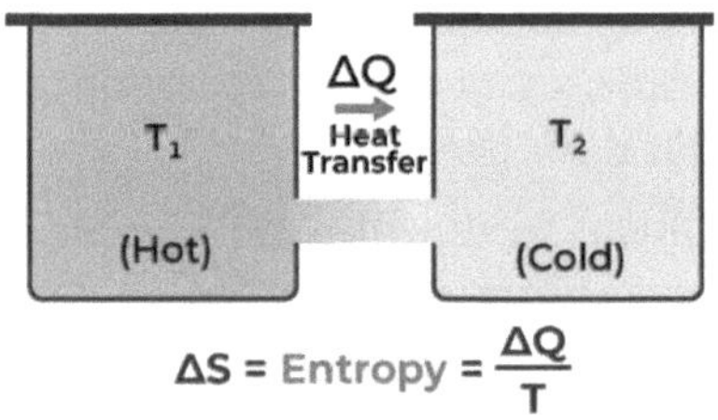

Plate 546: Second law of thermodynamics illustration

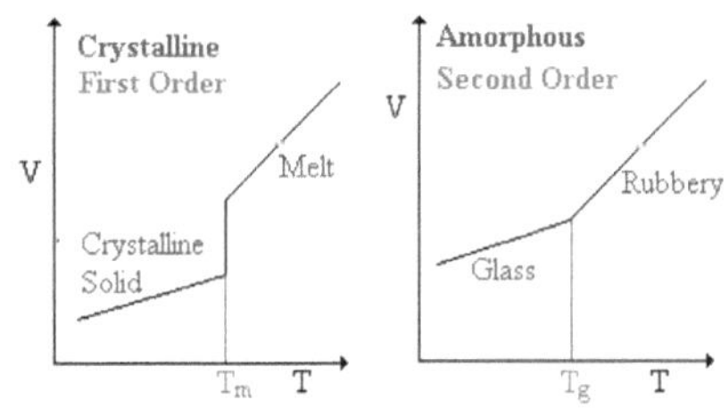

Plate 547: Second order transition illustration for glass (crystalline vs amorphous polymer)

Secondary colour: Violet, green and orange colours are known as secondary colour. These are made by combination of two primary colours. Combination of blue and yellow, yellow and red, and blue and red results green, orange and violet colour respectively.

Secondary container: A secondary container is an outer box, case or wrapper that holds or utilizes several cans, jars or pouches together but does not contact the food directly (Plate 548). Secondary containers are a necessary part of food packaging. It would not be possible to distribute products in glass jars, for example without corrugated secondary carton to protect against breakage. Secondary container has fewer but no less important functions. Secondary containers must also protect the primary containers form damage during shipment and storage. They must also prevent dirt & contamination from soiling the primary containers. Corrugated fibre board is most commonly used to make secondary shipping cartons. There are strict standards on the construction and use of the secondary containers to ship products. The size and strength of carton used must be selected for each type of product to be shipped. Cartons come in different designs, each of which is intended for different types or forms of products.

Plate 548: Secondary containers used for holding tetrapacks

Secondary organism: This is the organism which is recovered as a result of by-product of fermentation.

Secondary processing: It involves elementary modification of the natural foods by packing, waxing, pulping, juice extraction, preparation and preservation of the semi-finished products for later use.

Secondary settling tank: It is a part of effluent treatment plant. Secondary settling tank is used for removal of excess biomass from aeration basin. The design of the unit is similar to primary clarifier. The over flow from this tank is out drained and the excess sludge is then dried in sludge drying beds built adjacently 9 Plate 549).

Plate 549: Secondary settling tank

Secondary wholesale market: These are generally located at districts headquarters or important trade centres near railway stations, where produce is handled in large quantity. These are also known as mandis and are situated generally at district or taluka headquaters. Small merchants purchase from primary wholesale market and sale in this markets. Some cultivators directly sell their produce in these markets. Each market comprises area with a 10-20 miles' radius e.g. District and taluka market. Here the transaction is between wholesalers or between wholesaler and retailers.

Secondary yeasts: Secondary yeasts are recovered as a by-product of an alcoholic fermentation is strains of brewer's yeasts, *Saccharomyces cerevisiae* or *S. carlsbergensis* or of distiller's yeast, *S. cerevisiae var ellipsoids*.

Secular markets: These markets are of permanent nature. The commodity in these markets can be stored for many years e.g. machinery items, processed products.

Seed cake: A cake or cookie containing aromatic seeds (sesame or caraway) is called seed cake.

Seedy-cut: It is a blackish blemish sometimes found in pig meat.

Self cooling cans: These cans are self cooling cans and can be cooled by the consumer only after buying and just before consumption. These cans consist of three compartments. One compartment contains the food which is to be cooled. The two compartments contain different chemicals. One compartment contains gel and another one contain desiccant in vacuum. These two compartments are separated from each other by a thin membrane. When ever the consumer is interested to cool the can, they press the button provided. The pressing of the button pierces of the membrane between two compartments and results in evaporation of water from the gel. It results in endothermic reaction i.e. absorption of energy / heat. This absorption of heat results in cooling of the food or beverages. These are also called IC can (Instant Cooling Can). These cans are 100 % safe and environment friendly, self refrigerating process that cools using single water evaporation. No pressure and no gas is required for cooling and even is toxin free. In cans the product is cooled within 3 minutes. However, the cooling is uneven as it is by conduction process (Plate 550).

Self heating cans: These cans are self heating cans and can be heated by the consumer only after buying and just before consumption. These cans consist of three compartments. One compartment contains the food which is to be heated. The two compartments contain different chemicals. One compartment contains CaO (quicklime) and another one contains water. These two compartments are separated from each other by a thin membrane. When ever the consumer is interested to heat the can, they press the button provided. The pressing of the button pierces of the membrane between two compartments and results in mixing of quicklime and water. It results in exothermic reaction i.e. energy / heat is released. This released heat result in heating of the food or beverages. $CaO + H_2O \rightarrow Ca(OH)_2$, $CuSO_4 + Zn \rightarrow ZnSO_4 + Cu$. These are also called IH can (Instant Heating Can). These cans are 100 % safe and environment friendly and no need of the microwave heating. In cans the product is heated within 3 minutes. However the heating is uneven as it is by conduction process (Plate 551).

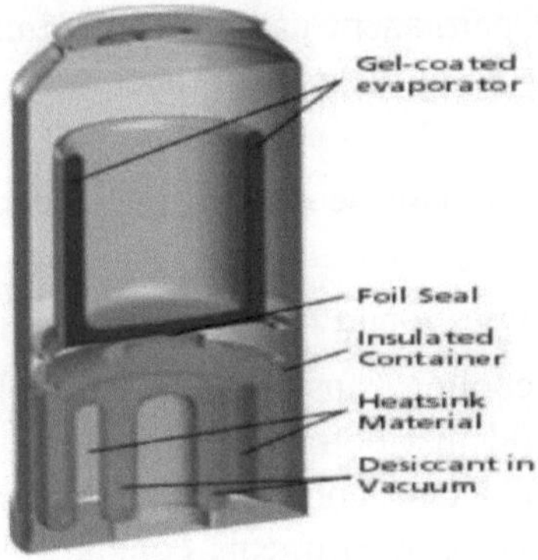

Plate 550: Shelf cooling can mechanism

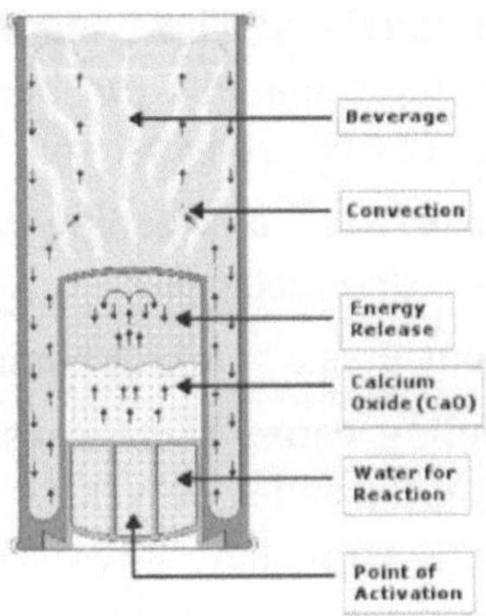

Plate 551: Shelf heating container

Selling climax: A sharp decline in stock prices for a short time on very heavy trading volume followed by a rally.

Selling point: An aspect or detail of something that is emphasized in selling or promotion.

SEM: It stands for scanning electron microscope / scanning electron microscopy. The SEM is used to generate high resolution two dimensional image of the shape of objects that display the special variation of properties. The important components of the SEM are the electro source (gun), electron lenses, sample stage, detectors and data out put devices (Plate 552).

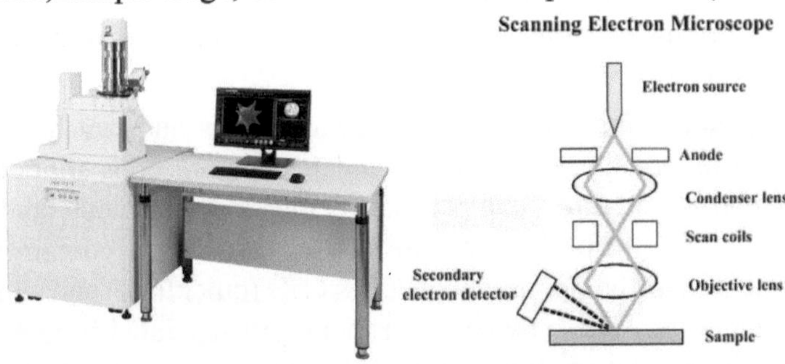

Plate 552: Scanning electron microscope

Semi automatic bottling machine: This is a machine which is used for filling the bottles with juice / RTS / beverages semi-automatically. Machines are available with 2-6 heads and so can fill 2-6 bottles at a time. This type of machine can fill 12 – 40 bottles per minute (Plate 553).

Plate 553: Semi automatic bottling machine

Semi fluid: A substance having the qualities of both fluid and solid e.g. lubricants. Fluid is viscous in nature.

Semi liquid: A substance having the qualities of both liquid and solid **e.g.** manure.

Semi-dry: A moderately dry wine is called semi dry.

Semi-drying: It is a process that dries imperfectly or slowly. It is used in drying of some oils as cottonseed oil.

Semi logarithmic: It is having one scale logarithmic and the other arithmetic. It is used to graph paper.

Semi permeable membrane: A membrane which is partially but not freely or wholly permeable is known as semi permeable. The semi permeable membrane is permeable usually to small molecules but not to larger particles.

Semolina: The fine particles of hard wheat which do not pass into flour during milling. It is also called dalia.

Senescence: Senescence is an irreversible process of gradual degeneration and disintegration of plant tissue eventually leading to the death of the organism. Or The process that follows physiological maturity or horticultural maturity and lead to the death of the tissue.

Sensible heat: The thermal energy whose transfer to or from a substance results in a change of temperature is called sensible heat.

Sensible heating: Heating of air without any change in its specific humidity is known as sensible heating.

Sensitivity / Threshold Tests: Sensitivity / threshold tests are used to evaluate the ability of the individual to smell, taste or feel specific attributes in foods. These tests are mostly used in the selection of individuals / panelists for sensory evaluation of food. The intensity of colour, flavor, taste like sourness, sweetness, bitterness can be measured by adopting this test.

Sensory: It is relating to sensation or to the senses by stimulation.

Sensory analysis: Examination of organoleptic attributes of a product by the sense organs.

Sensory evaluation: It is a process of conveying nerve impulses from the sense organs to the nerve centers. It is the process of checking the quality of the food on the basis of the organoleptic properties of the food.

Separated milk: The milk from which the cream has been removed by centrifuging is called separated milk. It has about half the energy value of whole milk but a relatively higher content of protein and minerals. It had a dry matter content of about 9%. The removal of fat means that milk is deficient in fat soluble vitamins like vitamin A, D, and F (essential fatty acids). It is skimmed milk.

Separator: It is a device for separating liquids of different specific gravities (as cream from milk) or liquids from solids.

Sepsis: A toxic condition resulting from the spread of bacteria or their toxic products from a point of infection.

Sequestrants: These are the chelating agents that combine with metals such as iron and copper and remove them from solution. These are not antioxidants, yet they play a most valuable role in stabilizing foods. Most commonly used chelating agents are citric acid & its salts, phosphates and salt of ethylenediamine tetra acetic acid (EDTA). These agents help to remove discolouration, rancidity, turbidity and flavour changes in foods. Calcium chloride, calcium phosphate, tartaric acid and citric acid are some examples of sequestrants.

Serine: A crystalline nonessential amino acid $C_3H_7NO_3$ that occurs as a residue in many proteins.

Serum acidity: The acid content of fat free portion of the cream or milk is called serum acidity.

Serum concentration process: It is an alternate method for concentration of pulpy fruit juices. In this method the fruit juice is centrifuged to separate the solid phase (pulp) and the liquid phase (serum) is then concentrated in an evaporator before mixing with the pulp.

Setting temperature: It is the temperature at which the setting of the jelly or jam takes place.

Setting time: It is the time taken for a jelly from the time of filling and occurrence of set at standard temperature. Setting is an important criterion in the manufacture of jam and jelly.

Sewage: It is a refuse of a liquid or a solid nature of domestic and industrial waste.

Shake and catch method: Shake and catch method of harvesting depends on the principle of inertia of tree shaking. This method is used for harvesting of mangoes.

Shape factor: Shape factor is a geometrical factor which is determined by the shapes and relative location of the two surfaces.

Sharp freezing: Freezing of the food at slow freezing rate is known as sharp freezing. Such freezing is done in still air at a temperature between -15 to -29°C and forms large ice crystals. It refers to the process where by the desired temperature -20°C (-15 to -29°C) is achieved with in 3 to 72 hours. This is essentially the type of freezing method utilized in the home freezer. It is also known as slow freezing.

Shear press: It is an objective testing machine that measures compressibility in extrusion and shear of food sample.

Sheet test: It is test used to determine the end point of jam and jellies. In this test, a small portion of the jam/jelly is taken with a large spoon / wooden ladle, cooled slightly and then allow to drop off keeping the spoon or ladle in horizontally inclined position. If the jam/jelly drop like syrup, then further concentration is needed. If it is in the form of flake or sheet, then the end point has reached (Plate 554).

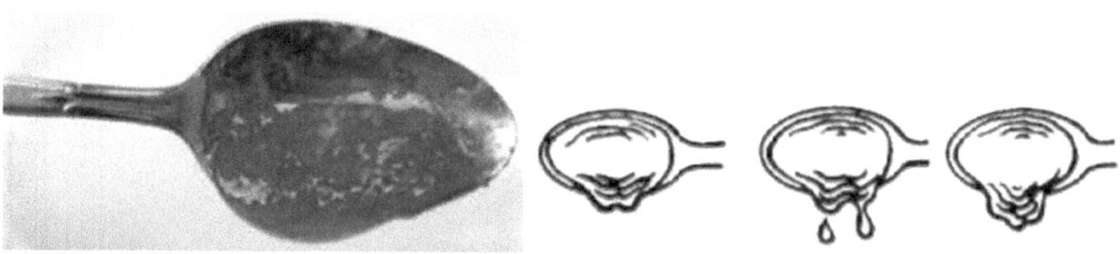

Plate 554: Test used to check end point of jam and jellies

Shelf life: It refers to the span up to which the perishable commodities can be kept at consumable and marketable stage after harvest. The time after processing during which a product remains suitable for human consumption, especially it is the time for which a food remains palatable and acceptable to consumers.

Shell fish: An aquatic invertebrate animal with a shell especially with an edible mollusk or crustacean.

Shell fishery: A commercially exploited population of shellfish is called shell fishery.

Sherbet: Clean sugar syrup which has been artificially favoured. It contains not less than 65% TSS and 10% juice. The contain SO_2 @ 350 ppm or benzoic acid @ 600 ppm.

Sherry: A Spanish fortified wine with a distinctive nutty flavor is called sherry. Sherry wine produced by special technique containing 18-21% alcohol and could be dry or sweet. The technique used is secondary fermentation by special type of 'Flor yeast' or by baking yeast.

Short chain fatty acids: These fatty acids having 10 or fewer carbon atoms in one molecule.

Short pack: It is normally used for only one serving e.g. jam and fruity.

Short period market: In these markets commodities are perishable and can be traded for some time. These commodities are like food grains and oilseeds.

Short ton: One short ton is equal to 2,000 lbs.

Shortening: An edible fat used to shorten baked goods is called shortening. It is commercially prepared edible fat used in frying, baking, cooking and ingredients in filling, icing and other confectionary items.

Shortometer: This is a device used to measure the tenderness of idly, pastry and cookies (Plate 555).

Plate 555: Shortometer

Shoulder: It is that portion of the glass container in which the maximum cross section or body area decrease to join the neck or finish area.

Shrikhand: It is a very popular product from milk based curds.

Shrink wrap film: Shrink wrap films are the films which are over drawn on the commodity by use of direct or indirect heat and cause the film to shrink. It is used for packaging of warm bread to increase shelf life. It comprises of 10 μm thick 3 layer having oxygen barrier property. A special machine is required for their application. Shrink films become more economic and are environmental friendly.

Shrink wrapping machinery: The shrink wrapping essentially consists of wrapping a film around a package, sealing it, and subsequently heating the film to a desired shrinkage as to make it tight fit. The heating of the film is accomplished in two ways i.e. immersion of the package in hot water and by blowing hot air over the package. The ability of a film to draw down over the content of a package when it is heated is very useful attribute in shrink wrapping (Plate 556).

Shrinkage ratio: The ratio of weight of raw material to the weight of the prepackaged finished product.

Sigmoidal Growth: This type of the growth is depicted by a curve which is curved in two directions like the letter S (Plate 557).

Plate 556: Shrink wrapping

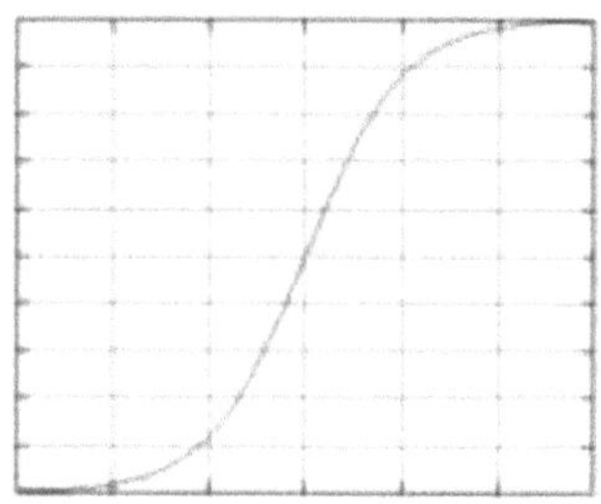

Plate 557: Sigmoidal growth

Silica coated films: Silica coated plastic also referred to as glass coated and have been found to have several benefits such as clear barrier materials, microwaveable, retortable and recyclable. Silica coating has been applied to PET containers polyesters, LDPE and PP films.

Silicates: These are salt or ester derived from a silicic acid especially numerous insoluble often complex metal salts that contain silicon and oxygen in the anion, constitute the largest class of minerals, and are used in building materials (cement, bricks, and glass).

Silo: It is a pit or airtight chamber (tall and cylindrical) designed for storing grains or for ensilage (Plate 558).

Plate 558: Siloes used for stotage of grains

Silver skin: The fine skin of the coffee bean is called silver skin.

Simmering: The gentle boiling of the products at a temperature of about 100°C is called simmering.

Simple protein: It is a protein (globulin) that yields amino acids as products of complete hydrolysis.

Simple sugars: These sugars are not decomposable into simpler sugars by hydrolysis. Simple sugars are either an aldose or ketose, and contain one or more hydroxyl groups per molecule. These sugars are also called monosaccharides.

Simulation: The act or process of simulating something (making fraud or adulterated product) or to counterfeit i.e. to make imitation of something else with intention to deceive consumer is called simulation.

Single cell protein (SCP): The term single cell protein (SCP) first time was coined in 1967 at Massuchusetts institute of technology (MIT), USA. SCP is the dried microbial biomass grown for its protein content which in term may be of bacterial, yeast, algal or fungal origin. Since, majority of the microorganisms used for SCP production are unicellular, the protein content form them is called SCP. In addition to their high protein content the microbial cells also contain some other important nutrients like vitamins, fats, carbohydrates, minerals etc.

Single effect evaporator: The evaporator which operated at single stage and has only one heat exchanger is called single effect evaporator (Plate 559).

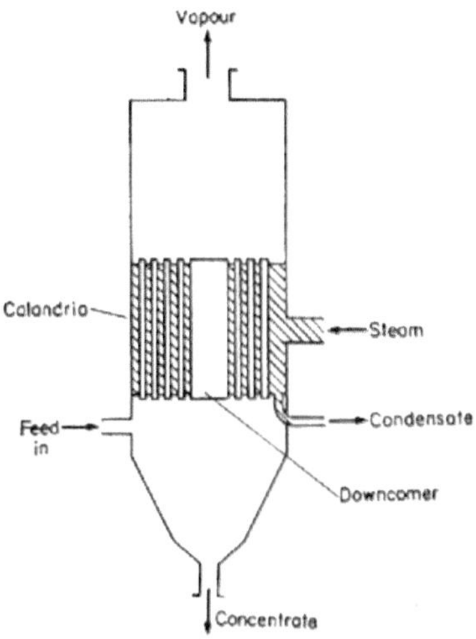

Plate 559: Single Effect Evaporator

Single sample: It is a procedure or act of presenting one sample early in an experimental project to determine acceptability and to aid in the decision on future development of the product.

Single sample test: This test is also known as monadic test. It is used for testing such foods which carryover / possesses an after-taste or flavour. Such foods preclude / avoid the testing of second sample in the same session.

Sinigrin: It is an isothiocyanate found in cabbage, can be broken down by myrosinase, an enzyme to yield a highly pungent compound Allyl isothiocyanate. or Sinigrin is a glucoside present in all the vegetable under genus Brassica which on hydrolysis under the influence of enzyme myrosinase produces allyl-150-thiocyanate which gives a typical pungent flavor.

Sinki: Sinki is a sour pickle prepared from radish taproots. It is consumed traditionally in India, Nepal and parts of Bhutan. It is used as a soup or eaten as a pickle.

Sinking pulp: This pulp is made up of smaller or finer pulp particles that form sediment on standing because of the juice saturation or higher density than the juice itself. It is also called spin down pulp. Such pulp results opaque appearance of the juice and it become turbid but impart desirable mouth feel. It can be overcome by using opaque container / labels.

Size enlargement: It is the process of production of large particles from smaller ones.

Size reduction: It is the process in which average size of the solid pieces of the food is reduced by application of grinding, compression or impact forces (Plate 560).

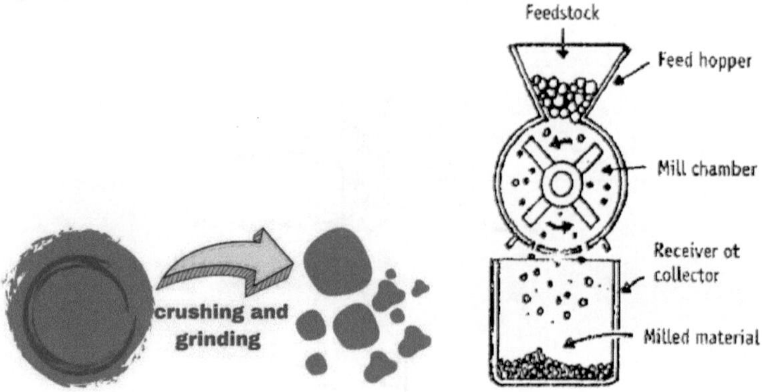

Plate 560: Size reduction process

Skewer: It is a thin, long metal or wood stick / pin used to hold or suspend fod during cooking.

Skiff: It is a specially designed boat used for setting nets and picking fish.

Skim: It is the process to remove the floating matter from the surface e.g. removal of the cream from milk.

Skim milk: Milk from which the cream has been taken is called skim / skimmed milk. It contains all other constituents of the milk except fat and fat soluble vitamin A and D.

Skimmer: A person that removes the skims or a flat perforated scoop or spoon used for skimming is called skimmer.

Skimming: It is a process to clear (liquid) scum or floating substance from the boiling syrup or to remove film or scum from the surface of a liquid or to remove cream from boiling milk.

Slaughter: The killing or the butchering of livestock for market or to kill (animals) for food is termed as slaughter.

Slaughter house: An establishment where animals are hygienically butchered and the carcases are prepared for sale to the public for human consumption is called Slaughter house. All carcases are examined by the qualified inspectors employed by the Local Authority and only those free from the disease are allowed for sale. It is also known as abattoir (Plate 561).

Slaughtering: The act of killing or butchering of livestock for market or to kill (animals) for food is known as slaughtering.

Sleepy cream: The semi-fluid state of the cream (even after churning) is called sleepy cream.

Sling psychrometer: It is used to measure the dry and wet bulb temperature. It contains two separate thermometers; the bulb of the one thermometer is covered with wet cloth (Plate 562).

Plate 561: Slaughter house

Plate 562: Sling psychrometer

Slip point: Slip point of the fat is the temperature at which the amount of the solid phase in the melting fat has become so low that an air bubble is forced upward in an open capillary tube filled with the fat.

Slipperiness: Degree to which the product slides over the tongue.

Slivovitz: A dry usually colorless plum brandy made in the Balkan countries is called Slivovitz.

Sloe gin: It is a gin made be extracting the flavour from sloe berry.

Slot machine: A machine whose operation is begun by dropping a coin into a slot.

Slow freezing: It refers to the process where by the desired temperature -20^{0}C (-15 to -29°C) is achieved with in 3 to 72 hours. This is essentially the type of freezing method utilized in the home freezer. It is also known as sharp freezing.

Slow freezing: Slow freezing rate is also known as sharp freezing.

Slow set pectin: Pectin having jelly grade of 150 is known as slow set pectin. Slow set pectins are more tolerant to excess acid level.

Slush: It is partly melted or watery snow or loose ice crystals formed during the early stages of freezing of salt water / refuse grease and fat from cooking on shipboard / paper pulp in water suspension.

Slush cooling: Cooling / freezing of the food commodities by using melted snow or loose ice crystals.

Slush drying: It is almost similar to freeze concentration.

Smelling salts: It is usually a scented aromatic preparation of ammonium carbonate and ammonia water used as a stimulant and restorative.

Smiley scale: It is scale based on very happy to very unhappy faces and used in evaluation of the food products when the respondents are unable to use to language easily.

Smoke point: It is the temperature at which the breakdown products of the fats become visible as smoke.

Smoking: Smoking is aprocess of flavouring, cooking or preserving food by exposing it to the smoke from burning or smoldering plant materials, most often wood (Plate 563). It is exposure of cured meat and fish products to smoke for the purpose of preserving them to increase their palatability by adding flavour and imparting a ric brown colour. The drying action of the smoke tends to preserve the meats, though many chemical present in the wood smoke e.g. formaldehyde and certain alcohol are natural preservative as well.

Smoothie: A creamy beverage made of fruit blended with juice, milk, or yogurt is called smoothie or smoothy (Plate 564).

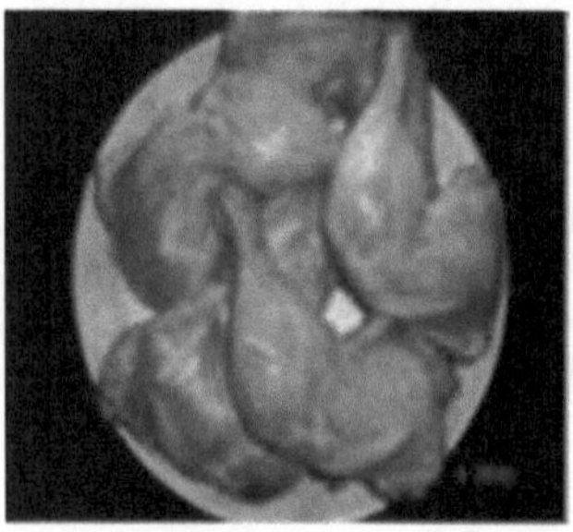

Plate 563: Smoked chicken leg

Plate 564: Smoothie prepared by using fruits and milk

Snack bar: A public eating place where snacks are served at a counter.

Snack food: A light meal or food eaten between regular meals is called snack food.

Snake root poisoning: This type of poisoning results from drinking milk of cow that have eaten snake root plants.

SO_2 generator: It is a paper pad containing KMS and citric acid in two different compartments. It releases / generate sulphur dioxide in the box containing grapes when combines with the moisture and thus protect the grapes from spoilage and increases the shelf life. Therefore it is called SO_2 generator / grape guard. Grape guard suppresses fungal growth, reduce physiological loss, berry decay can be controlled, berry drop reduced, and maintain quality. Grape guard is packed in tissue paper/ butter papers.

Soda pop: A beverage consisting of soda water, flavoring, and sweet syrup is called soda pop.

Sodium benzoate: A crystalline or granular salt $C_7H_5O_2Na$ used chiefly as a food preservative. This preservative is used in such food which contains water soluble colours usually dark colour.

Sodium bicarbonate: A white crystalline weakly alkaline salt $NaHCO_3$ used in baking powders and fire extinguishers and in medicine as an antacid is called sodium bicarbonate. It is also called baking soda or bicarbonate of soda.

Sodium carbonate: It is a sodium salt of carbonic acid used in making soaps and chemicals, in water softening, in cleaning and bleaching, and in photography. It is hygroscopic crystalline anhydrous strongly alkaline salt Na_2CO_3 and the transparent crystalline hydrate $Na_2CO_3 \cdot 10H_2O$ found in nature as natron. It is a starch modifying agent.

Sodium chloride: It is an ionic crystalline chemical compound consisting of equal numbers of sodium and chlorine atoms and is abundant in nature, and is used to season or preserve food. It is also called common salt.

Sodium citrate: A crystalline salt $Na_3C_6H_5O_7$ used chiefly as a buffering agent, as an emulsifier, and in medicine as an expectorant, chelator, alkalinizing agent, and blood anticoagulant.

Sodium fluoride: A poisonous crystalline salt NaF that is used in trace amounts in the fluoridation of drinking water and toothpastes, in metallurgy, as a flux (continuous flowing of fluid from the body i.e. diarrhea or dysentery), and as a pesticide.

Sodium free food: Food having less than 0.5 mg of the sodium per serving or per 100 gm.

Sodium hypochlorite: It is an unstable salt NaOCl produced in aqueous solution and used as a bleaching and disinfecting agent.

Sodium nitrate: It is a deliquescent crystalline salt $NaNO_3$ used as a fertilizer and an oxidizing agent and in curing meat.

Sodium nitrite: It is a salt $NaNO_2$ used in dye manufacture and as a meat preservative.

Sodium tripolyphosphate (STPP): STPP is a preservative for sea food, meats, poultry and animal feeds. It is used to retain moisture. When added to food it is denoted by E451.

Soft drink: It is usually carbonated nonalcoholic beverage especially soda pop.

Soft nose: Soft nose is a disorder of mango due to Ca deficiency. It causes breakdown of the fruit flesh at the fruit apex (Plate 565).

Plate 565: Soft nosein mango

Soft swell: A can with budged and but the gas pressure inside the can is low enough to permit the ends to be dented by pressure of fingers and with the release of pressure they will get back to their convex position.

Soft water: Water that readily produces leather with soap solution is called soft water. E.g. distilled water and demineralized water.

Soft wheat: Soft wheat is used for making cakes and cookies.

Softening enzymes: Pectin esterase (PE), polygalacturonic (PG) enzymes are fruit softening enzymes. These help in extraction and clarification of juices.

Sol: It is a colloidal dispersion of a solid dispersed in a liquid.

Solanine / solanin: A bitter poisonous crystalline alkaloid $C_{45}H_{73}NO_{15}$ found in the parts (tubers and fruits) of several plants (potatoes and tomatoes) of the nightshade family. Solanine alkaloid is present in potato which has got antibiotic activity.

Solgel: It is a state between sol and gel.

Solid waste: It refers to non-liquid, insoluble material including sewage sludge, agricultural refuse, garbage and industrial waste, demolition waste and mining residues that contain complex hazardous materials.

Solid waste: Solid waste is a non liquid insoluble material including sewage sludge, agricultural refuse, garbage and industrial waste that may contain complex hazardous substances.

Solute: Substance dissolves in a liquid to form a true solution.

Solution: It is a liquid in which the solute is dissolve to form a true solution.

Solvent: Liquid in which the solute is dissolved to form a true solution.

Sonic drying: Sound energy (sound) along with heat is used for drying food commodities but causes noise pollution problem.

Sonicator: It is an instrument using sound waves for the preservation of the food. However, during this operation heat is produced in the products which need to be controlled (Plate 566 and 567).

Plate 566: Sonicator

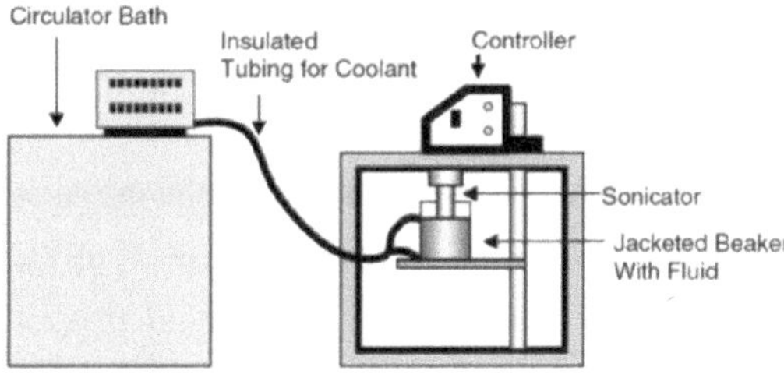

Plate 567: Sonicator working setup

Sorbic acid: It is a crystalline acid $C_6H_8O_2$ obtained from the unripe fruits of the mountain ash or synthesized and used as a fungicide and food preservative. This is effective against mold and yeast and used in cheese spread @ 1000-1500 ppm.

Sorbitol / glucitol / alditol: It is a faintly sweet alcohol $C_6H_{14}O_6$ that occurs in some fruits, is made synthetically, and is used as a humectant and softener and in making ascorbic acid. Sorbitol is a white crystalline alcohol found in certain fruits and berries manufactured by the catalytic hydrogenation of sucrose, use as sweetener and in the manufacture of ascorbic acid.

Sorption: It is the process or state of sorbing a solution.

Sorption-isotherm: It is a plot of free moisture content (expressed on dry weight basis) of a food verses relative vapour pressure (or water activity at low pressure) at constant temperature is known as a moisture sorption isotherm. If the water content of the food is plotted against the water activity at constant temperature, a sigmoid curve usually results. This curve is known as sorption isotherm for that product. Soption moisture anylyser can be used for measurement of sorption isotherm (Plate 568 and 569).

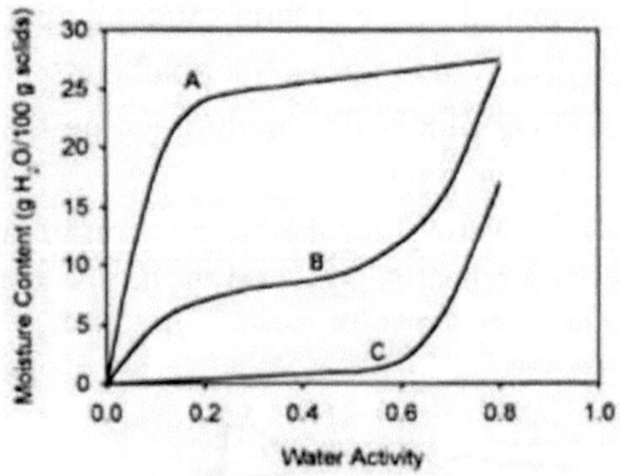

Plate 568: Sorption isotherm curve

Plate 569: Moisture sorption analyser

Sorting: It is the process of separation of foods into different categories on the basis of measureable physical properties. It also helps in removing of unwanted material or diseased portions.

Souchong: A tea made from the larger older leaves of the shoot is called souchong.

Sour dough: Leavened dough in which fermentation is active is called sour dough.

Sousvide Products: The food after preparation is vacuum package, pasteurized and refrigerated. They are reheated prior to consumption and possess better nutritional and sensory qualities.

Sousvide technology: It is a vacuum cooking/ packaging technology. The term sousvide means under vacuum. In this method the foods are cooked in sealed evacuated heat stable pouches such that flavor and nutritional quality retained. This technology has emerged in response to consumer demands for ready to eat micro-waveable convenience foods with extended shelf life that retain fresh like characteristics.

Soxhtec: It is an apparatus used for the estimation of the oil in the food sample. In this apparatus a small amount of the sample is required as compared to soxhlet apparatus. Petroleum ether is used as solvent during estimation of oil content (Plate 570).

Soya / soy: A brown liquid sauce made by subjecting beans (soybeans) to long fermentation and to digestion in brine.

Soya lecithin: It is an emulsifier, antioxidant, and antispattering agent for biscuits, confectioneries etc.

Soya milk: The milk obtained from the soybeans is called soya milk (Plate 571). It resembles dairy milk and is an excellent alternate to lactose intolerant. One kg of the soybean yielded 6-8 litre of the soy milk nutritional compared to cow milk. This milk is available in the market in different flavours like rose, pineapple and chocolate.

Plate 570: Soxhtec for oil estimation

Plate 571: Soy milk

Soybean: It is a hairy annual Asian legume (Glycine max) widely grown for its oil rich proteinaceous seeds and for forage and soil improvement.

Soybean Oil: A pale yellow drying or semidrying oil that is obtained from soybeans and is used chiefly as a food, in paints, varnishes, linoleum, printing ink, and soap, and as a source of phospholipids, fatty acids, and sterols is called soybean oil. It is also called soya oil.

Space food: IMF foods are termed as space food (Plate 572).

Spaghetti: The pasta made in thin solid strings is called spaghetti.

Spaghetti squash: An oval winter squash (*Cucurbita pepo*) with flesh that after cooking is similar in texture to spaghetti.

Spaghettini: A pasta thinner than spaghetti but thicker than vermicelli is called spaghettini.

Spam: The tinned meat made from ham is known as spam. It is a trademark used for a canned meat product particularly spiced ham (Plate 573).

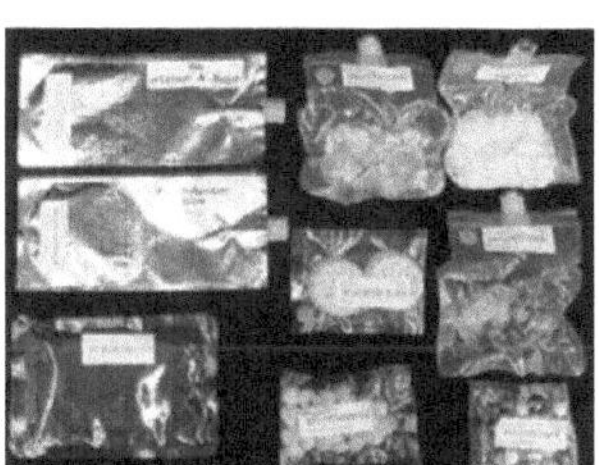

Plate 572: Space food

Plate 573: Spam

Sparger: It is an apparatus / equipment used for washing of mash.

Sparkling wine: Sparkling wine contains CO_2 produced during fermentation process i.e. secondary fermentation in bottle e.g. Champagne sparkling wine. CO_2 in sparkling wine is equal 5 g/l at 20°C.

Special property tests: These tests are frequently used in the tinplate industry to evaluate corrosion resistance of tin plates. These includes the Pickle Lag (PL) Test, the Iron solution value (ISV), the Tin Crystal Size (TCS) and the Alloy Tin Couple (ATC) tests.

Specialty food: It is a food product of a special kind having a distinctive quality or special excellence.

Specific gravity: The ratio of the density of a substance (food) to the density of pure water taken as a standard when both densities are obtained by weighing in air. Specific gravity of the food is the relative gravity / weight of solid or liquid as compared to pure distilled water at 62°F (16.7°C). Specific gravity is the relative gravity/ weight of solid or liquid compared to pure distilled water at 62°F (16.7°C). It bears no units as it is the relative quantity.

Specific heat: The heat in calories required to raise the temperature of one gram of a substance one degree Celsius is called specific heat. Or Specific heat is the ratio of the quantity of heat required to raise the temperature of certain mass of a substance through certain range of temperature to the quantity of heat required to raise the temperature of the same mass of water through the same range of temperature. The specific heat of a food can be estimated as specific heat = 0.008 x (% water in food) + 0.20.

Specification: It is a written description of an invention for which a patent is sought and is a statement of legal particulars.

Spectrophotometers: Spectrophotometers works on principle when light falls upon a homogeneous medium, a portion of the incident light is reflected, a portion is absorbed within the medium and the remainder is transmitted. UV-Visible spectrophotometers are double beam with a holographically ruled single grating monochromator. The spectral range is 195-900nm with a 0.1 nm resolution and a photometric range of -0.3 to 3.0 absorbance units. These instruments have automatic baseline correction, spectral substraction, axis expansion, differentiation and multi-wavelength measurements.

Spice: It is one of the aromatic vegetable products (e.g. pepper or nutmeg) used to season or flavor foods.

Spiced vinegar: The vinegar prepared by steeping the leaves or spices in the ordinary vinegar is known as spiced vinegar.

Spin: It is the process to form a thread by extruding a viscous rapidly hardening fluid.

Spindown pulp: This pulp is made up of smaller or finer pulp particles that form sediment on standing because of the juice saturation or higher density than the juice itself. It is also called sinking pulp. Such pulp results opaque appearance of the juice and it become turbid but impart desirable mouth feel. It can be overcome by using opaque container / labels.

Spinning: A method of fishing in which a lure is casted by use of a light flexible rod, a spinning reel, and a light line.

Spirit: The liquid containing ethyl alcohol and water that is distilled from an alcoholic liquid or mash is called spirit.

Spirit gum: It is a solution (gum Arabic in ether) used for attaching false hair to the skin is called spirit gum.

Spirit of wine: Alcohol is known as spirit of wine.

Spirit vinegar: The vinegar prepared by acidous fermentation of a distilled alcoholic fluid which in turn produced by aerobic fermentation is known as spirit vinegar.

Spoilage: The phenomenon in which a substrate becomes unsuitable for use. The physical and chemical changes in the food are brought by action of microorganisms, enzymes.

Spongy tissue: Spongy tissue is a physiological disorder in mango var alphonso. Spongy tissue is due to heat of convection from soil.

Spore: It is a primitive unicellular often environmentally resistant dormant or reproductive body produced by plants, fungi, and some microorganisms and capable to develop into a new individual either directly or after fusion with another spore. It is dormant stage of the microorganisms.

Sporicidal: Any agent which has ability to kill the spores is called sporicidal.

Spray drying: It is a process where liquid droplets are rapidly dried as it comes in contact with a stream of hot air. Spray drying is the fastest method of drying due to larger surface area of droplets produced in the spray mechanism. Milk, fruit juice or liquid foods can be dried by using spray drying methodology (Plate 574 and 575).

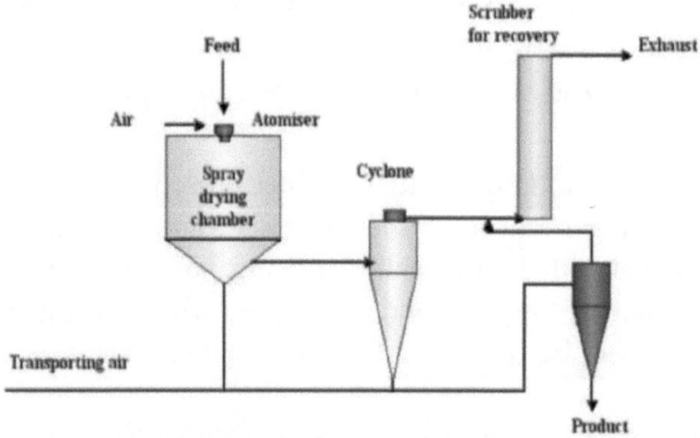

Plate 574: Spray dryer line diagram

Plate 575: Spray dryer used for spray drying

SPS: It is Sanitary and Phyto-Sanitary Agreement of the World Trade Organization (WTO)

SQC: Statistical quality control (SQC) employs statistical principles and methods to assess the magnitude of variation and to detect cause of that variation. The variation is based on the law of probability. SQC will indicate the limits beyond which these variations in the product should not go with out correction. Thus, SQC is really sampling of the product, determining the quality variation of the sample, and relating the finding to the entire lot under consideration.

Squash: It is unfermented fruit juice beverage which consists essentially of strained juice containing moderate quantities of fruit pulp to which cane sugar is added for sweetening. It should contain 25% juice and 40% TSS (Plate 576).

Plate 576: Squash bottles

Srikhand: It is milk base curd.

SSJ: It stands for single strength juice. The juice obtained by first pressing is called single strength juice (SSJ).

Stabilizers/Gelling agents/Thickeners: These substances helps to improve texture of foods, inhibits crystallization of sugar, formation of ice, stabilize emulsions and foams. These are used to keep products in a set state, such as jellies, jams and baby foods. They are also used to keep ice cream creamy. They generally improve consistency and will affect the appearance and texture of foods. The common ones are modified food starch and vegetable gum. They combine with water to form gels. Gum arabic, agar-agar, alginic acid, starch and its derivatives, gelatin, pectin, amylase, CMC, are examples of such additives. Gravies, pie fillings, cake toppings, jellies, puddings and salad dressings are some foods contain stabilizers & thickeners.

Stack burning: If the filled cans after processing are not rapidly cooled and are stacked in warehouses, the heat inside the can will cause the burning of food constituents and also helps to multiply the heat resistant microorganisms leading to the loss of flavor and nutrition and spoilages of canned foods, called stack burning.

Stalagmometer: An instrument used for measuring surface tension by determining the exact number of drops in a given quantity of a liquid. The lower the surface tension, the small will be the size of each drop and more drops will be in a given volume. It is also called stactometer. It is one of the volumetric tensiometer. The weight of the drop is in equilibrium state with the surface tension when it falls from the instrument. So, based on Tates law; $mg=2\pi r\sigma$, where $2\pi r$ is the circumference and σ is the surface tension, mg is the mass of the drop (equivalent to force) (Plate 577, 578 and 579).

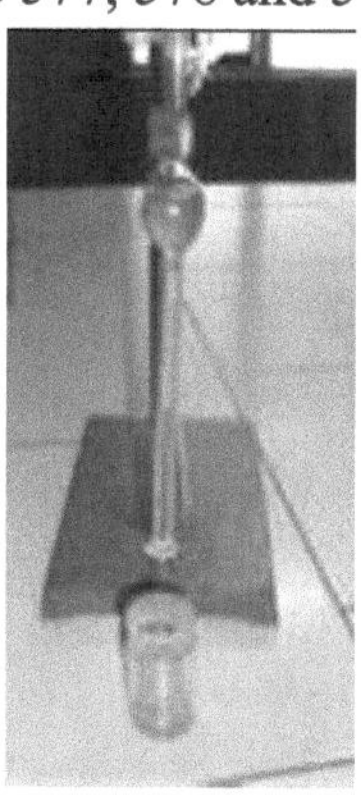

Plate 577: Stalagmometer

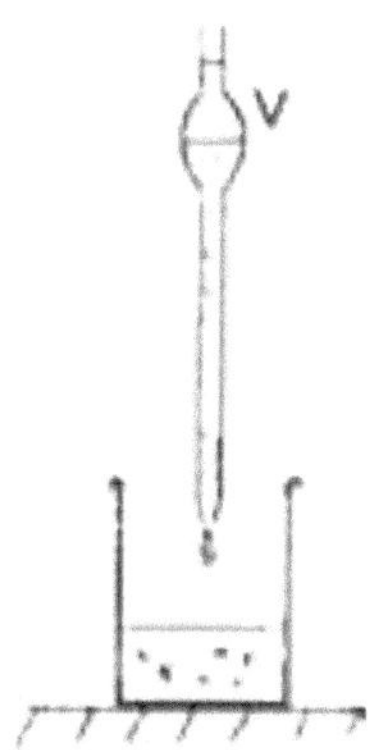

Plate 578: Drop measurements using stalagmometer

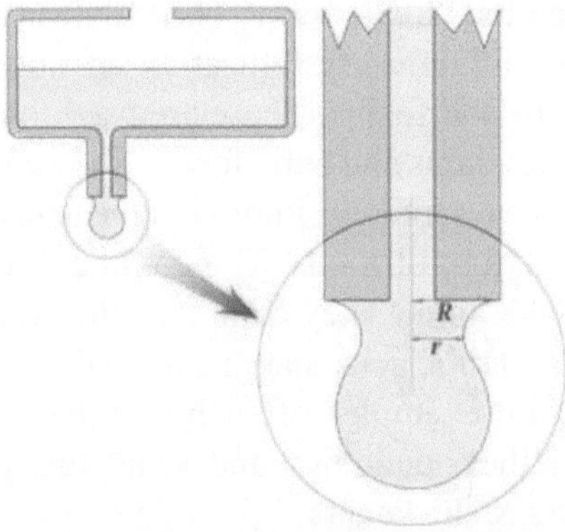

Plate 579: Measurement of surface tension based on Totes law

Stale: It refers to tasteless or unpalatable bread.

Staling: It is change in texture and flavour of starch based products during storage.

Stand - up – pouch: Stand up pouches is also known as stand bags or stand up bags. These flexible packages are with or without zipper system. It is a new introduction. This concept has been widely accepted in the products like Idli mix, Dosa mix, and soup mix etc.

Stand time: It is the period of the length for which a food is allowed to stand in a microwave oven without operation of magnetron tube so that the residual heat in the food can be transferred by conduction.

Standard: Standard in relation to any food article of food means the standards notified by the Food Authority.

Standard setting body: Government or private bodies that establish standards which may be the subject of a certification programme.

Standard solution: It is a solution of known concentration or known normality.

Standardization: Standardization means adjustment of food recipe etc by raising or lowering of a specific component in order to improve the quality and acceptability of the product.

Staphylococcus aureus: It is nonmotile gram positive spherical bacteria that occur singly, in pairs or tetrads, or in irregular clusters and include causative agents of various diseases (skin infections, food poisoning, and endocarditis). The source of this pathogen is human being in itself i.e. it is available in the skin, nose, throat, cuts and sores of the human. Consequently it is easily transmitted to any foods by handling and poor hygienic practices.

Staple food: The foods that are eaten to provide the energy e.g. rice, roti, potato etc.

Starch: It is a white odourless, tasteless, granular or powdery complex carbohydrate $(C_6H_{10}O_5)_x$ that is the chief storage form of carbohydrate in plants, is an important foodstuff, and is used also in adhesives and sizes, in laundering, and in pharmacy and medicine (Plate 580). It is an insoluble crystalline substance in the natural form. Starch is a long chain molecule of glucose polymer. Starch is major carbohydrate in fruit and vegetables. More the starch more is glucose and fructose on degradation. Starch give indication of TSS. Starch is stored in specialized storage structure called plastids (Amyloplast). Chloroplast present in the leaves synthesis carbohydrate. Starch is a straight chain molecule of amylose having 200-10000 subunits of glucose & amyl pectin. Starch is also seen in branched form i.e., minimum of 2000-200,000 glucose units. The ratio between the amylase and amyl pectin in a particular species /type is genetically controlled by amyl pectin comprising of 60-95%. Amyl pectin is more important in plant because pectin forms the framework of plants and so amyl pectin is large in quantity in plants e.g. developing fruit of apple, pear, mango, banana, guava, sputa have more of starch than that of fruit sugars.

Plate 580: Potato starch

Starch equivalent: This is the term used for calculation of the livestock ration to indicate the feeding value of feeding stuff.

Starch test: It is a common maturity test in fruits and is found by cutting the fruit in half along the equator and placing cut surface in the iodine solution.

Starchy foods: A food containing, consisting of, or resembling starch is called starchy food.

Statistics: A branch of mathematics dealing with the collection, analysis, interpretation, and presentation of masses of numerical data is called statistics.

Statistical quality control (SQC): It is one of the most useful tools that can be of great value for interpretation of data pertaining to all reports/ research publications and to improve manufacturing efficiency. Statistical quality control mainly based on or employing the principles of statistics.

Steady state: It is a state or condition of a system or process (as one of the energy states of an atom) that does not change in time. Broadly, it is a condition that changes only negligibly over a specified time. It is the state in which the temperature distribution a body remains constant with time.

Steady state process: In total mass balance, inflow = outflow + accumulation. If accumulation is '0' inflow = outflow and the process is called steady state process.

Steak: It is a slice of meat cut from a fleshy part of a beef carcass or is a similar slice of a specified meat other than beef i.e. ham or it is a cross-section slice of a large fish e.g. swordfish or ground beef prepared for cooking or for serving in the manner of a steak e.g. hamburger.

Steak house: It is a restaurant whose specialty is beefsteak.

Steak tartare: The highly seasoned ground beef eaten raw is called steak tartare.

Steam: A vapor arising from a heated substance or the invisible vapor into which water is converted when heated to the boiling point or it is the mist formed by the condensation on cooling of water vapor or it is water vapor kept under pressure so as to supply energy for heating, cooking, or mechanical work.

Steam based kettle: The steam based double walled steel jacketed kettle is used in the food processing unit for preparation of jam, jelly etc. The hot steam is passed through the double wall of the kettle which helps in uniform boiling and heating of the product. Better quality product can be prepared by using this apparatus (Plate 581).

Plate 581: Double walled steel jacketed kettle

Steam boiler: The boiler used for producing steam is called steam boiler.

Steam economy: It is ratio of quantity evaporated per hour to steam required per hour. In single effect evaporator it is less than one, while in multi effect evaporator, it is greater than one.

Stearate: It is a salt or ester of stearic acid.

Stearic acid: A white crystalline fatty acid $C_{18}H_{36}O_2$ obtained by saponifying tallow or other hard fats containing stearin is called stearic acid. It is a commercial mixture of stearic and palmitic acids.

Stearin: An ester of glycerol and stearic acid is called stearin.

Steriflame process: In steriflame process, cans are heated rapidly by rotating above direct flame.

Sterilised milk: The homogenized milk is bottled, sealed and then heated to 100°C for upto an hour. This results in partial caramalization of lactose and gives the milk a brownish tinge and slightly cooked flavour. The process destroys most of the microorganisms' along with reduction in 1/3rd of thiamine, half of the vitamin C and folic acid and almost all vitamin B_{12}. This milk can be kept in sealed packs upto 3 months.

Sterilization: Sterilization is the process of destruction of all infectious agents from an environment or from the food at a temperature above 100°C. This includes all living cells, viable spores, viruses and viroids.

Sterilization by direct heating: It can be achieved either by steam injection or steam infusing. Sterilization tank with steam injection is presented in Plate 582.

Plate 582: Sterilization tank (steam based)

Sterilization by indirect heating: It can be achieved by using suitable heat exchangers (Plate, tubular, or scraped surface heat exchanger) or hot water.

Sterilization by ohmic heating: It is novel sterilization process in which heat is generated within a food by the passage of electric current, making it possible for particles to heat at the same or faster rate than fluids (an ideal ohmic process) whereas in practical scenario, heating rates depends on the electrical and thermal properties of both particle and liquid. This technique utilizes the principle that heat can be generated throughout a material by the passage of electric current, using the inherent electrical resistance of the material.

Sterilizing value: It is also called lethal rate. Log (N/N_0) is called the sterilizing value. It is the number of decimal reduction required. It is also expressed as ratio of F-value and D – value.

Steroids: It is one of the compounds containing a 17 carbon, 4 ring systems and including the sterols and numerous hormones (e.g. anabolic steroids or corticosteroids) and glycosides.

Sterol: It is one of the various solid steroid alcohols (cholesterol) widely distributed in animal and plant lipids

Steroplast: End product of gram negative bacteria is Steroplast.

Stickers on fruits / produce: The four- or five-digit numbers identify the produce, indicating size, growing method, type of food (apple or orange for instance) and variety (such as a Honeycrisp or Golden Delicious apple). The voluntary labels tell you whether you are purchasing organic or conventionally grown produce. A four-digit code denotes conventional cultivation. These fruits undergo growth using traditional farming practices, often involving pesticides and chemical fertilizers, potentially affecting soil health. A five-digit code starting with the number 8 indicates genetic modification. If it has 5 digits, and begins with a 9, it means that the produce was organically grown. A 5-digit code that begins with an 8 indicates that the produce was genetically modified (Plate 583).

Plate 583: Illustration of Sticker on fruit

Still wine: Still wine contains no CO_2 produced during the fermentation process.

Stir–fry: It is a process to fry quickly over high heat in a lightly oiled pan while stirring continuously.

Stir–fry vegetables: A dish of vegetables that is stir-fried.

Storage: Storage is the holding of produce from time of production until needed by the consumers. Storing creates time utility. Storage helps to spread out market supply. Some products are stored for short period whereas fresh fruits, vegetables require cold storage.

Stormer viscometer: This type of instrument is used to measure the resistance (R) to flow.

Stowage: It is an act / process / place of stowing goods in storage or it refers to storage capacity.

Strain: The deformation of a material body under the action of applied forces. It causes a change of form or size by application of external force.

Strain gauge/ extensometer: An instrument for measuring minute deformations of test specimens caused by tension, compression, bending, or twisting is called strain gauge / extensometer.

Strainer: A device (as a sieve) to retain solid pieces while a liquid passes through / any of various devices for stretching or tightening something.

Strass: Strass is dough that contains a considerable proportion of fat and is used for making pastry crust / fancy rolls. Or Strass is a confection made by evaporating fruit with sugar.

Strecher degradation: It is degradation of amino acids to aldehyde with one carbon atom less.

Stress: Force per unit area is called stress. Stress is applied on any food articles to get strain (deformation).

Stress metabolites: These are known as Phytoalexins. These are produces due to physiological stresses, microbial invasion or due to chilling injury.

Stretch cling film: It is a film used fro stretch wrapping of the food. Its chemical composition provides high tensile strength and provide a rubber band effect.

Stroke: It is the sudden diminution or loss of consciousness, sensation, and voluntary motion caused by rupture or obstruction of a blood vessel (by a clot) of the brain. It is also called apoplexy / cerebrovascular accident.

Strong wine: A wine having alcohol strength of 16-21% is called strong wine.

Stun: It is the effect of something that stuns.

Stunning: It is the process to make a person or animal senseless, groggy, or dizzy.

Stunning agents: The agents which causes senseless, groggy or dizziness is called stunning agents.

Styrene: It is a fragrant liquid unsaturated hydrocarbon C_8H_8 used in making synthetic rubber, resins, and plastics.

Styrofoam: It is an expanded rigid polystyrene plastic.

Sub threshold level: It is the concentration of the taste compound at a level that is non-detectable but is capable of influencing other taste perceptions.

Subclinical deficiency: This type of the deficiency is due to inadequate intake of a nutrient leading to a deficiency which cannot be detected in a clinical test.

Suberin: A complex fatty substance found in the cell walls of cork is called suberin.

Suberization: It is the process of developing of secondary skin for protecting potatoes against rotting/ microbial spoilage. It is process of conversion of the cell walls into corky tissue by infiltration with suberin.

Subjective Methods of quality evaluation: In this method, individual is required to give his opinion about qualitative/ quantitative values. This method is also referred as sensory method. It is by experience of the individual. Different subjective methods are used for estimation like: 9 - point Hedonic Scale, Triangular Test or Composite Test.

Sublimation: The change of phase directly from solid phase to vapour phase without passing through the liquid phase is called sublimation.

Substrate phosphorylation: When high energy substrate is converted to lower energy substrate with the formation of ATP.

Substrate: The substance on which the enzyme act is termed as Substrate.

Subtilin: It is an antibiotic obtained from *B. subtilis* used in the preservation of asparagus, corn and peas. It is effective against gram positive bacteria.

Succulometer: This instrument is used for determination of maturity of corns and apples. This instrument measure volume of the extractable juice under controlled conditions of time and pressure.

Sucralose: It is a sweetner and is derived from table sugar (sucrose). It closely resembles table sugar in taste and is highly water soluble. It is not absorbed in the digestive tract and so adds no calories to food.

Sucrase: It is an enzyme that catalyses the hydrolysis of sucrose into glucose and fructose. It is also called invertase.

Sucrose: It is a sweet crystalline dextrorotatory disaccharide sugar $C_{12}H_{22}O_{11}$ that occurs naturally in most plants and is obtained commercially from sugarcane or sugar beets. It is 12-carbon sugar.

SUFU: SUFU is known as Chinese cheese / bean cake made from cakes of soyabean curd (Tofu) by the action of a mold.

Sugar: A sweet crystallizable material that consists wholly or essentially of sucrose, is colorless or white when pure tending to brown when less refined, is obtained commercially from sugarcane or sugar beet and less extensively from sorghum, maples, and palms, and is important as a source of dietary carbohydrate and as a sweetener and preservative of other foods is called sugar. It is water soluble compound that vary widely in sweetness, include the monosaccharides and oligosaccharides, and typically are optically active.

Sugar (Control) Order, 1966: Indigenous production of raw sugar and its quality are controlled under this order. The Directorate of Sugar, Ministry of Agriculture and Rural Development is the operating authority of this order.

Sugar acids: These are oxidized derivatives of a monosaccharide such as aldonic acids and uronic acids.

Sugar alcohol: It is monosaccharide in which aldehyde group is reduced to alcohol except glycerol others are found in the most of plants.

Sugar apple: It is a tropical American tree (*Annona squamosa*) of the custard apple family and has edible sweet pulpy fruit with thick green scaly rind and shining black seeds.

Sugar beet: A white rooted beet grown for the sugar in its roots is sugar beet.

Sugar bush: A wood in which sugar maples predominate is sugar bush.

Sugar cane juice: Canes are harvested when TSS content reaches to about 18-19% and are concentrated to 30°Brix.

Sugar free food: Food having less than 0.5 gm of sugar per serving or per 100 gm.

Sugar off: The complete process of boiling down the syrup in making maple sugar until it is thick enough to crystallize is called sugar off. It results in reaching the state of granulation.

Sugarcane: A stout tall perennial grass (*Saccharum officinarum*) native to tropical Southeast Asia that has a large terminal panicle and is widely grown in warm regions as a source of sugar.

Sugarhouse: It is a building where sugar is made or refined. It is a place where maple sap is boiled and maple syrup and maple sugar are made.

Sugaring off: The process of converting maple syrup into sugar is called sugaring off.

Sugarloaf: The refined sugar molded into a cone is called sugarloaf.

Sugarplum: A small candy that has the shape of a ball or disk is called sugar plum. It is also called sweetmeat.

Sugars properties: Glucose contain aldose sugars whereas mannose, galactose, fructose contain ketone sugars. Reducing sugars are glucose, fructose, galactose, mannose, ribose, xylose and glycogen while non-reducing sugars are sucrose and raff nose.

Sugary starch: The starch which contains higher percentage of the amylose than normal starch is called normal starch. Sugary starch have rough surface of the granules.

Sulfur dioxide: A heavy pungent toxic gas SO_2 that is easily condensed to a colorless liquid, is used in making sulfuric acid, in bleaching, as a preservative, and as a refrigerant, and is a major air pollutant in industrial areas.

Sulfur stinker spoilage: Spoilage of low acid canned foods by anaerobic thermiphile (*Clostridium nigrificans*). Here hydrogen sulfide gas is evolved due to decomposition of proteins containing sulphur and rotten egg odour and blackening of food (due to FeS formation) are imparted.

Sulfuric acid: It is a heavy corrosive oily dibasic strong acid H_2SO_4 that is colorless when pure and is a vigorous oxidizing and dehydrating agent.

Sulfuring: Exposure of the prepared fruit pieces to burnt sulphur fumes before drying which helps to maintain an attractive light colour, conserve vitamin C, A and makes drying process easier by softening cell wall, displaces air from the tissues, destroys enzymes that cause darkening of cut surface and kills many of the microorganisms present. It is a fumigation process.

Sulphide spoilage: It is due to *Desulfomaculum* (*Clostridium*) *nigrificans* in low acid foods. The spores are heat resistant and their presence is an indication of under processing. It is also called sulphur stinker.

Sulphitation: Treatment of the prepared and blanched vegetable with sulfide solution either by dipping or spraying for sufficient time. Generally 0.5% KMS solution helps to achieve all benefits of sulphuring along with better retention of colour of vegetable.

Sulphur resistant (SR) cans: These are gold enamel cans and are coated with C- enamel. They are meant for non-acid foods.

Sulphur stinker: It is due to *Desulfomaculum* (*Clostridium*) *nigrificans* in low acid foods. The spores are heat resistant and their presence is an indication of under processing. It is also called sulphide spoilage.

Sump and flash mixer tank: It is a part of effluent treatment plant. Here, alum and lime are mixed properly with the help of agitator and mixers.

Sun drying: Removal of moisture from food by exposure to the sun rays is called sun drying.

Sunki: Sunki is a non-salted and fermented vegetable product prepared from the leaves of Otaki-turnip in Japan. *Lactobacillus plantarum* is used for fermentation.

Sunset Yellow FCF: It is also known as Orange Yellow S. It is a synthetic yellow azo dye, manufactured from aromatic hydrocarbons from petroleum. When added to food it is denoted by E110.

Super saturated solution: True solution which contains more solute than theoretically can be dissolved at that temperature, a situation created by cooling a heated saturated solution carefully is called super saturated solution.

Supercooling: It is process of going below the freezing point without the formation of the ice crystals. Or It refers to cooling below the freezing point without solidification.

Supercritical fluid extraction (SFE): SFE is the process of separating one component from another using supercritical fluid as the extracting solvents. The extraction is usually from the solid matrix but can also be done from liquid. SFE can be used for preparation of sample for analytical purpose, for removal of the unwanted material from the product (decaffeination) or for collection of the essential oils. Carbon dioxide is the most commonly used super critical fluid that has a critical temperature of 31°Cand critical pressure of 74 bars. No residue of CO_2 is left after SFE. Ethanol and methanol can also be used for extraction purposes. As carbon dioxide is non polar cannot be used for extraction of polar solvents. In such cases ethanol can be used. Maintenance of the high pressure is the necessity of SFE. It is alternative for liquid extraction by using solvent like hexane or dichloromethane (Plate 584).

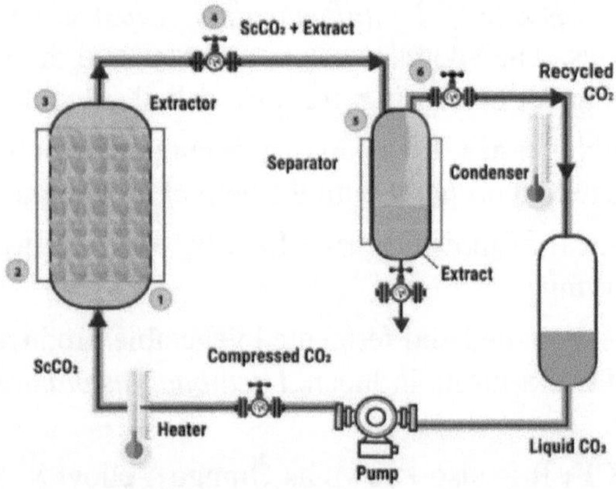

Plate 584: Supercritical fluid extraction system

Supercritical fluid: A state of fluid having a temperature above a critical temperature is called supercritical.

Superficial scald: Diffused brown discolouration of the fruit skin, somewhat rough in severe cases. It is developed after several months of cold storage which becomes more expensive after removal to warm temperature and during ripening after storage.

Supersaturation: A state in which a compound remains in a saturated solution without being crystallized.

Surface tension: The energy required to increase the unit surface area is the surface tension. The force per unit length that tends to contract the surface of a liquid is known as surface tension (Plate 585). It may also be defining as the force in Newton acting at right angle along the surface of a liquid one meter in length. Stalagmometer is used for measuring surface tension and is expressed as M/m is SI system or dynes/cm in cgs system. Milk has surface tension of 50 dyne per cm at 20°C.

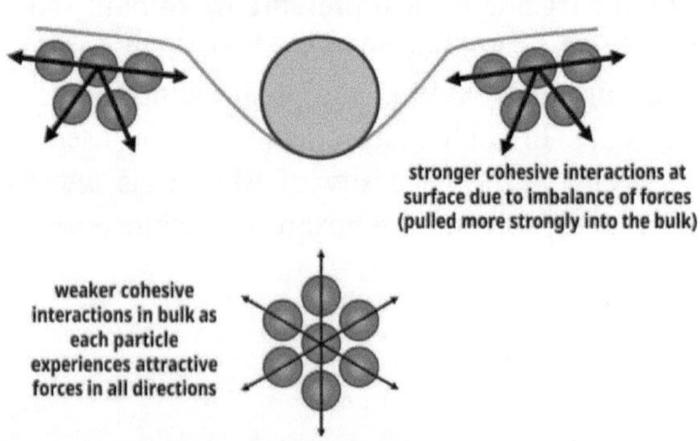

Plate 585: Surface tension illustration

Surface-active agents: These additives are used to stabilize oil-in-water, water-in -oil, gas-in-liquid and gas-in-solid emulsions. or Surface active agent is the additive that promotes the formation of an emulsion. These additives are used to stabilize oil-in-water, water-in -oil, gas-in-liquid and gas-in-solid emulsions. The emulsifying agents used in food production are amphophilic substances; they are able to stabilize a system which is naturally unstable in oil or water. Natural origin emulsifier is Lecithin and synthetic is mono and triglycerides, bile acid, fatty acids & their derivatives can be used as emulsifying agents. Synthetic surface active agents include defoaming compounds and detergents e.g., propylene glycol monostearate and monosodium phosphate. These compounds contain both polar and non-polar groups and thus are drawn to the interface between the two phases of an emulsion to coat the surface of the droplets. These are also called as Emulsifiers (Plate 586).

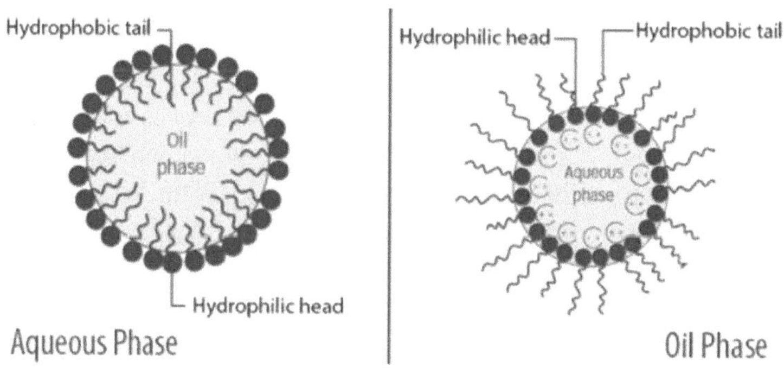

Plate 586: Surface-active agents

Surfactant: These are the substances that lower the surface tension of the liquid and help in wetting of powders. These are also called emulsifiers.

Suspensoid: It is the colloidal dispersion of the gas in the solid.

Sweating: The process for equalization of the moisture content of the dried products is called sweating. Sweating is the equalization of moisture to a desired level of the dried fruit and nuts before packaging.

Sweet flag: A perennial marsh herb (*Acorus calamus*) of the arum family with long narrow sword-shaped leaves and an aromatic rootstock. It is also called calamus. It is prohibited food additive.

Sweet juice: It is a liquid product which contains at least 85% juice and 10% TSS.

Sweet meat: Food rich in sugar or a candied confection or crystallized fruit is called sweet meat.

Sweet wine: Sweet wine contains unfermented sugar or is added later on and wine alcohol content 11-14%.

Swell: A can with tightly bulged ends due to formation of CO_2 and other gases inside the can as result of microbial decomposition of foods is called swell. This food has an offensive and sour odour and is generally discoloured and unfit for consumption. It is also called Blower.

Swellfish: It is puffer fish of Tetraodontidae family.

Swelling power of starch: Swelling power of waxy starch > normal starch> sugary starch.

Swept surface heat exchanger: It is a piece of the equipment used in food processing wherein the product is continually swept off the heat transfer surface thereby constantly exposing it to unheated or less heated product mass thus accelerating the process of heat transfer and accomplishing it much faster and within a much shorter time. It is also called scraped surface heat exchanger. Scraped Surface Heat Exchangers are used for difficult heat transfer applications, with high viscosities and/or where fouling becomes a problem, the preferred option is a scraped surface heat exchanger. Scraped surface heat exchangers keep heat transfer high and the heat transfer surface constantly clean (Plate 587).

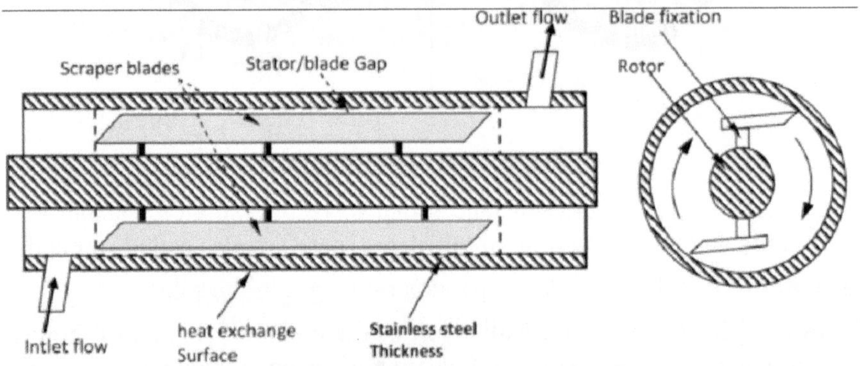

Plate 587: Scraped surface heat exchanger

Swiss chard: A beet (*Beta vulgaris cicla*) having large leaves and succulent stalks often cooked as a vegetable is called Swiss chard.

Swiss cheese: A hard cheese characterized by elastic texture, mild nutlike flavor, and large holes that form during ripening. It is also known as Emmental cheese.

Swiss steak: A slice of steak pounded with flour and braised usually with vegetables and seasonings is called swish steak.

SWOT: It stands for strength, weekness, opportunities and thrust area.

Syconium: The multiple fleshy fruit of a fig in which the ovaries are borne within an enlarged succulent concave or hollow receptacle is called syconium.

Synbiotics: Synbiotics is the word coined for the combined administration of specific probiotics to provide definite health benefits by synergistic action. A synbiotic is a product containing probiotic (fibre, pectin etc) and prebiotic (microbes) in which the prebiotic compound specifically favors for probiotic compound.

Syneresis jelly: A condition of spontaneous exudation of fluid from a jelly, caused by excess of acid leading to break down of jelly structure through hydrolysis or decomposition of pectin, low concentration of sugar, insufficient pectin, premature gelation and fermentation of jelly. It is also called weeping jelly.

Syneresis: It is loss of fluid from jellies or separation of liquid from the gel. It occurs if the acidity of the juice is very high e.g. grape jelly. Or Syneresis is the separation of liquid from a gel caused by contraction.

Synergism: The interaction of discrete agencies (industrial firms), agents (drugs), or conditions such that the total effect is greater than the sum of the individual effects is called synergism.

Synergistic: The process having the capacity to act in synergism is called synergistic.

Synthetic beverage: A non alcoholic beverage or syrup that contains no fruit juice but having an artificial flavour or colour resembling any fruit.

Syntrophy: The interdependence of living organisms (photosynthetic cells) and heterotrophy on each other is known as syntrophy. They depend on carbon and oxygen on each other.

Syrup: It is a fruit beverage containing atleast 25% fruit juice and 65% TSS along with 1.3 – 1.5% acid. Or A thick sticky solution of sugar and water is also called syrup.

Syruping: It is the process covering the fruit or their pieces with the hot strained sugar syrup having > 20-25°Brix strength. It is done prior to seaming and exhausting.

SWOT: It stands for strengths, weaknesses, opportunities and threats.

[illegible]: [illegible] the fruit of [illegible] which the [illegible] which on entering [illegible] is called [illegible].

Syndication: [illegible] used in the [illegible] of [illegible] to providing [illegible] by [illegible] [illegible] among the [illegible] the problem [illegible] specifically [illegible].

Synergy: [illegible] simultaneous [illegible] [illegible] of [illegible] [illegible] [illegible].

[illegible]

the separation of liquid from a gel caused by contraction.

Synergism: The interaction of [illegible] or other [illegible] so that the total effect is greater than the sum of the individual [illegible] called synergism.

[illegible]: [illegible]

[illegible]

[illegible]

Syrup: It is a [illegible] along with [illegible] also called [illegible].

Systemic: [illegible]

T

Table cream: It is also called coffee cream, single cream or homogenized cream. It is used in coffee and tests about 20% butter fat.

Table oil expeller: It is a type of the machine that is used for extraction of the oil from the oil seeds (Plate 588).

Plate 588: Table oil expeller

Table ready meat: Cooked and smoked sausages are called table ready meat.

Tal prolong: This wax is esters of fatty acid glycosides NaCMC.

TAME: It stands for p-Toluene sulfonyl L- arginine methyl ester. It is used in isolation of Trypsin inhibitor.

Tandoor: The clay oven in India (Hindustani) is called Tandoor.

Tandoori: A food spiced and cooked over charcoal in a tandoor e.g. tandoori chicken and tandoori roti.

Tankard: The tall beer mug with a handle is called tankard.

Tannin: These are soluble astringent complex phenolic substances of plant origin used in tanning leather and dyeing fabric, manufacturing ink, clarifying wine and beer, and in medicine. Tannins bind with iron and interfere with iron absorption. Black and red varieties of beans are high in tannins. Tannins are widely spread in nature and almost present in all the plant materials. These are poly-phenolic compounds and mainly divided into two groups i) hydrolysable tannins and condensed tannins

Tapka: It is a maturity index for mango harvesting indicated by fall of ripe fruits naturally from the birds.

Tapping: The process of harvesting latex from trunk of rubber plant, fruit of papaya etc is known as tapping.

Tapster: One whose business is to draw ale is called tapster.

Tartrazine: It is a synthetic lemon yellow azo dye and is used as food colouring agent. It is water soluble. When added to food it is denoted by E142.

Taste: Taste is due to sensation felt by tongue. Taste may be sweet, salty, sour & bitter.

TASTE: Thermally Accelerated Short Time Evaporator. It is a high temperature evaporator with very short holding times (Plate 589).

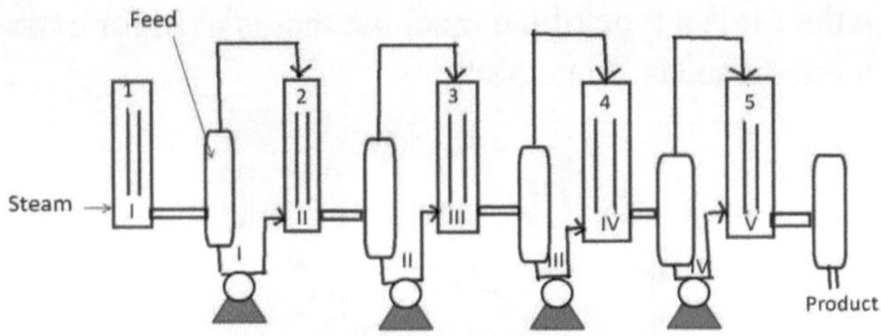

Plate 589: Multi effect evaporators (TASTE)

TBA value: It is also called peroxide or Lea value.

TBT: It refers to Technical Barriers to Trade Agreement of the World Trade Organization (WTO). It deals with product specification and packaging. TBT is a category of NTB to trade and used to regulate the markets, protect their consumers, and preserve natural resources by protecting domestic industries. TBTs have great potential for being misused by the importing countries as they are non transparent obstacles to trade.

TDT curve: A thermal death time (TDT) curve can be plotted by collating D-values at different temperatures (Plate 590).

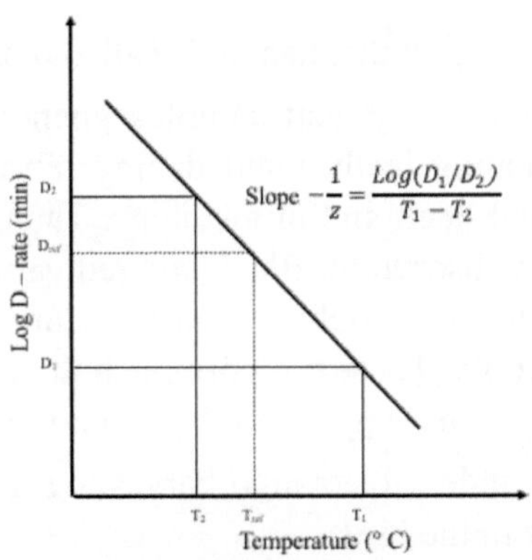

Plate 590: TDT Curve

Tea: It is a shrub (*Camellia sinensis* of the family Theaceae, the tea family) cultivated in China, Japan, and the East Indies. The leaves, leaf buds, and internodes of the tea plant prepared and cured for the market, classed according to method of manufacture into one set of types (green tea, black tea, or oolong), and graded according to leaf size into another (orange pekoe, pekoe, or souchong). It is an aromatic beverage prepared from tea leaves by infusion with boiling water.

Tea bag: A bag usually of filter paper holding enough tea for an individual serving is called tea bag.

Tea ball: A perforated metal ball that holds tea leaves and is used in brewing tea in a pot or cup.

Tea cake: A small flat cake usually made with raisins is called tea cake.

Teaberry: The use of checkerberry leaves as a substitute for tea is called teaberry.

Teapot: A vessel with a spout and a handle in which tea is brewed and from which it is served.

Teapoy: A small box, can, or chest used especially to keep tea in is called teapoy.

Teat dipping: It is a method of applying disinfectant to a cow's teats as a precaution against the spread of mastitis, involving the dipping of each teat into a cup containing the disinfectant.

Technical enzyme: The separated and purified enzymes are termed as technical enzymes.

TEM: It stands for transmission electron microscope / microscopy.

Tempeh: An Asian food prepared by fermenting soybeans with a rhizopus is called tempeh.

Temper fresh: This wax is sucrose polyester and unsaturated FA.

Temperature: The degree of hotness or coldness measured on a definite scale or it is the degree of heat that is natural to the body of a living being or abnormally high body heat is called temperature.

Temperature inversion: An increase of temperature with height through a layer of air is called temperature inversion.

Temperature quotient: The number of times a reaction rate change with a 10°C change in temperature is called Q_{10} value. If reaction rate doubles with 10°C change in temperature, $Q_{10} = 2$. It is also called Q_{10} value.

Tempering: It is the process of bringing a product to desired moisture content for processing. Tempering is controlled humidification of wheat to adjust moisture upto 15-16%. It is also called conditioning. Or It is the process of the removal of the heat resulting from crystallization of fats and maintenance of a selected temperature to promote the formation of stable, and desirable crystals. Or It is the process of softening a substance (hardened steel or cast iron) or food by reheating at a lower temperature. The process of annealling or toughening of glass by gradually heating and cooling is also called tempering.

Tempoyak: It is fermented pulp of a durian fruit (*Durio zibethinus*). It is popular product of Malaysia. It is prepared by mixing durian pulp with salt and placing in asealed container. Fermentation takes about 7 days.

Tempura: The seafood or vegetables dipped in batter and fried in deep fat is called tempura.

Tenderay process: It is an air accelerated aging process. It is used for beef ageing. The beef is aged in 2 or 3 days by combining high humidity and air circulation with a temperature of about 18°C. However, this speeds up the microbial growth which can be checked by UV light.

Tenderfrank process: It is used for meat. In this case no casing is used and all operations of forming, cooking and smoking is done in a same way as that for frankfurters.

Tenderization: The act to make meat or meat products tender by applying a process or substance that breaks down connective tissue is called tenderization.

Tenderometer: An instrument by which toughness of the seed coat and firmness of pulp is determined and is mostly used to determine the seed quality of pea. Tenderometer is used to measure the compression and shearing of peas for preservation purpose.

Tensile strength: The greatest longitudinal stress a substance can bear without tearing apart is called tensile strength.

Tensiometer: It is a device used for measuring tension of structural material or an instrument for measuring the surface tension of liquids.

Tequila: Distilled pulque (distilled agave wine) is called Tequila. It is produced in Mexico.

Terminal markets: These are the markets in which the produce is either finally disposes off, direct to consumer or processors or assemble for shipment to foreign countries or assembled for export. These are located in Metropolitan cities like Mumbai, Madras and Calcutta. These markets are the parts where warehouses and storages are available/ and cover a wide area e.g. Bombay terminal market. It is a hub and spoke model, where hub is the main terminal market and spokes are the collection centres. This market has forward linkages with wholesalers, processors, exporters, traders, retailers etc. and backward linkages with farmers, farmer associations, cooperatives societies, self help group (SHG) etc.

Termination stage in rancidity: The chain reaction in the formation of the free radicals and hydro peroxides continue until either the whole of the unsaturated fatty acids have been exhausted or the free radical has inactivated each other. This is called termination stage.

Terpene / terpenoides: These are isomeric hydrocarbons $C_{10}H_{16}$ found present in essential oils (from conifers) and used as solvents and in organic synthesis. Broadly it is one of the hydrocarbons $(C_5H_8)_n$ found in essential oils, resins, and balsams.

Tertiary colour: Combinations of three primary colours give rise to tertiary colours.

Tertiary containers: Tertiary containers group several secondary cartons together into pallet loads or shipping units. The objective is to aid in the automated handling of larger amount of products. Typically, a forklift truck or simpler equipment is used to move and transport these tertiary loads.

Tertiary processing: It involves high level of the modifications of the natural produce by preparation of different types of the value added products like ketchup, RTS beverages and ice cream etc.

Test and cylinder: Jute is also called test and cylinder.

Tetra Pack: Tetra pack concept is a paper board foil plastic laminate and is receiving widest acceptance at present. This laminated structure consist of as many as six layers of materials i.e. polypropylene (PP), surlyn, aluminium foil, polyethylene (PE), paper board and polyethylene as inner food contact layer (Plate 591). Other combinations that can be used include laminated having EVOH, PE, PS or a metalized polyester. Aluminium foil is the most commonly used heat sealing and food contact material. Paper board besides providing rigidity protects aluminium foil from mechanical damage.

Plate 591: Tetra Pack containers

Tetra pack 11 Slim Carton: This is a new carton with re-cap opening used by Smithkline Beecham for their Ribena Black Currant and Ribena light. The Re-cap opening consists of a reclosable plastic cap placed over a folded foil strip which is temper evident. A pear shaped opening is revealed when the foil strip is pulled back. The device gives easier open-ability, better pouring and reclose-ability. The carton itself has a 13% larger display area compared to standard 11 packs, but requires 14% less shelf space.

Texture: Texture is the composite of those properties which arise from the structural elements of a product, and the manner in which this composite registers with the physiological senses. Or it is overall assessment of the feeling by mouth and hand or it is sense of touch by hand and mouth.

Texture analyser: It is an apparatus used for the measurement of the texture of the fruits, vegetables, cereals based food, meat etc (Plate 232). It is physical method of measurement of the texture attribures like hardness, cohesiveness and crushability.and thus by measuring textures various quality attributes can be predicted like gumminess, brittleness, chewiness etc of foods. It is based on the principle of force require to disintegrate the product. It can be monitored by the computer or manually. Different types of the probes are used for different products depending upon their characteristics.

Texturized vegetable protein: The protein from some vegetables especially soybeans used as a substitute for or added to meat is called texturized vegetable proteins.

Texturometer: It is a simulation device that measures physical properties of food such as hardness, cohesiveness and crushability. It is also called texture analyser (Plate 592).

Thawing: The process of melting a frozen food to a liquid state is called thawing (Plate 593).

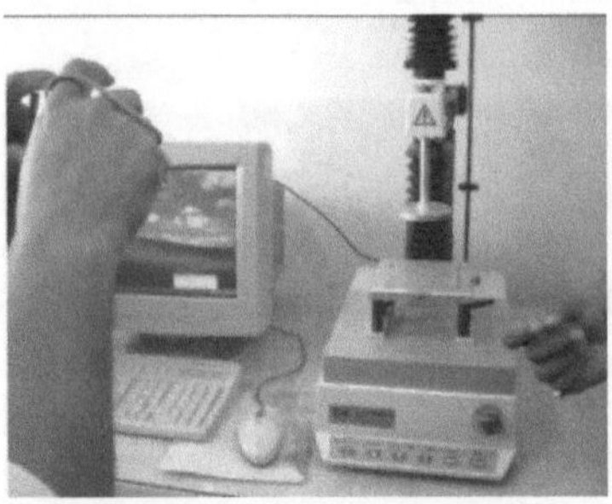

Plate 592: Texture analyzer

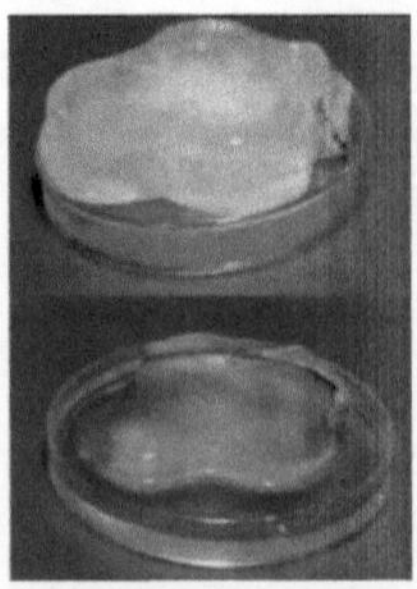

Plate 593: Thawing of ice

The Coffee Act, 1942: It provides rules and regulations regarding curing, assessment of quality and enforcement of quality control measures at different stages of marketing coffee within the country and abroad.

The Insecticide Act, 1968: This Act is regulated and governed by directorate of Plant Protection, Quarantine and Storage under the Ministry of Agriculture. This Act describes the safe use of insecticides to ensure that residual levels of chemicals in food do not pose a health hazard.

Therapeutic diet: It is related to that diet which is used for the treatment of disease or disorders by remedial agents or methods rather than a diagnostic specialty and thus assisting in a curing disease or disorder.

Therapeutic index: It is a measure of the relative desirability of a drug for the attaining of a particular medical end that is usually expressed as the ratio of the largest dose producing no toxic symptoms to the smallest dose routinely producing cures.

Therapeutic nutrition: The nutritional composition of a particular diet which is used for treatment of disease or disorders by remedial agents or methods rather than a diagnostic specialty and thus assisting in a curing disease or disorder is called therapeutic nutrition.

Thermal conductivity (k): The quantity of heat transmitted per unit time per unit area per unit temperature gradient. It is the ability of the food or package to conduct a heat.

Thermal conductivity detectors (TCD): TCD is working on phenomenon "if you blow air over a hot wire, it will be cooled". A thin filament of wire is placed at the end of GC column and heated by passing a current through it. The carrier gas molecules strike the hot wire, as each molecules hits wire, it takes away some heat from wire and as wire is cooled, its resistance changes. Sensitive to all organic compounds (10^{-7}g/sec.). TDC is sensitive to flow rate and temperature changes, detector temperature limits is 500^{0}C and have carrier gases like N_2, H_2, He, Co_2 and Ar.

Thermal death point (TDP): It is the lowest temperature necessary to kill all the microorganisms in 10 min. Different microorganisms will respond differently to high temperatures e.g., endospore-formers such as C. botulinum) being more heat tolerant.

Thermal death time (TDT): TDT is defined as the time required at a given temperature to kill a stated number of organisms under specified conditions or the short time required at certain temperature to kill a stated number of micro-organisms (spores) under specified conditions. TDT determines experimentally on a logarithmic scale and the corresponding temperature on a linear scale. Unless the original number of the micro-organisms involved in known, the TDT value is meaningless because different population results in different curves. It the value of Z is large; the temperature has a lesser effect on the TDT than when Z is a small number. Most resistant bacterial spores exhibit Z values within range of 18 to 27°F (10-15°C). It is the heat resistance of micro-organisms in 10 minutes at a given temperature. The thermal resistance of the spores is co-related with dipicolinic acid (DPA) content which is 15% of the dry weight of spores. The heat resistant of microorganisms is usually expressed in terms of their TDT (Plate 594).

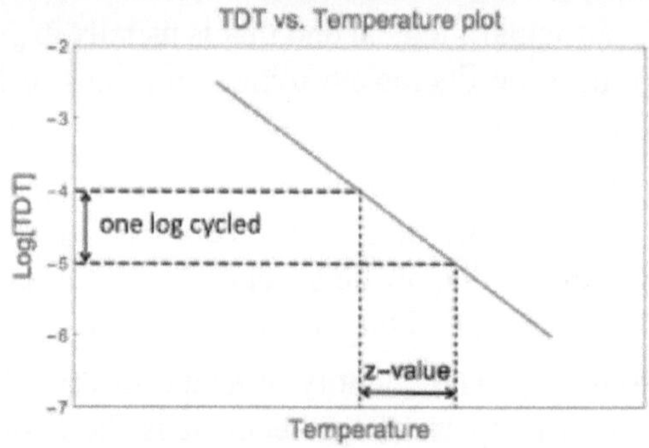

Plate 594: Thermal death time curve

Thermal diffusivity: It determines the rate at which a non uniform temperature distribution approaches equilibrium conditions.

Thermal fogging: The volatile chemicals can be applied as a vapour which is circulated in a confined space containing the crop. Some chemicals need to be heated to aid circulation and this process is thermal fogging. CIPC also called chloropropham is applied as thermal fog.

Thermal pollution: The discharge of heated liquid (as wastewater from a factory) into natural waters at a temperature harmful to the environment is called thermal pollution.

Thermal process: It is the application of the heat to food before and after sealing in the container for a certain period of time and a specific temperature, scientifically determined to achieve commercial sterility. The use of the heat is done to kill the micro-organisms. It involves combination of time and temperature. A particular microorganism which is to be killed is called thermal indicator microorganisms for food processing e.g. *Closteridium botulinum*. It is presumed that if *Closteridium botulinum* is killed, all other are also killed.

Thermal radiation: It is the transfer of energy by that part of the electromagnetic spectrum with wavelengths between 10^{-7} and 10^{-4} m.

Thermal resistant constant: It is similar to z-value. Z value is equal to the reciprocal of the slope of the thermal death time (TDT) curve; hence Z-value provides information on the relative resistance of microorganisms to different destructive temperature.

Thermal resistant curve: A plot of decimal reduction time against temperature on a semi log scale is called Thermal resistant curve.

Thermization: The treatment of milk at 63-65°C for 15 second is called thermization.

Thermo plasticity: It is related to dehydrated fruit/vegetable. Fruits are high in sugars and soften and melt at the elevated temperature of drying. They will also stick to the pan and would be difficult to remove. Even all the water is removed; its give indication that it contains water and on cooling the thermoplastic solids hardens into a crystalline or amorphous gloss form.

Thermoduric organisms: The micro-organisms which can survive at relatively high temperature but do not necessarily grow at this temperature are known as thermoduric organisms e.g. *Streptococcus, Lactobacillus*.

Thermoduric: Microbes able to survive high temperatures specifically at a temperature of pasteurization are called thermodurics.

Thermoduric: The resistant to heat as a mechanical agent is called thermoduric.

Thermodynamics: It is the branch of the physics that deals with the mechanical action or relations of heat. The science of the relations between heat and other forms of energy is called thermodynamics. Or It is deals with the relation between heat and work. It predicts different relationships between properties of the matter.

Thermodynamic process: The energetic evolution of a thermodynamic system proceeding from an initial state to a final state is called thermodynamic process.

Thermodynamic wet bulb temperature: It is the temperature at which the air can be brought to a saturation state adiabatically by the evaporation of water into the flowing air. It is also called adiabatic saturation temperature.

Thermoelectric: It is the branch of the physics that is producing electricity by a difference of temperatures.

Thermogenesis: It is the process or mechanism of production of heat (by burning stored body fats) in an animal body by physiological process.

Thermo-hygrograph: It is an instrument combining hygrograph and thermograph to record simultaneously on the same chart. It measures atmospheric temperature and humidity.

Thermolabile: The substances which become unstable when heated at a temperature above 55°C (131°F) e.g. many immune bodies, enzymes, and vitamins are called thermolabile.

Thermometer: It is an instrument for measuring temperature. It is a graduated glass tube containing mercury or alcohol.

Thermophillic Acid (TA) Spoilage: The spoilage of low and medium acid canned foods by *Clostridium thermosacharolyticum* which breaks sugars and form acid and gas (moisture of H_2 and CO_2) resulted in swelling and in severe cases even bursting of can may occur and the spoiled food usually has a sour odour.

Thermophillic organisms: The micro-organisms which not only survive at relatively high temperature but also require high temperature (above 45^{0}C) for their growth and metabolic activity are known as thermophillic organisms e.g. *Bacillus* spp., *Clostridium* spp. However, spore forming bacteria found in tomato juice require heating at a higher temperature of 88^{0}C for much longer duration for their growth.

Thermopile: An apparatus consisting of thermocouples combined so as to multiply the effect and used for generating electric currents or for determining intensities of radiation is called thermopile.

Thermoplastic: The substance that becomes plastic (capable of softening or fusing) on heating and hardens on cooling is called thermoplastic. Thermoplastics can be heated and cooled repeatedly without any loss of mechanical and physical properties. **Or** The thermo plastics are those materials which can be heated and cooled repeatedly without appreciable loss of mechanical and physical properties. Polyethylene (PE), Polypropylene (PP), Polystyrene (PS), Polyvinyl chloride (PVC) Nylon also known as amides, Polyester also polyethylene terephthalate (PET), polyvinylidene chloride (PVDC), Ionomers etc are thermoplastics. A thermoplastic, or thermosoftening plastic, is any plastic polymer material that becomes pliable or moldable at a certain elevated temperature and solidifies upon cooling. Most thermoplastics have a high molecular weight.

Thermosets: These are the material which soften on application of heat but harden on cooling. Material which soften once on application of sufficient heat but harden on cooling and rarely used for packaging e.g.

polyester, epoxy, furan, ionomer. Thermosets are formed when two or more components chemically react with each other under ambient conditions or when induced by radiation or heat to form a highly cross-linked network. The formation of thermosets is an irreversible process. Thermosets are typically hard and rigid.

Thermosetting: It is the state of a substance which is capable of becoming permanently rigid when heated or cured.

Thermostat: It is an automatic device for regulating temperature by controlling the supply of gas or electricity to a heating apparatus. Or Thermostat is a device that automatically regulates or responds to temperature or thermostat is a device that is used to provide or control the temperature.

Thermovinification: Thermovinification is heating of intact / crushed grape to about 75^0C to denature the skin enzymes and cooled to 30^0C prior to preparation of red wine. Thermovinificatiion technique is useful in better extraction of red colour from red grapes.

Thiamine \ thiamin: A vitamin $C_{12}H_{17}N_4OSCl$ of the vitamin B complex that is essential to normal metabolism and nerve function and is widespread in plants and animals is called thiamine. It is also called vitamin B_1. Thiamine is stable to heat in acid food but less stable in neutral and alkaline food and is also sensitive to SO_2. It is found in unrefined cereals, beans, and liver, a deficiency of which causes beriberi. The synonyms are vitamin B_1, aneurin, antineuritic factor, antiberiberi factor.

Thin boiling starch: The starches produced by HCl treatment are termed as thin boiling starch.

Thin layer drying: It refers to the drying of grains which are entirely exposed to the air and air is moving through the grains.

Thixotropic fluids: These are the fluids for which the dynamic viscosity decreases with time for which shearing forces are applied. The fluids go from gel to sol and then back to gel. Apposite to the thixotropic is the rheology.

Thixotropic: It is the ability of a gel to become more fluid with increasing shears and then to regain previous viscosity after shear rate is slowed.

Thixotropy: It is the property of various gels of becoming fluid when disturbed as by shaking (Plate 595).

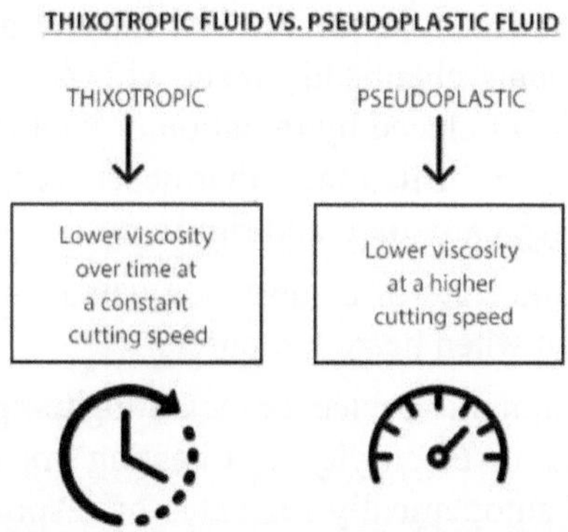

Plate 595: Thixotropic vs pseudoplastic fluids

Three stage dehydration: It is the process of drying of the food commodities at three different temperatures (higher followed by lower temperature). At first the temperature is kept at 75°C for 2 hours, then at 68°C for 4-5 hours followed by dehydration at 60°C till drying upto desired moisture. This type of the dehydration process results better dehydration of the products with acceptable quality.

Threonine: A colorless crystalline essential amino acid $C_4H_9NO_3$ is called threonine.

Thresh \thrash: It is the act to separate seed from a harvested plant mechanically.

Thresher: It is a threshing machine. It is a power driven machine commonly used for threshing of cereals.

Threshing: It is the process of removing the grains from the earheads.

Threshing machine: A machine for separating grain crops into grain or seeds and straw is called threshing machine.

Threshold level: It is the concentration of a taste compound at a barely detectable level.

Threshold: It is the point at which a physiological or psychological effect begins to be produced. Or It is a level, point above which something will take place and below which it is not or will not.

Throttling valve: Expansion value used in refrigeration is also known as throttling valve.

Tie elements: Tie substance is components of a system that can be used to relate the quantity of one process to another. The tie substances do no change during a process. These are also called tie substances.

Tin: It is a soft faintly bluish white lustrous low melting crystalline metallic element that is malleable and ductile at ordinary temperatures and that is used in containers, as a protective coating, in tinfoil, and in soft solders and alloys. Or it is a box, can, pan, vessel, or a sheet made of tinplate.

Tin plate: Common material used for metal cans which is a rigid and impervious material, consisting of a thin sheet of low carbon steel coated on both sides with a very thin layer of tin.

Tirage: It is a beverage containing 2.5% sugar and 2% ethanol as brandy.

Titrant: A substance (as a reagent solution of precisely known concentration) that is added in titration.

Titrate: The solution whose concentration is to be determined by titration is called titrate.

Titration: A method of analyzing the composition of a solution by adding known amounts of a standardized solution until a given reaction (colour change, precipitation or conductivity change) is produced. Or Titration is the process of chemical analysis in which the quantity of some constituent of a dissolved sample is determined by adding an exactly measured quantity of another dissolved substance with which it reacts in a definite, known proportion. The solution of known concentration is gradually added to the unknown solution from a burette (a long measuring tube with a valve at the bottom) until the equivalence point (end point) is reached. The amount of the unknown substance can then be calculated. The equivalence point is determined by a detectable change, for instance, of colour in an indicator substance (e.g., litmus) or in an electrical property. Reactions used in titration include acid-base reactions, precipitations, formation of complexes, and oxidation-reduction reactions. or

Titration: It is a method or the process of determining the concentration of a dissolved substance in terms of the smallest amount of a reagent of known concentration required to bring about a given effect in reaction with a known volume of the test solution.

Titrometer: It is a semi-automatic apparatus used for titration purpose. It automatically titrate the prepared samples. The titration stops automatically upon neutralization of the samples (Plate 596).

TLC: It stands for thin layer chromatography. Thin Layer Chromatography is a technique used to isolate non-volatile mixtures (Plate 597). The experiment is conducted on a sheet of aluminium foil, plastic, or glass which is coated with a thin layer of adsorbent material. The material usually used is aluminium oxide, cellulose, or silica gel.

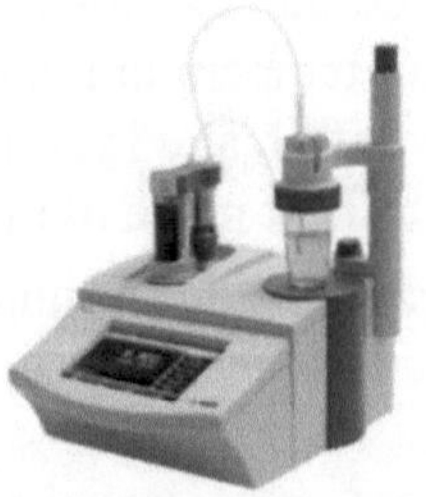

Plate 596: Titrometer

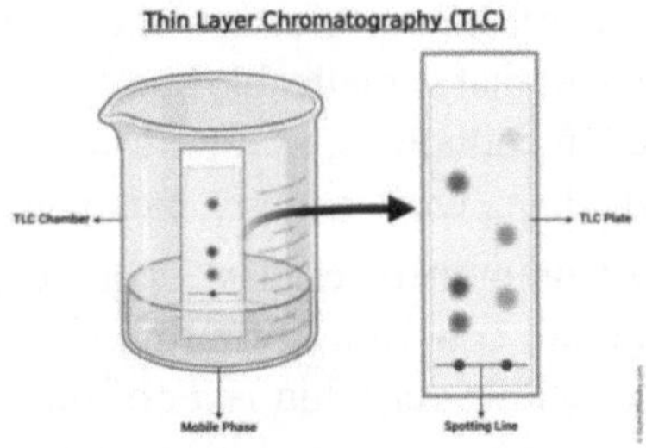

Plate 597: TLC

Toast: The sliced bread browned on both sides by heating on toaster is called toast and such food is called toasted bread. In other words toast means to make bread crisp, hot, and brown by heat.

Toaster oven: It is a small electrical appliance that can function as an oven and is used for toasting breads or other food item.

Tocopherol: It is one of the fat soluble oily phenolic compounds with varying degrees of antioxidant vitamin E activity e.g. α-tocopherol.

Toddy: A sugar containing juice which is obtained by tapping the unopened spadix of the coconut palm. Or The fresh or fermented sap of various chiefly Asian palms is called toddy. Or Toddy is a hot drink consisting of liquor (rum), water, sugar, and spices.

Toffee / Toffy: A candy of brittle but tender texture made by boiling sugar and butter together is called toffee.

Tofu: The soyabean curd (Tofu) made by the action of a mold is called Tofu (Plate 598). It is a soft food product prepared by treating soybean milk with coagulants (magnesium chloride or diluted acids). It is also called bean curd or soya panner.

Plate 598: Tofu

Tokay: Fortified wine from Hungary is known as Tokay. It is a naturally sweet wine from the area around Tokaj, Hungary while it is a blend of Angelica, port, and sherry made in California.

Tomatine: A steroidal glycoalkaloid present in *Lycopersicon* and *Solanum* spp which exhibits an antibiotic activity against a number of organisms (Plate 599).

Tomato paste: It is concentrated and strained tomato product and contain not less than 25 % tomato solids.

Tomato puree: It is concentrated and strained tomato product but thinner than the paste and contain not less than 9 % salt free tomato solids.

Tomato sauce: Tomato sauce is also known as red sauce.

Tomography: It is a method of producing a three-dimensional image of the internal structures of a solid object (food) by the observation and recording of the differences in the effects on the passage of waves of energy impinging on those structures. It is a non-destructive method (Plate 600).

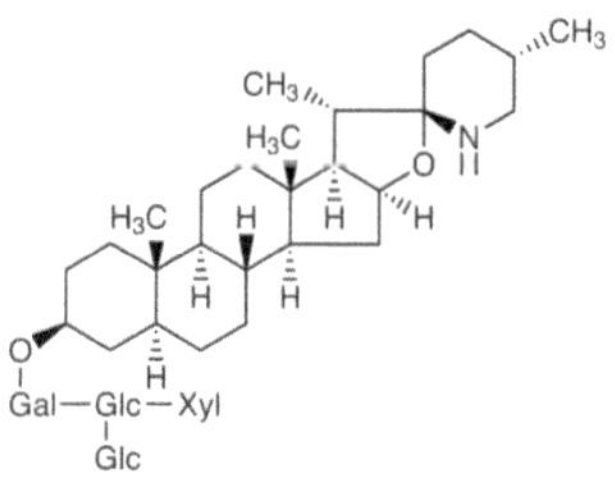

Plate 599: Tomatine structure

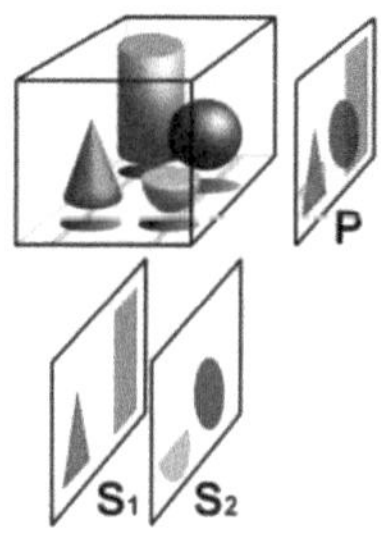

Plate 600: Three-dimensional image by tomograpgy

Tomotsu: Cathecol based sachets is called Tomotsu. It is used for absorption of oxygen in the active packaging.

Ton of refrigeration: The standard unit of refrigeration capacity which is derived on the basis of the removal of the latent heat from 1 ton of water at 32^0F in order to produce 1 ton of the ice at the same temperature in 24 hrs. Since the latent heat of Fusion for water is 144 BTU /lb, a ton of refrigeration is 144 times 2000 lb or 288000 BTU. It is the amount of the refrigeration effect produced by uniform melting of 1 tonnes of ice form 0°C in 24 hours.

Toned milk: It refers to milk obtained by addition of water and skim milk powder to the whole milk. The toned milk should contain a minimum of 1.5% and 9% SNF in double toned milk while 3% fat and 8.5% SNF in single toned milk.

Top fermented beers: Top fermented beers include ale, stout, and porter; characterized by a prominent released carbon dioxide, a sharper and more strongly hopped flavour than lagers, and an alcohol content of 4–6.5% by volume.

Toppan C: This is ascorbic acid based sachets used for O_2 absorption are marketed as Toppan C (when combined with iron).

Torr: The unit of the pressure is called Torr. 1 Torr = 1 mm of Hg column = 133.3 N/m^2. The torr is defined as 1/760 of one standard atmosphere, while the atmosphere is defined as 101325 pascals. Therefore, 1 Torr is equal to 101325/760 Pa (Plate 601).

atm	Pa	mmHg	Torr	Psi
1	101325	760	760	14.7

Plate 601: Illustration of different presre units

Torrid: It is a food parched with heat of the sun or dried with intense heat.

Total acid: Organic acids in the fruit juice such as malic in apple, tartaric in tamarind, citric in lemon which usually measure by titration with standard alkali.

Total mass balance: In total mass balance, inflow = outflow + accumulation. If accumulation is '0' inflow = outflow and the process is called steady state process. If the accumulation is not 'O' then the conc. of components in the system could change with time and the process is called an unsteady state process.

Total quality control: It is a system that integrates quality development, maintenance and improvement of different parts of an organization.

Total quality management (TQM): TQM is a management approach for an organization focused on quality based on the participation of all its members and aiming at long term success through customer satisfaction and benefits to all members of organization and society. It is basically a strategy derived from internal /external customer and supplier wants /needs that have been determined through daily management. TQM is a method through which management and employees can get involved in the continuous improvements of the production of food products. It is a combination of quality and management tools aimed at increasing business and reducing losses. So, TQM is the foundation for activities of all personnel in manufacturing, marketing, engineering, research and development (R & D), sales, purchasing, human resource management (HRD) etc. in relation to food industry. Or TQM is a management approach to long term success through customer satisfaction. It is based on the participation of all members of an organization in improving processes, products, services and the culture in which they work.

Total Soluble Solids / sugars (TSS): TSS is defined as the amount of the water soluble sugars, acids, vitamins (B & C), proteins present in the fruits and vegetables. TSS is measured in term of percent sucrose. Total soluble solids include sugars, organic acids and other soluble component in the juice which is measured in terms of percentage of pure sucrose as the major constituent.

Tote: It is a large plastic bin used to deliver the fish to the processor.

Toxic: A substance containing poisonous material especially when capable of causing death or serious debilitation. The consumption of insecticides exhibit symptoms of toxicosis in the patients.

Toxic metals: Metals can enter food from a number of source and the most significant sources of toxic metals are environmental pollution, soil in which food stuffs grown, equipments, utensil and containers for cooking, processing and storage, food processing water etc.

Toxin: A poisonous substance that is a specific product of the metabolic activities of a living organism and is usually very unstable, toxic when introduced into the tissues, and typically capable of inducing antibody formation.

TQC: It stands for Total Quality Control.

TQM: It stands for Total Quality management.

Trace element: A chemical element present in minute quantities and used by organisms and held essential to their physiology.

Traceability: The ability to trace the history, application or location of an item and like items or activities by means of recorded identifications. It is the ability to follow the movement of a food through specified stage (s) of production, processing and distribution.

Trade agreement: An international agreement on conditions of trade in goods and services or an agreement resulting from collective bargaining is called trade agreement.

Trade and Merchandise Marks Act: The Trade and Merchandise Marks Act, 1958 acts as safeguard to help the buyer of quality as well as to protect the genuine manufacturer. Under this act, any person who deceptively uses as a registered trade mark rights of the genuine manufacturer and is liable for civil as well as criminal proceedings in the court of law.

Traditional foods: These are the foods which are prepared traditionally and are based on the custom and social attitudes of the peoples.

Trans fat: A fat containing trans fatty acids is called trans fat.

Trans fatty acid: An unsaturated fatty acid characterized by a trans arrangement of alkyl chains that is formed especially during the hydrogenation of vegetable oils and has been linked to an increase in blood cholesterol (Plate 602).

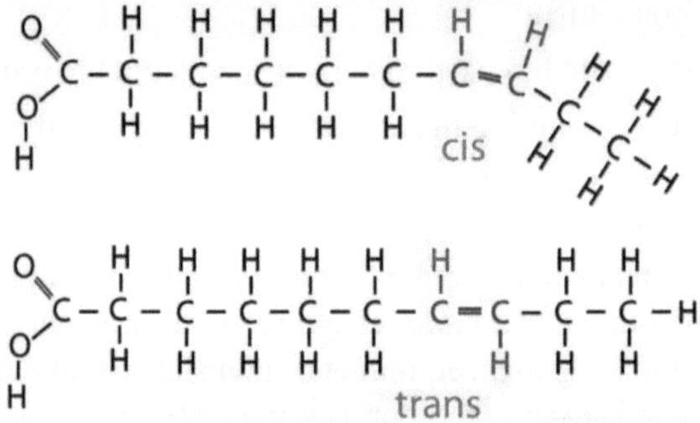

Plate 602: Cis vs Trans fatty acids

Transient heat conduction: It similar to unsteady heat transfer. It is the state in which the temperature distribution in a body does not remains constant with time.

Transmembrane potential: It is based on Dielectric rupture theory. Based on this theory, an external electric field can induce an electric potential difference across cell membranes. This is called Transmembrane potential.

Transmissivity: The ratio of the transmitted energy to the incident energy is called transmissivity.

Transmittance (T): The radiation entering the cell has an intensity I_0 and energy radiation has a value of I. The ratio of I/I_0 is called transmittance (T) and I/I_0 x 100 is per cent transmittance (%T).

Transpiration: A physiological process of giving off water vapour from the internal tissues of living plants through the aerial parts under the influence of sun light which is regulated to some extent by the protoplasm.

Transpiration pull: A capillary pull arising out of a fall in pressure due to transpiration at the top of the plant is called transpiration pull.

Transpiration ratio: The amount of water (g) transpired per gram of dry matter produced which depends on soil moisture, environment variety, and growth stage and growth duration.

Transport Agency: This agency assists in the movement of the produce from one market to another e.g. railways, trucks, bullock carts, camel carts, tractor trolleys.

Transportation: Physical movement of produce from the place of production to the final consumer is called transportation.

Trehalose: A crystalline disaccharide $C_{12}H_{22}O_{11}$ that is found in various organisms (fungi and insects), is about half as sweet as sucrose, and is sometimes used as a sweetener in commercially prepared foods is called trehalose (Plate 603).

Triangle test: In this test, the panelists are provided with three samples. Out of these three samples, two samples are identical and the other one is different. Here all the three samples are given to the panelists simultaneously and panelists are asked to find out the odd sample.

Trichinosis: It is the illness caused by *Trichinella spiralis*, a worm which infests pork (Plate 604).

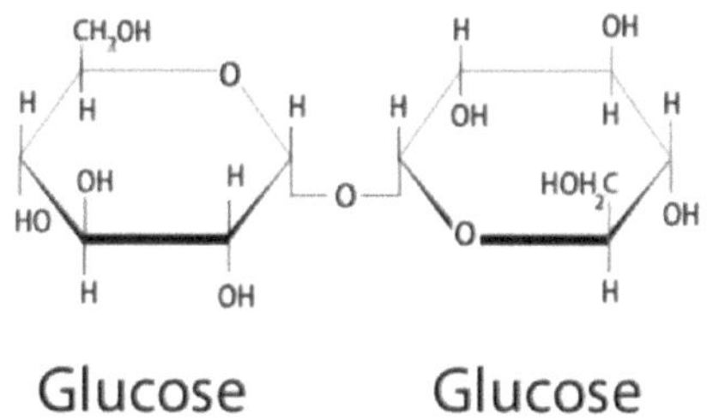

Plate 603: Trehalose structure

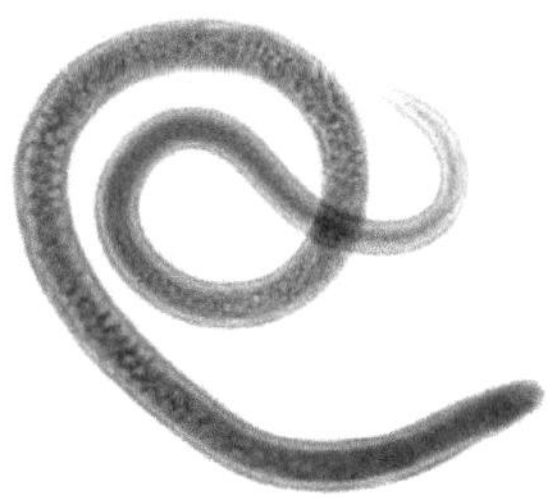

Plate 604: Trichinosis

Triple point of water: It is the intersection of three phase boundaries of water at 0°C and pressure of 4.7 mm of Hg. The triple point of the water shows relationship between temperature and pressure indicating how freeze drying is possible. At triple point water can remain either in solid, liquid or vapour form at 0°C by the variation of pressure only. At 4.7 mm Hg pressure the water from the solid phase can be converted directly into vapour phase at 0°C. It is the point where all the 3 stages like solid, liquid and vapour exist. It is the eutectic temperature at which eutectic mixture occurs (Plate 605).

Trisaccharide: A sugar that yields on complete hydrolysis three monosaccharide molecules is called trisaccharide. Trisaccharides sugars are raffinose, manninotriose. Sugars have free ketone/ potentially free aldehyde groups are reducing sugars based on their ability to act as reducing agent (Plate 606).

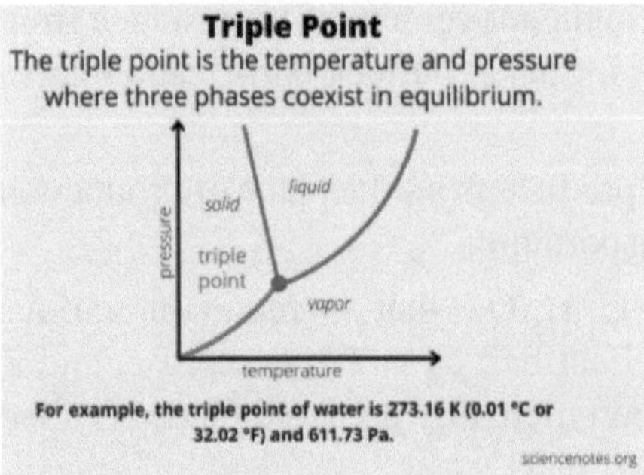

Plate 605: Triple point illustration of water

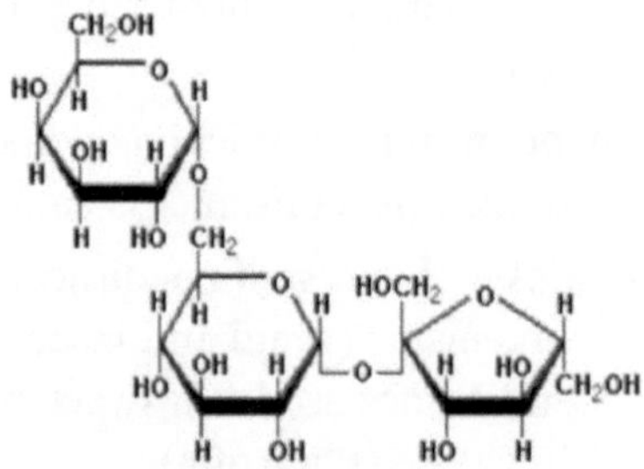

Plate 606: Raffinose structure

Truck crops: These are bulky crops that yield in tones and are grown for distant markets where these are usually transported in trucks without packing in wooden boxes or bags e.g. onion, potato etc.

Truck gardening: The perishables (cabbage and cauliflower) are grown in bulk for selling in distant markets and usually the transport of these commodities is done by means of trucks.

True solution: Dispersion in which ions or molecules not larger than one milli micron are dissolved in a liquid (usually water) is called true solution (Plate 607).

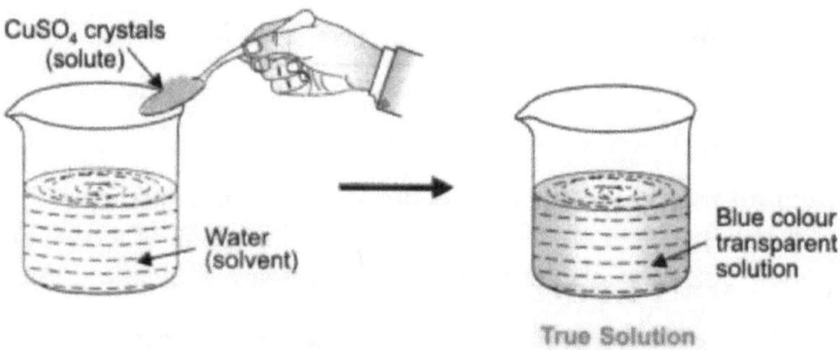

Plate 607: True solution

Truss: It is the process to prepare a bird for cooking after killing and plucking, by removing the head and viscera and tying or skewering.

Trypsin: A proteolytic enzyme that is secreted in the pancreatic juice in the form of trypsinogen and is activated in the duodenum, and is most active in a slightly alkaline medium. Ovoinhibitor @ 0.1% inhibits trypsin.

Tryptophan: It is a crystalline essential amino acid $C_{11}H_{12}N_2O_2$ that is widely distributed in proteins.

TTI: It stands for Time temperature indicator. The enzymes in blanching or pasteurization act as TTI.

Tun: Vessel used for mashing of malt is known as tun. Tun a large cask used especially for wine making and is a unit of liquid capacity which is equal to 252 gallons (Plate 608).

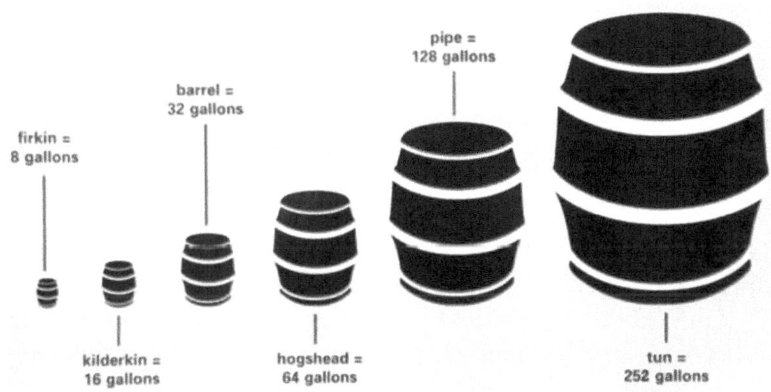

Plate 608: TUN illustration

Turbidity test: This is a statutory test and the sterilized milk has to pass this test before selling to the public consumption. The main principle of this test is that when milk is boiled, all of the albumen in it is precipitated. It involves the addition of the ammonium sulphate to precipitate other substances e.g. casein, which are then filtered out. The filtrate is heated and any albumen in the milk is revealed by turbidity. If the milk is adequately sterilized by heating above 100°C, then all albumen would get precipitated during the sterilization and the test would reveal no turbidity (Plate 609).

Plate 609: Turbidity test illustration

Turbulent flow: A fluid flow in which the velocity at a given point varies erratically in magnitude and direction (Plate 610).

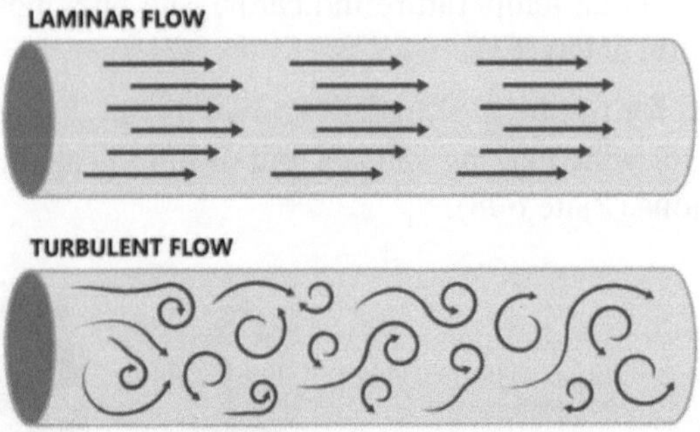

Plate 610: Turbulent vs laminar flow

Tuti fruity: It is a candy like product prepared from bits of raw papaya after peeling.

Tuxford's index: This is a formula for relating height of the children to weight of the children. Children having weight more than the average have an index greater than one and having weight lighter than the average have an index lesser than one. The Tuxford's index formula for boys = W/H x (386-m)/270 and for girls = W/H x (308-m)/235. W is weight in pound, H is height in inches, m is age in months.

Twelve - D (12-D) process: 12-D process this refers to be combination of time and temperature necessary to reduce a colony of *C. botulinum* spore from 10^{12} to 1.

Two sample difference / rating tests: These tests are used to test two samples of same food product having some differences in their qualities. In this test, the panelist is served with four pairs of samples. Here, out of four pairs, the two pairs contain test samples which are mixed or duplicated with the reference or standard samples. Other two pairs contain only the test samples. Here the panelists are asked to evaluate each pair independently to find the degree of difference between the test and standard samples. The scale may be used indicate no difference to large difference in the samples.

Two stage dehydration: It is the process of drying of the food commodities at two different temperatures.

Tyndel effect: The scattering of the light by the particles of colloidal solution in the path of the light is called tyndel effect (Plate 611 and 612).

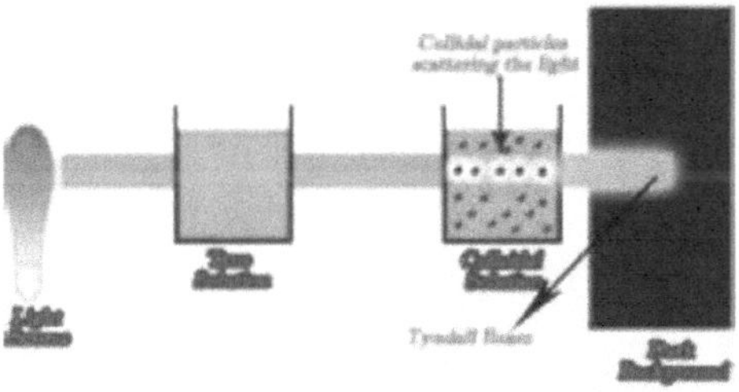

Plate 611: Pictorial representation of Tyndel effect

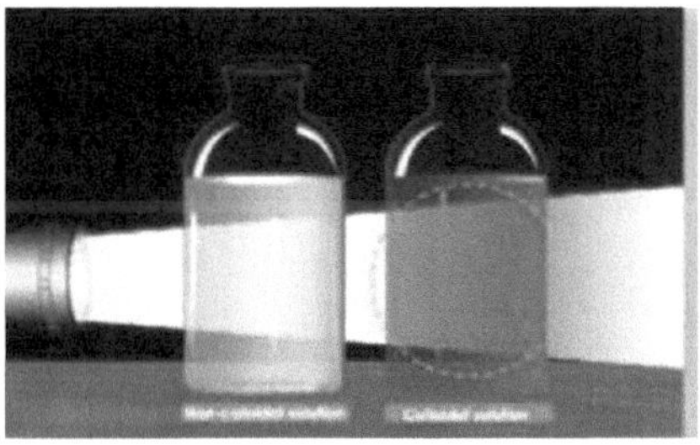

Plate 612: Scattering of light due to Tyndel effect

Tyndalization: Tyndallization is a process from the nineteenth century for sterilizing substances, usually food, named after its inventor John Tyndall, that can be used to kill heat-resistant endospores.Tyndallization, also called fractional sterilization and discontinuous heating or intermittent sterilization, is a form of sterilization that involves boiling the goods to be sterilized in their cans or jars at 100 degrees Centigrade for about 15 to 20 minutes a day, for three days in a succession. After each heating, the resting period will allow spores that have survived to germinate into bacterial cells; these cells will be killed by the next day's heating. During the resting periods the substance being sterilized is kept in a moist environment at a warm room temperature, conducive to germination of the spores. When the environment is favourable for bacteria, it is conducive to the germination of cells from spores, and spores do not form from cells in this environment (Plate 613).

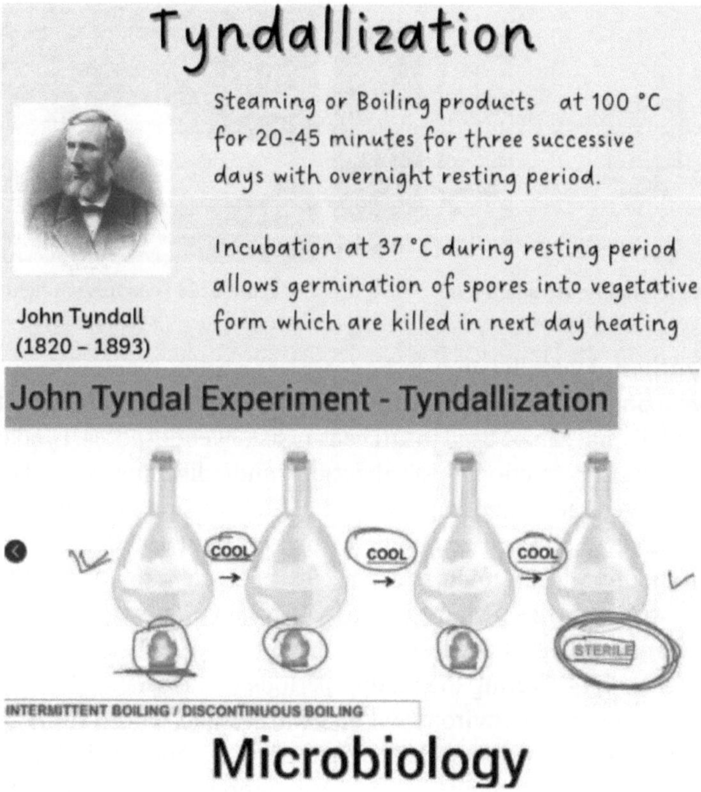

Plate 613: John Tyndall concept

Type I Jellies: These are standardized jellies prepared from fruit juices of a single variety of fruit.

Type II Jellies: These are standardized jellies prepared from fruit juices of more than one variety of fruit.

Type III Jellies: These are unstandardized jellies prepared from fruit juices of a single variety of fruit.

Type IV Jellies: These are unstandardized jellies prepared from fruit juices of more than one variety of fruit.

Tyrosinase: A tyrosinase is a copper containing enzyme that promotes the oxidation of phenols (tyrosine) and is widespread in plants and animal.

Tyrosine: Tyrosine is a phenolic amino acid $C_9H_{11}NO_3$ that is a precursor of several important substances (epinephrine and melanin).

Tyrothricin: It is a polypeptide antibiotic mixture of gramicidin and tyrocidine, produced by Bacillus *brevis* and used as bactericide and fungicide.

U

U value: U value is a measure of the heat transmission through a building part (wall or window) or through a given thickness of a material (insulation). U values with lower numbers indicate better insulating properties (Plate 614).

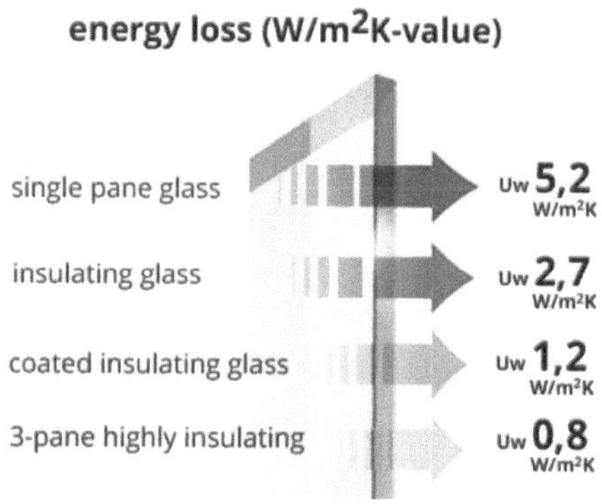

Plate 614: U value of biulding

UHT milk: This milk is homogenized, heated to atleast 135°C for about 2-3 second, cooled rapidly and then packed in sterilized containers aseptically. The milk in packs can be stored for several months without refrigeration. The nutrient loss is negligible. Used for making yogurt commercially as it does not need to be boiled again.

UHT pasteurization: The milk is heated to ~150°C for a fraction of seconds practically no holding period. It is also called ultra pasteurization.

UHT sterilization: It is method in which food is sterilized, cooled and then aseptically placed in previously sterilized containers which are subsequently sealed in an aseptic environment.

Ultimate analysis: It is used for determination of a particular element or a compound in a food material.

Ultra heat treated milk: It is designation of the milk sold for liquid consumption after subjection to ultra high temperature to sterilize the milk by killing bacteria and other microbes including spores.

Ultra-centrifuge: A high speed centrifuge able to sediment colloidal and other small particles and used especially in determining sizes of such particles and molecular weights of large molecules by sedimentation. It is also used in analytical work in the laboratory to precipitate desirable or undesirable molecules. Broadly it is also called centrifuge (Plate 615).

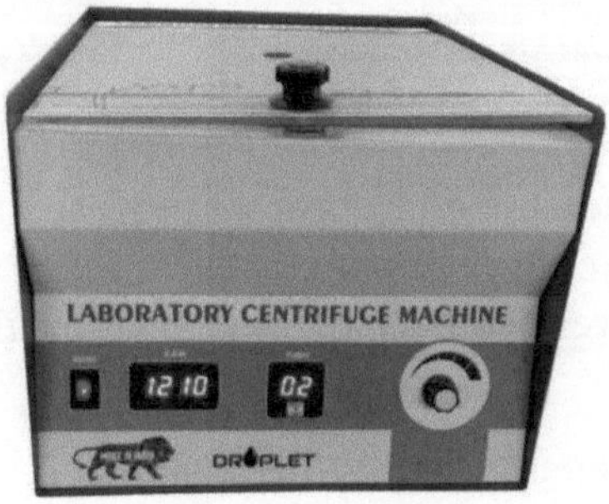

Plate 615: Centrifuge

Ultra-filtration: Ultra filtration is the filtration through a medium (semi permeable capillary wall) which allows small molecules (water and solute with small particle size) to pass but holds back larger ones (protein) (Plate 616). It is used for concentration of protein only. The membrane is less tight and it restrict only macromolecules i.e. proteins and allow to pass water, sugar, vitamin etc.

Ultra-microscope: It is an apparatus for making visible by scattered light particles too small to be perceived by an ordinary microscope.

Ultra-microtome: It is a microtome for cutting extremely thin sections for electron microscopy (Plate 617).

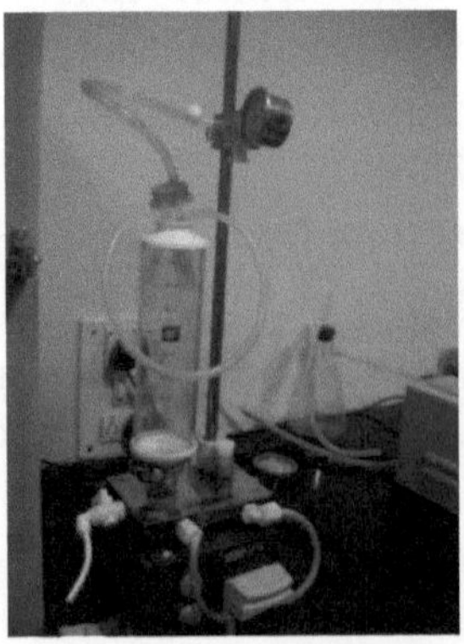

Plate 616: Ultra- filtration (UF) unit for concentration of proteins

Plate 617: Ultra-microtome for cutting extremely thin sections

Ultra-pasteurized: A food which is subjected to pasteurization at higher than normal temperatures specially to extend shelf life is called ultra-pasteurised.

Ultra-sonic: Waves and vibrations having a frequency above the human ear's audibility limit of about 20,000 hertz are called ultrasonic.

Ultra-sonication: The application of ultrasound waves in the preservation of the food is known as ultrasonication. Ultrasound comprises vibrations similar to sound waves, but of a frequency too high (18kHz-500MHz) to be heard by the human ear. Such sound waves in combination with pressure and heat have more effect than the individual. Such combinations include thermosonication (heat + ultrasound), manosonication (pressure + ultrasound) and manothermosonication (pressure + heat + ultrasound). These combinations are more effective against microorganisms and enzymes. The combined use of heat and ultrasound, or heat and ultrasound under pressure, markedly increases the lethality of heat treatments and consequent reductions in time and/or temperature of heat processes. Sonicator is used for generation of the ultrasounds. Sometime, it is also called as sonication (Plate 618).

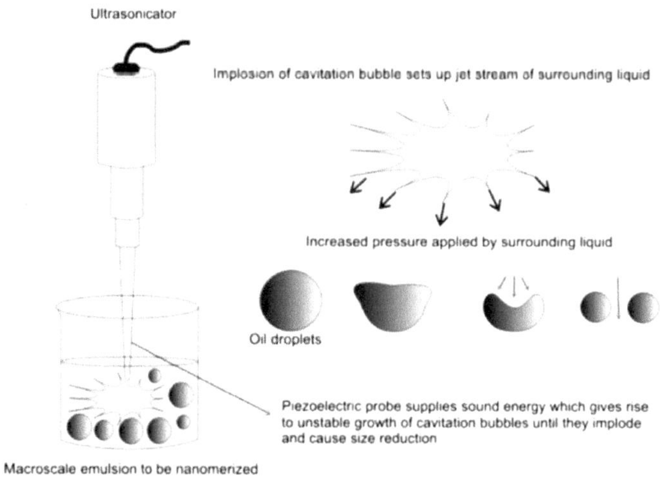

Plate 618: Ultra sonication

Ultra-structure: Biological structure and especially fine structure (as of a cell) not visible through an ordinary microscope

Ultra-violet (UV) rays: The waves situated beyond the visible spectrum at its violet end. Such radiations have wavelengths shorter than wavelengths of visible light and longer than those of X-rays. These have low degree of penetration (Plate 619).

Unassorted tin plate: Standard grade of tin plate also called as Unassorted (U/S or U/A) (Plate 620).

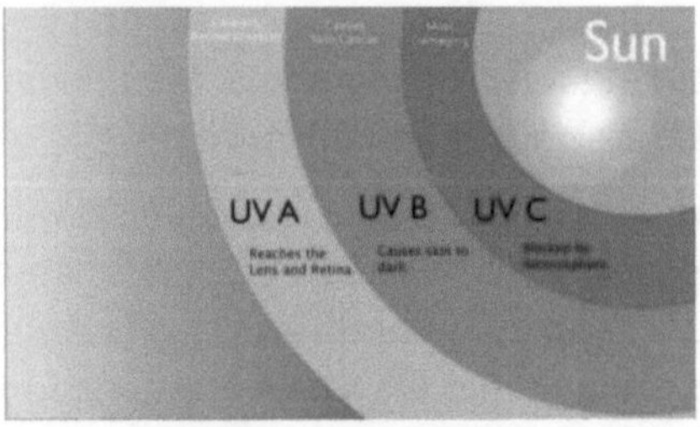

Plate 619: UV rays

Plate 620: Unassorted tin plate

Unavailable carbohydrate: Unavailable carbohydrate refers to pentosans, pectins, hemi-cellulose and cellulose which are not digested by monogastric animals but only by ruminants.

Unbound moisture: It is the water held by the material in excess of the equilibrium moisture content corresponding to saturation in the surrounding atmosphere.

Undecorticated seeds: The seeds which have not been decorticated by the decorticator are called undecorticated seeds.

Under-nutrition: Insufficient supply of the essential nutrient is called under – nutrition.

Unfermented beverages: The fruit juices which do not undergo alcoholic fermentation are called unfermented beverage.

Unintentional (indirect) additives: These substances are also known as non-intentional or incidental or indirect additives which have no intended function in the finished products, but become part of the food product during some phase of production or handling. Additives added in very small quantities as a result of a processing or packaging, are indirect additives.

Unit operation: The term unit operation has been restricted to those operations in which the changes are primarily physical.

Unit package: It is a package which constitute unit at the consumer level e.g. toothpaste, a bottle of jam.

Unit process: It is a processing job consists of series of unit operation.

Universal testing machine / Shear press: This machine is used to measure the tenderness of fruits and vegetables (Plate 621).

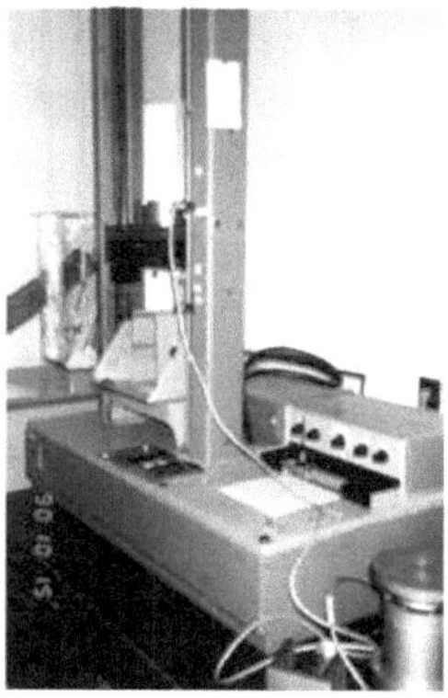

Plate 621: Universal testing machine / Shear press

Unregulated markets: Here business is conducted without any set of rules and regulations. Traders frame rules and conduct business. These markets suffer from various defects in functioning.

Unsaturated Fats: Unsaturated fats are oils and are liquid at RT. Unsaturated fats are obtained from vegetables and marine sources. These have -CH=CH- bonds in their fatty acid structure (Plate 622).

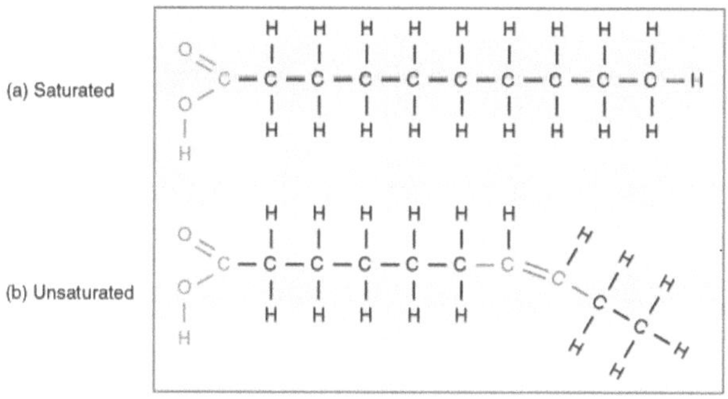

Plate 622: Saturated vs Unsaturated Fats

Unsaturated solution: True solution capable of dissolving additional solute at the temperature of the solution.

Unsteady state heat transfer: It is the state in which the temperature distribution in a body does not remains constant with time. Eg. heat transfer through a metal rod.

Unsteady state process: In total mass balance, inflow = outflow + accumulation. If the accumulation is not '0' then the concentration of components in the system could change with time and the process is called an unsteady state process. Unsteady-state flow is a flow that occurs while the pressures and/or rate changes with time.

Unsweetened juice: A natural juice pressed out of a fruit, strained and kept practically unaltered in its composition during its preparations and preservation is called unsweetened juice. Unsweetened juice is also called fruit juice.

Upperization process: It is the trade name for the sterilization of the milk and other liquids on a continuous flow principle which utilizes direct injection system. In this process, the food is preheated to 80°C, after which the temperature is raise to 150°C for 1-2 second by direct injection of the steam in the upperization unit followed by abrupt cooling under vacuum to remove the injected steam. The process can be used in conjunction with the aseptic processing in the tetra pack containers. The milk after upperization is homogenized to prevent the chalky flavour. It is similar to UHT process.

Ursolic acid: It is a pentacyclic alcohol ($C_{30}H_{48}O_3$) solution, found in leaves and berries of plants and used as emulsifying agent especially in pharmaceutical and food industry.

V

Vacreation: The pasteurization of the milk or cream under reduced pressure by direct steam is called vacreation. This process removes feed and other volatile flavors from cream and to pasteurize it for butter making. This is also called vacuum pasteurization (Plate 623).

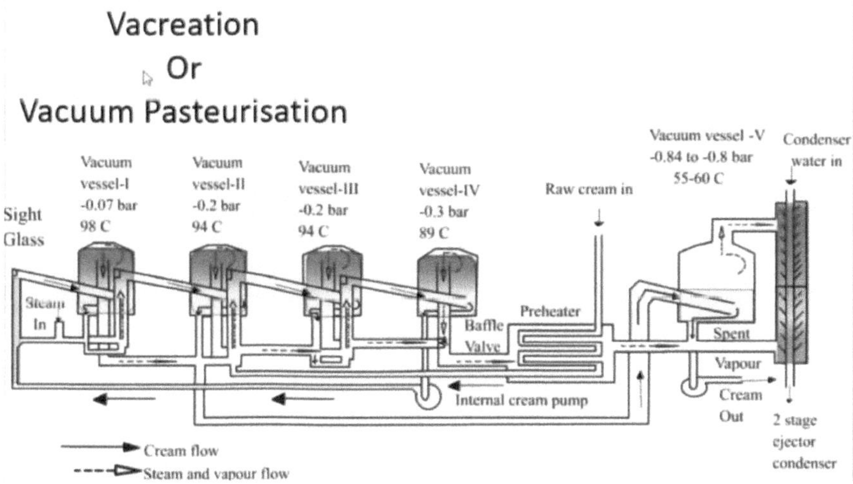

Plate 623: Vacreation or vacuum pasteurization process

Vacuum: The state of negative pressure. It is equal to atmospheric pressure minus absolute pressure.

Vacuum cooling: A technique of cooling vegetable (leafy vegetables, asparagus, brussels sprouts) having a high surface to volume ratio; rapidly and uniformly by boiling off some of their water at 1^{0}C and at low pressure (5 mm of Hg) into a sealed container. The produce is cooled by evaporation of water from the tissue surface and is more rapid than hydro cooling (Plate 624).

Vacuum drier: A drier used to dry heat sensitive products where low pressure lowers the temperature, helps to retain natural flavour and minimizes oxidation and browning is called vacuum drier (Plate 625).

Vacuum gauge: A gauge indicating degree of rarefaction (state or region of minimum pressure in a medium) below atmospheric pressure (Plate 626).

Vacuum packed: A pack having much of the air removed before being hermetically sealed is called vacuum pack (Plate 627).

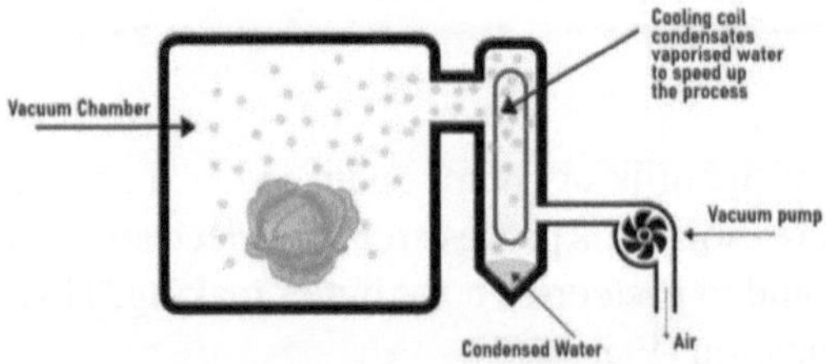

Plate 624 : Vacuum cooling

Plate 625: Vacuum oven/dryer

Plate 626: Vacuum gauge

Plate 627: Vacuum packed soybean

Vacuum pan: A tank with a vacuum pump for rapid evaporation and condensation (sugar syrup) by boiling at a low temperature is called vacuum pan.

Vacuum puff drying: It refers to dehydration of the liquid material in vacuum by evaporating water under pressure in excess of 1 mm of Hg so that the product puff during drying and get expanded. The puffed material has 20 times the volume of the liquid concentrate.

Vacuum pump: A pump for exhausting gas from an enclosed space is called vacuum pump.

Value: Differentiation / identification of black and white colour are called value (Plate 628).

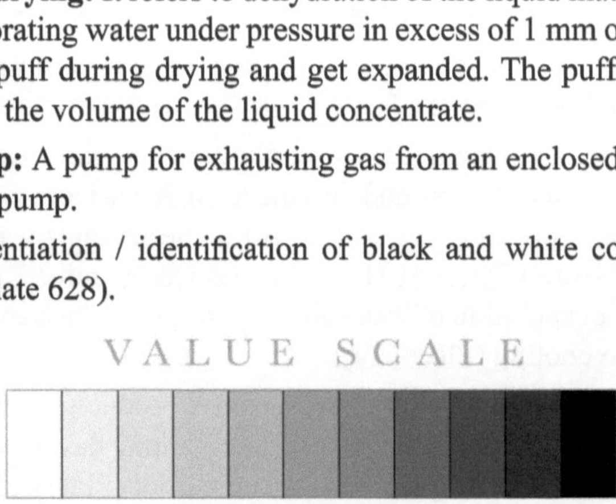

Plate 628: Value in colorimeter

Value addition: It is process by which a fresh commodity is altered in such a way that it offers great convenience for handling, transportation, storage and consumption. It may or may not involve processing or addition of various additives but must definitely add to its nutritional value and economical returns.

Value based pricing: It means price of the product or commodities based on the value addition like quality characteristics, grading etc. Price of the product depends upon values of its sensory, physical, chemical and biological qualities.

Vanilla: It is a commercially important extract of the vanilla bean that is used as a flavoring.

Vanilla bean: It is a long capsular fruit of a vanilla (*Vanilla planifolia*) that is an important article of commerce (Plate 629).

Plate 629: Vanilla beans

Vanilla flavoured pudding: If natural flavour of vanilla alongwith artificial vanilla flavour or only artificial vanilla flavour is present in a pudding then it is called as vanilla flavoured pudding.

Vanilla pudding: If natural flavour of vanilla is present in a pudding then it is called as vanilla pudding.

Vaporimeter: It is an instrument for measuring the vapour pressure.

Vaporizer: It is an instrument/ apparatus used for converting liquid into vapour and discharging it in fine spray.

Vapour compression cycle: It is a system in which low pressure vapour refrigerant from the evaporator is compressed and delivered to condenser at high temperature.

Vapour heat treatment (VHT): It is used to control fruit fly in fruits after harvesting. The most difficult stage of fruit fly to control by HWT (hot water treatment) or VHT is the larval stage, because at this stage larva tends to move into the interior from the fruit surface and thus fruits must be exposed to high temperature for shorter period. VHT for citrus, mangoes, papaya and pineapple is done at 43 °C for 8 hours. It is carried out in VHT plant on commercial scale (Plate 630).

Plate 630: Vapour heat treatment plants

Vapour pressure: It is pressure exerted by a molecule of a compound attempted to be by a gaseous rather than the liquid state. It can be measured Using the Clausius-Clapeyron Equation. This equation have the formula of In(P1/P2) = (ΔH_{vap}/R)((1/T2) - (1/T1)) where, ΔH_{vap}: The enthalpy of vaporization of the liquid, R: The real gas constant, or 8.314 J/(K × Mol), T1: The temperature at which the vapor pressure is known (or the starting temperature.), T2: The temperature at which the vapor pressure is to be found (or the final temperature.), P1 and P2: The vapor pressures at the temperatures T1 and T2, respectively. Other way to find vapour pressure is by Raoult's Law having formula $P_{solution} = P_{solvent} X_{solvent.}$ Where, $P_{solution:}$ The vapor pressure of the entire solution, $P_{solvent}$: The vapor pressure of the solvent, $X_{solvent}$: The mole fraction of the solvent (Plate 631 and 632).

Clausius-Clapeyron Equation

$$\ln\frac{P_1}{P_2} = \frac{\Delta H_{vap}}{R}\left(\frac{1}{T_2} - \frac{1}{T_1}\right)$$

Plate 631: Clausius-Clapeyron Equation

$$P_{solution} = P_{solvent} X_{solvent}$$

The vapor pressure of the entire solution; The vapor pressure of the solvent; The mole fraction of the solvent

Plate 632: Raoult's Law

Variety meat: An edible part (liver or tongue) of a slaughter animal other than skeletal muscle is called variety meat.

VAT: It stands for value added tax.

Veal: The flesh of young calf or kid used as food is called veal. It is slaughtered at less than 15 weeks old.

Vegan: One who does not eat even the animal produce like eggs, milk, butter, curd and cheese etc is called vegan. Or a strict vegetarian who consumes no animal food or dairy products is called vegan

Vegetable: An herbaceous plant (cabbage, bean, or potato) grown for an edible part that is eaten raw or cooked as part of a meal.

Vegetable oil: An oil of plant origin especially fatty oil from seeds or fruits in known as vegetable oil. It is the oil which is obtained by various plant parts leaves, stems, and seed and comprises of esters of fatty acids and glycerol.

Vegetable wax: A wax of plant origin secreted commonly in thin flakes by the walls of epidermal cells is called vegetable wax.

Venison: The flesh of the deer used as food is called venison. Or the edible flesh of a game animal and especially a deer is called venison. It gives red meat.

Very fast freezing rate: Very fast freezing rate is also known as cryogenic freezing.

Very hard durum wheat: Very hard durum wheat is used for making pasta products like Macroni and Noodles.

Very low sodium food: Food having 35 mg or less sodium per serving or per 100 gm.

Very short period markets: These are for few hours and are mostly for highly perishable commodities like fruits, vegetables, fish, milk, etc.

Veterinary residues: Hormones, growth regulators and antibiotics used in animal treatment can pass into food and can cause toxic responses when consumed. These are called veterinary residues

Vibration test: Vibration shocks occur during in plant handling and in shipment. Internal handling on conveyor and industrial trucks could produce damage to highly fragile products. A vibrating table is used for studying this type of hazard. The vibrating table consists of a bed which is driven by two eccentrics, one at each end. To the top of the vibrating bed a platform is attached and platform had circular harmonic type of vibration with amplitude of vibration of one inch and frequency is 120-360 cycle/mint. This test helps to study the package failure/content damages and the weakness that may develop during transit (Plate 633).

Plate 633: Vibration tester

Village markets: A market which is located in a small village, where major transactions take place among the buyers and sellers of a village is called a village market. Buying and selling activities are confined among buyers and sellers of the village or nearby villages mostly for perishable commodities.

Village Merchant: He is an important agent in the collection of produce and move so when the mandi is situated at a considerable distance from the village. He advances from his shop either on credit or for exchange of food grain or so price given for cultivator's produce. The quantities of agricultural produce so collected are either disposed off in the mandi or retained for resale in the village in the processed forms, such as rice, flour, oil etc.

Vinegar: It is a sour liquid obtained by fermentation of dilute alcoholic liquids and used as a condiment or preservative (Plate 634). Vinegar is an acidulous liquid prepared from sugar or starchy material by first alcoholic and subsequently acetic acid fermentation. It should contain at least 4% of acetic acid and corresponding quantity of flavouring, colouring and extracted substances, fruit acids, esters and inorganic salts, according to its origin. It is the product of acetones fermentation of dilute alcoholic liquors (less than 10-12% v/v of ethanol).

Plate 634: Vinegar in bottles

Vinegar eel: A minute free living nematode worm (*Turbatrix aceti*) often found in great numbers in acidic vegetable or vegetable derived fermenting matter (unpasteurized vinegar) is called vinegar eel (Plate 635).

Vinegar fly: The drosophila, a fly infesting vinegar is known as vinegar fly. It causes spoilage of pickles (Plate 636).

Virdigris: It is the formation of green rust on the copper (Plate 637).

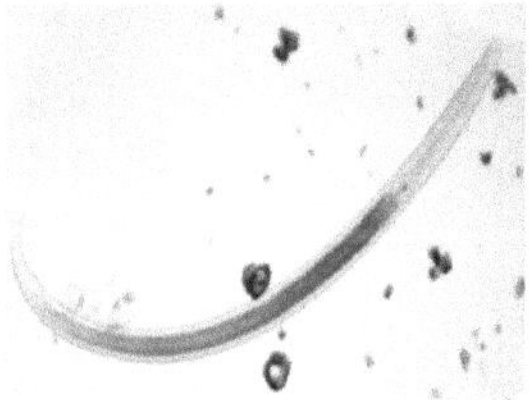

Plate 635 : Vinegar eel

Plate 636: Vinegar fly

Plate 637: Virdigris of copper pipe

Virescence: A disorder in which certain vegetative or floral structures become and remain abnormally vegetative and green. Or It is the condition of becoming green especially a condition due to the development of chloroplasts in plant organs (petals) which are normally white or colored.

Virgin Kraft: According to American paper Institute contain 80% new fibre and not more than 20% recycled fibre is called virgin kraft.

Virulent: Highly pathogenic organism having the strongest capacity for causing a disease is called virulent.

Viruliferous: An agent (insects) containing, producing, or conveying of infection of a virus is called viruliferous.

Viscoelastic: A material having appreciable and conjoint viscous and elastic properties and constituting the state of viscoelastic materials.

Viscometer: An instrument which is used to measure viscosity is called viscometer (Plate 638 and 639).

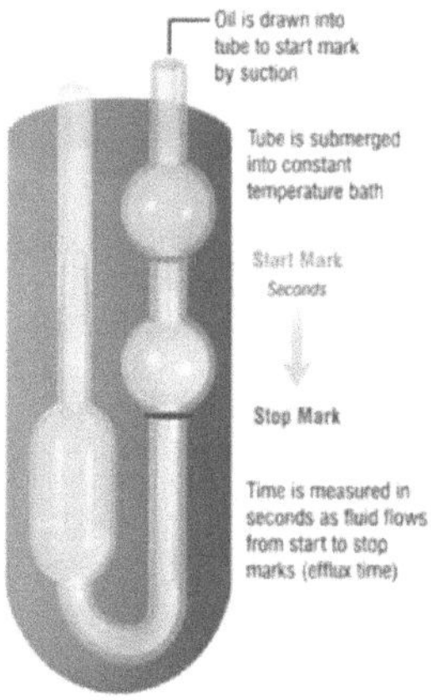

Plate 638: Ostwald viscometer

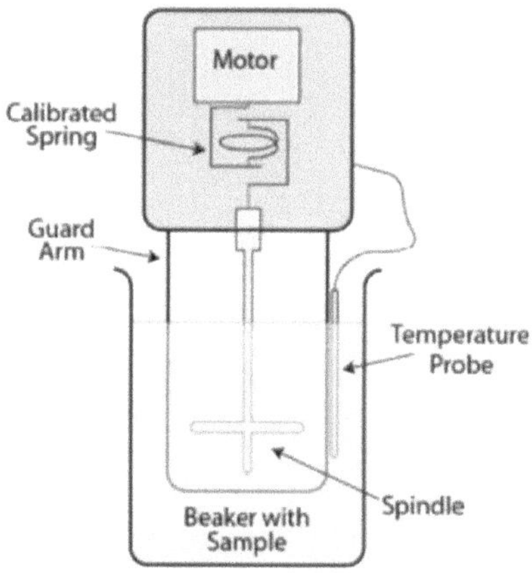

Plate 639: Working principle of rotational viscometer

Viscose: A viscous golden brown solution made by treating cellulose with caustic alkali solution and carbon disulfide and used in making rayon and films of regenerated cellulose is called viscose.

Viscosity: Force required for drawing a liquid from a spoon over the tongue. Viscosity is the rate of flow per unit force. It is the property of resistance to flow in a fluid or semifluid. Viscosity is generally used with Newtonian fluids. Dynamic viscosity is the resistance to movement of one layer of a fluid over another. The unit of dynamic viscosity is Pa s. Usually, it is measured in centipoise (cP). Kinematic viscosity is expressed as the ratio of fluid dynamic viscosity to its density. The unit of measurement of kinematic viscosity is m^2s^{-1}(Plate 640 and 641)

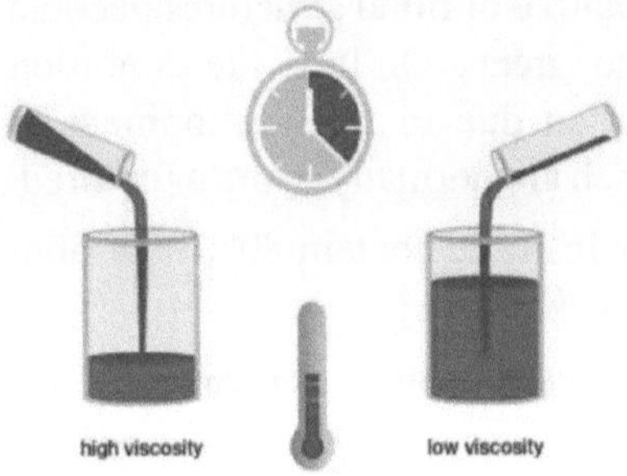

Plate 640: Viscocity illustration

$$\text{Dynamic viscosity } \nu \ (m^2/s) = \frac{\text{Viscosity } \mu}{\text{Density } \rho}$$

where, Viscosity μ (Pa · s) = (kg/(m · s)), Density ρ (kg/m^3)

Plate 641: Formula of viscosity

Viscosity index: An arbitrary number assigned as a measure of the constancy of the viscosity of lubricating oil with change of temperature. The indexes with higher numbers indicate that viscosities change little with temperature (Plate 642).

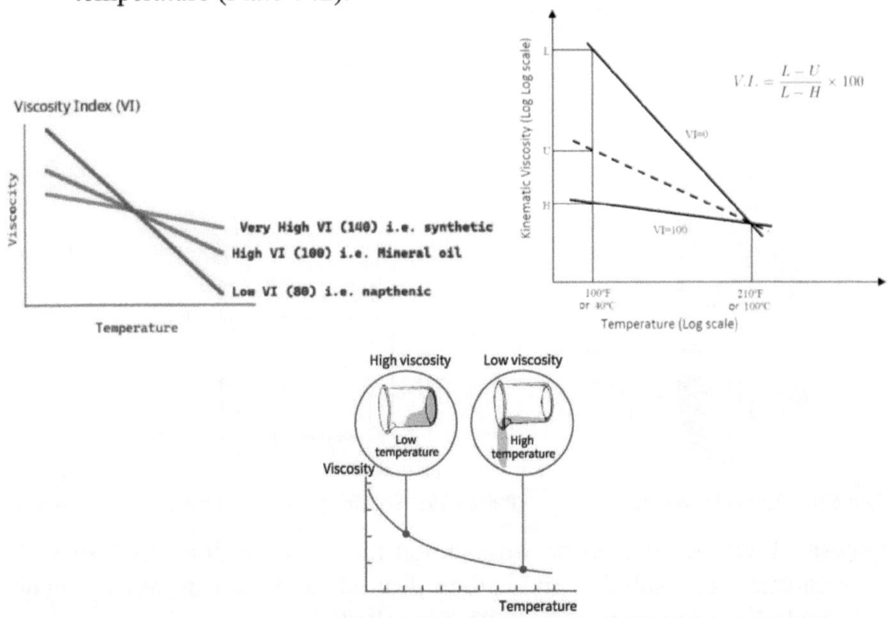

Plate 642: Viscosity index

Viscous elastic fluid: Some fluids have elastic properties, which allow them to spring back when a shear force is released e.g. egg white.

Vitabon GMA: This is ascorbic acid based sachets used for O_2 absorption are marketed as Vitabon GMA (when combined with iron).

Vitamers: These are the substances which are structurally related to vitamins but the biological activity is less than the true vitamin.

Vitamin: The various organic substances that are essential in minute quantities to the nutrition of most animals and some plants, act as coenzymes and precursors of coenzymes in the regulation of metabolic processes but do not provide energy or serve as building units, and are present in natural foodstuffs or sometimes produced within the body.

Vitamin A: Vitamin A also called retinol. Fat-soluble alcohol is most abundant in fatty fish and especially in fish-liver oils. It is not found in plants, but many vegetables and fruits contain beta-carotene, which is readily converted in the body to vitamin A. It functions directly in vision, especially night vision. A derivative, retinaldehyde, is a component of the visual pigments, including rhodopsin, in the retina. Humans require vitamin A in very small amounts. Unlike carotenes, it is toxic in large amounts and is readily destroyed by exposure to heat, light, or air. Vitamin A is fat-soluble vitamins (as retinol) found in animal products (egg yolk, milk, or fish liver oils) or a mixture of them whose lack in the animal body causes epithelial tissues to become keratinous (in the eye with resulting visual defects). It is an alcohol ($C_{20}H_{29}OH$) called retinol. Vitamin A_1 is found in mammals and salt water fish while Vitamin A_2 is found in fresh water fish.

Vitamin B complex: A group of water soluble vitamins that are found in yeast, seed germs, eggs, liver and flesh, and vegetables and that have varied metabolic functions and include coenzymes and growth factors is called vitamin B complex. It is also called B complex.

Vitamin B_1: It is a vitamin $C_{12}H_{17}N_4OSCl$ of the vitamin B complex that is essential to normal metabolism and nerve function and is widespread in plants and animals is called thiamine. It is also called vitamin B_1. Thiamine is stable to heat in acid food but less so in neutral and alkaline food and is also sensitive to SO_2. It is found in unrefined cereals, beans, and liver, a deficiency of which causes beriberi. The synonyms are vitamin B_1, aneurin, antineuritic factor, antiberiberi factor.

Vitamin B_2: The vitamin B_2 is also known as riboflavin. It is yellow crystalline compound $C_{17}H_{20}N_4O_6$ of the vitamin B complex that occurs both free (milk) and combined (liver) as a component of coenzymes (FMN) which are essential to normal metabolism. The product of photooxidation of this vitamin destroy vitamin C. Other synonyms are vitamin G and lactoflavin.

Vitamin B_6: The pyridoxine or a closely related compound found widely in combined form and considered essential to vertebrate nutrition is Vitamin $B_{6.}$ The pyridoxine is a crystalline phenolic alcohol $C_8H_{11}NO_3$ of the vitamin B_6 group found in cereals and convertible in the organism into pyridoxal and pyridoxamine. The synonyms are pyridoxine and antiacrodynia factor.

Vitamin Bc: It is related to folic acid.

Vitamin $B_{T:}$ It is related to essential dietary factor for the mealworm *Tenebriomolitor* and other certain spp and now known to be identical with carnitine which plays an important part in the fat synthesis by transferring acetyl across the mitochondria membrane but it is not a dietary essential. The carnitine originally called as Vitamin $B_{T.}$

Vitamin B_{12}: It is a complex of cobalt containing compound $C_{63}H_{88}CoN_{14}O_{14}P$ that occurs in liver and is essential to normal blood formation, neural function, and growth, and is used in treating pernicious and related anemia and in animal feed as a growth factor. It is also called cyanocobalamin. Vitamin B_{12} is synthesized by bacteria and mould and is commercial by product of antibiotic production. Vitamin B_{12} is also called antipernicious anemia factor. It is important in nucleic acid formation, in fat and CHO metabolism. Vitamin B_{12}, also called cyanocobalamin, is the largest vitamin molecule and contains cobalt in its structure.

Vitamin C: It is a water-soluble vitamin $C_6H_8O_6$ found in plants and in fruits and leafy vegetables or made synthetically and used in the prevention and treatment of scurvy and as an antioxidant for foods. It is also known as ascorbic acid, hexuronic acid, cevitamic acid and antiscorbic acid. Vitamin C or ascorbic acid is an enediol isomer of 2 – keto-L gluconolactone with a configuration similar to L- glucose. Vitamin C removes high level of cholesterol from blood of rats and prevents common cold in human. L-Dehydroascorbic acid is less stable than L- ascorbic acid but the both have the same nutritional effects. Human cannot synthesis vitamin C.

Vitamin D: It is a fat soluble vitamin chemically related to steroids, essential for normal bone and tooth structure, and found in fish liver oils, egg yolk, and milk. Vitamin D is formed in skin of humans/animal by activation of sterols by UV light from sun. Its synonym is antirachitic vitamin.

Vitamin D_2: It is chemically calciferol. It is an alcohol $C_{28}H_{43}OH$ prepared by irradiation of ergosterol and used as a dietary supplement in nutrition and medicinally used in the control of rickets and related disorders.

Vitamin D_3: It is chemically cholecalciferol. It is a sterol $C_{27}H_{43}OH$ that is a natural form of vitamin D found in fish, egg yolks, and fish liver oils and is formed in the skin on exposure to sunlight or ultraviolet rays.

Vitamin E: It is one of the fat soluble vitamins that are chemically tocopherols and are essential in the nutrition of various vertebrates in which their absence is associated with infertility, degenerative changes in muscle, or vascular abnormalities. Vitamin E is found in leaves and in seed germ oils, and used chiefly in animal feeds and as antioxidants. Vitamin E is an anti-sterility factor in rats and is essential for normal muscle tone in dogs and other animals. Vitamin E is a storage antioxidant. Its synonyms are alpha tocopherol and anti-sterility vitamin.

Vitamin F: The essential fatty acids (polyunsaturates) are known as vitamin F (Plate 643).

Essential Fatty Acids with Numeric Formula

COMMON NAME (Abbreviation)	NUMERIC FORMULA (chain length: double bonds)	LOCATION OF FIRST DOUBLE BOND (relative to omega end of chain)
Linoleic acid (LA)	18:2	Omega-6
Alpha-linolenic acid (ALA)	18:3	Omega-3
Arachidonic acid (AA)	20:4	Omega-6
Eicosapentaenoic acid (EPA)	20:5	Omega-3
Docosahexaenoic acid (DHA)	22:6	Omega-3

Plate 643: Vitamin F

Vitamin G: Earlier the riboflavin is called vitamin G. It is now called Vitamin B_2. Lactoflavin is also called vitamin G. It is obsolute name of the Vitamin $B_{12.}$

Vitamin H: It is a colorless crystalline growth vitamin $C_{10}H_{16}N_2O_3S$ of the vitamin B complex found in yeast, liver, and egg yolk. Biotin is known as Vitamin H.

Vitamin K: These are two naturally occurring fat-soluble vitamins $C_{31}H_{46}O_2$ (vitamin K_1) and $C_{41}H_{56}O_2$ (vitamin K_2) essential for the clotting of blood because of their role in the production of prothrombin. Vitamin K is essential for blood clotting. Green vegetable, spinach and cabbage are the source of vitamin K. It is also synthesized by bacteria in human intestinal tracts. Its synonym is antihemorrhagic vitamin.

Vitamin L: This is not established vitamin and it is usually found in yeast. It is essential for lactation.

Vitamin P: The group of plant flavonoids is known as vitamin P. These are water soluble pigments. It is also called permeabiliate vitamin. It affects the strength of the wall of the blood capillaries. It is also as bioflavonoids.

Vitamin T: It is a mixture of folic acid, niacin and deoxyribosides; and it is found in insect cuticle, mould mycelia and yeast fermentation liquor.

Vodka: A colorless liquor of neutral spirits distilled from a unmalted mash (rye or wheat) and specially prepared form potatoes is called vodka (Plate 644). It is popular in U.S.S.R. It contains 40-60% alcohol by volume.

Plate 644: Vodka in bottle

Volatile oil: The essential oil is called volatile oil.

Volatile: A compound is volatile when it evaporates or vaporizes at ordinary temperature on exposure to the air

Volumeter: An apparatus for measuring the volume of the baked product is known as volumeter (Plate 645).

Volumetric analysis: It is the quantitative analysis for accurate measurement of volumes of food products or liquids (Plate 646).

Vomitoxin: It is also known as deoxynivalenol (or DON) and has been found in grain crops worldwide. It is produced by *Fusarium graminearum* and its presence is promoted by wet weather conditions (Plate 647).

Plate 645: Volumeter for baked products

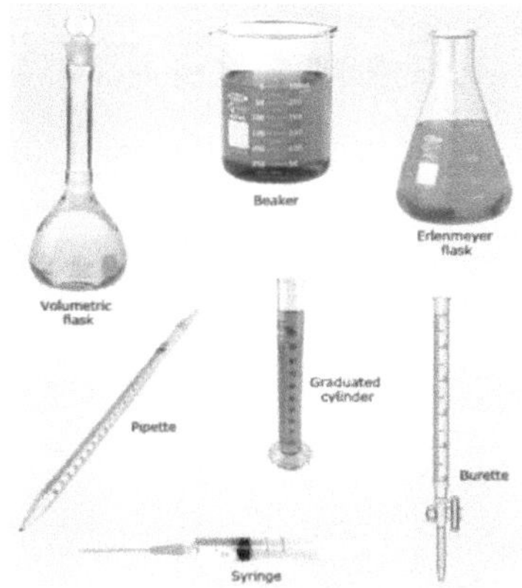

Plate 646: Volumetric analysis

Plate 647: Vomitoxin on corn

W

Wafer: A thin crisp cake, candy, cracker or chips is known as wafer.

Waffle: A crisp cake of batter baked in a waffle iron is called waffle

Warner-Bartzler Shear: It is one of the oldest but commonly used devices used to measure the tenderness of meat. It is used to test the quality in the cooked meat (Plate 648).

Plate 648: Different models of Warner Bartzler shear apparatus

Wastage: Wastage is a term that cannot be precisely defined by even moral value / judgment and defined on the context / situation on which it is used.

Waste water: Water that has been used in a manufacturing process and goes waste is called waste water.

Water: Water (H_2O) is a clear transparent, colourless, odourless and tasteless liquid; neutral in reaction; boils at 100°C, freezes at 0°C and has maximum density (1 g/cm^3) at 4°C. It exists as snow, ponds, lake, rivers, seas, oceans etc., also occur as rain and is there in the form of moisture in atmosphere. Pure water is bad conductor of heat and electricity. Water is an essential part of the body structure as a carrier of nutrients and regulator of a number of body functions. Fruit and vegetables contain > 80% water and sometime upto 95% water e.g. cucumber, lettuce, marrow, lettuce, melon. The starchy tubers and seeds, for example yam, cassava and cork, contain less water, but even > 50%. So water content must be optimum at the harvesting time. The human body comprises of about 65 % water, while that of fruits and vegetables 80-90% water.

Water / effluent treatment plant (W/ETP): Water /Effluent Treatment Plant (W/ETP) in the food processing industries is an important component for treatment of polluted water, which otherwise awakened the environment protection agencies; as this polluted water/effluent affects the surrounding environment and causes nuisance if not handled properly (Plate 649).

Plate 649: Effluent Treatment Plant

Water activity (a_w): It is defined as the ratio of the vapour pressure of water in a material to the vapour pressure of pure water at the same temperature. Water activity (a_w) may be expressed in terms of available water for growth of microorganisms (depends on kind of microorganisms) which is the ratio of vapor pressure of the solution (of solute in water in most food) to the capillary pressure of solvent (usually water), for pure water $a_w = 1.00$. RH = a_w x100 = 1.00x 100 = 100%. Water activity is a dimensionless ratio aw= $P_{H2O\ (food)} / P_{H2O\ (Water)}$. Water activity is based on Raoults law. Water activity is the measure of the ERH. It is the percent of the RH of an atmosphere in contact with the product where no sorption or desorption of the water occur.

Water bath canning: Heat processing of food in containers immersed in water at atmospheric pressure is called water bath canning (Plate 650).

Water binding capacity of starch: Water binding capacity of waxy starch > normal starch> sugary starch.

Water biscuit: A cracker of flour and water and sometimes fat is called water biscuit (Plate 651).

Water colour: The most common cause of water color is the presence minerals. Red and brown colors are due to iron; black to manganese or organic matter; and yellow to dissolved organic matter such as tannins. Iron and manganese are common, at least in small amounts, in most rocks and sediments.

Plate 650: Water bath canning

Plate 651: Water biscuits

Water solubility Index (WSI): It is the amount of the dried solid recovered by evaporating the supernatant from the WAI test.

Watt: It is power of an agent working at the rate of 1 joule or 10^{-7} ergs per second.

Wave number: The number of waves per cm is called wave number.

Wavelength: It is the distance between two peaks of waves (cm).

Waxes: Waxes are esters of higher fatty acid with monohydric alcohols and hydrocarbons and some free fatty acids. There are solvent waxes (alliphalic hydrocarbons), Water waxes (shellac, natural gum) and paste or oil waxes (paraffins). Waxes according to use are storage wax, pack out wax, high shine wax. Waxes are a diverse class of organic compounds that are lipophilic, malleable solids near ambient temperatures. They include higher alkanes and lipids, typically with melting points above about 40 °C (104 °F), melting to give low viscosity liquids. Waxes are insoluble in water but soluble in nonpolar organic solvents such as hexane, benzene and chloroform. Natural waxes of different types are produced by plants and animals and occur in petroleum (Plate 652).

Normal Non-normal

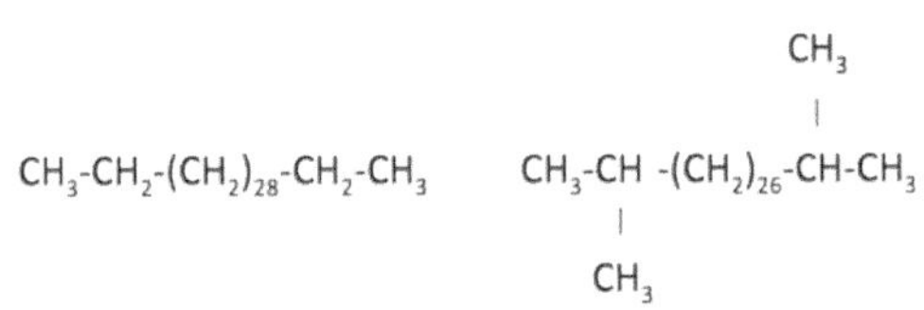

Plate 652: Paraffin wax

Waxing: It is a short term storage technique of fresh fruit and vegetable under ambient conditions by applying wax emulsion containing paraffin wax matrix ethanol and oleic acid which provides a thin discontinuous layer on the fruit surface thus curtails the respiration and transpiration (Plate 653). It results in increase in shelf life. It also helps to keep away the microbes when fungicides like, bentonite are used in wax emulsion.

Waxy starch: The starch which contains 100% of the amylopectin is called normal starch. Waxy starch have smooth surface of the granules (Plate 654).

Plate 653: Kinnow fruits coated with wax

Properties	Amylose	Amylopectin
Water solubility	Less soluble	More soluble
Viscosity (thickener, stabilizer)	Low	High
Retrogradation (gelling)	High	Low
Syneresis (releasing water)	High	Low

Plate 654: Waxing starch characteristics

Weating: It is a term used for all wheat offals other than bran containing not more than 6% fibre and used as feed stuff.

Wet and dry bulb thermometer: It is an instrument or a pair of thermometers hold side by side. Out of these, bulb of one being covered with piece of muslin cloth that is kept constantly wet; used for recording the humidity of the atmosphere; the difference between the readings of the two thermometers indicated the amount of the water vapour in the air (Plate 655).

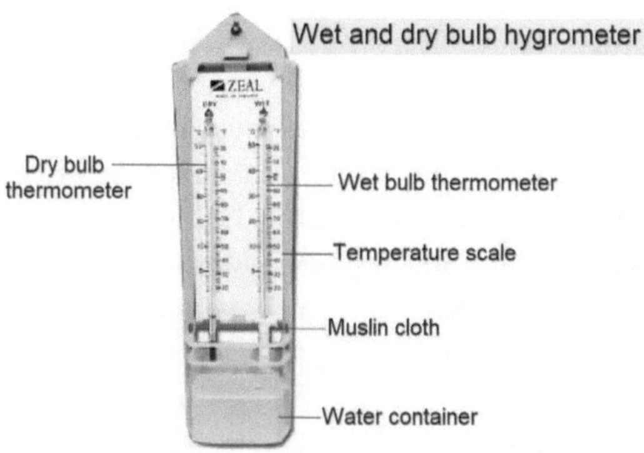

Plate 655: Wet and dry bulb thermometer

Wet bulb depression: It is the difference between the two simultaneous reading of a wet and dry bulb thermometer is called wet bulb thermometer.

Wet bulb temperature: It is the temperature when large amount of unsaturated air is brought in contact with ordinary thermometer with at least 5m /s velocity whose bulb is kept moist by a wet wick snugy wrapped around it.

Wet dog aroma: The off aroma commonly occurring in the citrus juices due to excessive heat treatment during recovery of the essence oil is called wet dog aroma.

Wetness: Amount of moisture perceived on product's surface.

Wettability: The ability of a cake or other food to absorb moisture during a controlled period of time is called wettability.

Wetting agents: A substance which lowers the surface tension of water so that it spread out rather than remains in droplet form when sprayed.

Whetting: Process of making knife sharp is called whetting.

Whey: It is the watery part of milk that is separated from the coagulable part or curd in the process of making cheese and that is rich in lactose, minerals, and vitamins and contains lactalbumin and traces of fat. It is by-product of cheese (Plate 656).

Plate 656: Whey

Whey protein: It is a protein obtain be concentrating whey (Plate 657 and 658).

Whipping: The beating of the cream to produce froth and foam is called whipping. Whipped cream is a special foam possessing remarkable stability and is used in cakes, ice cream etc for decorative purpose (Plate 659).

Plate 657: Whey protein powder

Plate 658: Commercial packing of whey protein

Plate 659: Whipping

Whiskey: It is alcoholic product obtained from distillation of beer or liquor distilled from fermented wort which is obtained from rye, corn, or barley mash.

Whiskey sour: A cocktail consisting of whiskey, sugar, and lemon juice shaken with ice is called whiskey sour.

White meat: White meat is light coloured meat as that of poultry, game-birds and even veal, rabbit and pig (pork), though pig's meat is often classified as red meat (Plate 660).

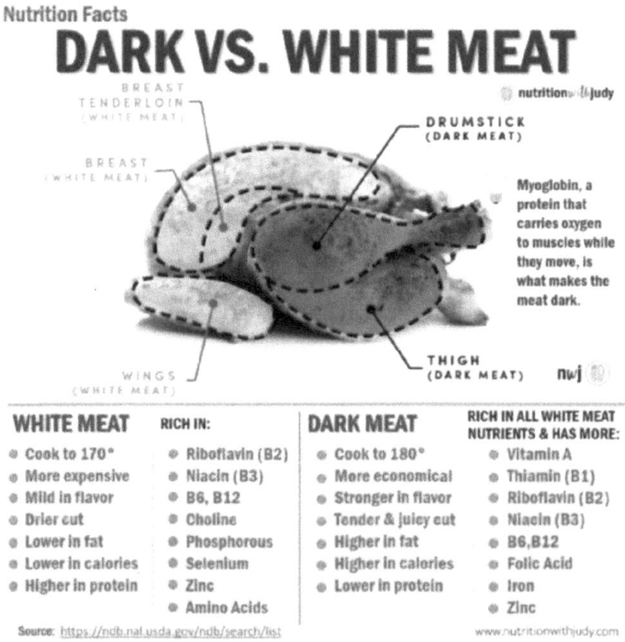

Plate 660: White meat vs red (dark) meat

White wine: White wine is produced by extraction of free run juice from white grapes. Maceration of the skin is avoided for white wine.

Whole food: A natural food and especially an unprocessed one (vegetable or fruit) is called whole food.

Whole wheat: Flour made of ground entire wheat kernels is called whole wheat.

Wholesale markets: Here commodities are brought by and sold in large lots or in bulks. Transaction takes place generally between traders.

Wholesalers: Wholesalers are those traders who sell and purchase the goods in very large quantity. Whole selling is the one that covers activities of all individuals or businessmen, which sell to or negotiate sales with customers, who buy for resale or industrial use. His position is that of an intermediary between manufacturer and retailer. Wholesaler is a merchant or middleman who sells chiefly to retailers, other merchants, or industrial, institutional, and commercial users mainly for resale.

Windrowing: The windrowing is traditional drying of onions in the field for 1-2 weeks. In this practice the dry leaves of one bulb covers the adjacent bulb to reduce the exposed surface to sun. It is not used due to its problematic nature to diseases.

Wind break: It is a traditional way of storing onions.

Wine: The alcoholic fermented juice of fresh grapes used as a beverage is called wine. Or wine is undistilled liquor produced by alcoholic fermentation of the juice of ripe grape or other fruit by saccharomyces cerevisiae var ellipsoids yeast culture. Traditionally wine mean fermented grape juice. Or Wine is the product made by the alcoholic fermentation of grape or grape juice unless otherwise specified by yeast and subsequent aging process (Plate 661 and 662).

Wine cellar: It is a room for storing wines. A stock of wines is also called wine cellar (Plate 663).

Plate 661: Red wine

Plate 662: White wine

Plate 663: Wine Cellar

Wine cooler: A vessel or container in which wine is cooled is called wine cooler. A carbonated beverage that contains a mixture of wine and fruit juice is also called wine cooler.

Wine flower: When unfermented juice is exposed to air, a film of yeast is formed on the exposed surface. It is called wine flower.

Wine grower: A person who cultivates a vineyard and makes wine is called wine grower.

Wine maker: A person who makes wine or one who supervises the wine making process at a winery is called wine maker.

Wine press: A vat in which juice is expressed from grapes by treading or by means of a plunger is called wine press (Plate 664).

Wine sap: An apple with deep red skin and juicy somewhat tart flesh is called wine sap.

Wine skin: A bag that is made from the skin of an animal (as a goat) and that is used for holding wine is called wine skin (Plate 665).

Plate 664: Wine Press

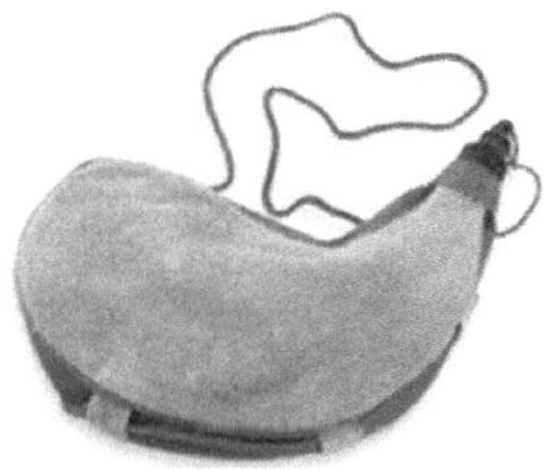

Plate 665: Wine skin

Wine taster: A person who tastes and evaluates wine especially professionally is called wine taster.

Wine yeast: Saccharomyces cerevisiae var ellipsoideus.

Winery: A wine -making establishment is called winery.

Winey or winy: A food product having the taste or qualities of wine is called winey.

Winnower: It is a power operated machine used for winnowing. The machine is provided with one or more sieves and fan (Plate 666).

Plate 666: Winnower

Winnowing: It is the process of separating grains from the mixture of the grain and chaff by using the natural or artificial air. or The treatment of grains by exposure to a current of air so that waste matter is eliminated or to removes as chaff is called winnowing.

Winterization: Winterization is the process in which the higher melting triglycerides are crystallized. The oil is filtered to remove these crystals. The remaining oil has a lower melting point and does not crystallize at refrigeration temperature. It is referred to a Salad oil (Plate 667).

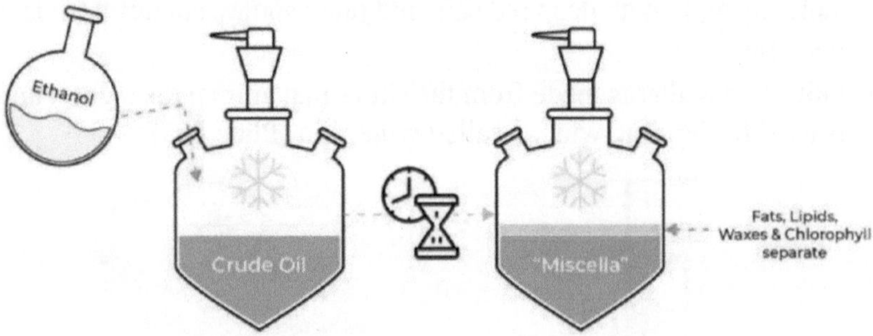

Plate 667: Winterization for oil and fats

Withering: This is a process carried out by spreading tea leaves thinly on the racks or shelves, to dry the leaves partially (Plate 668). It lowers the water content of leaves by about 40%.

Wood sugar: The xylose is known as wood sugar. It is a crystalline aldose sugar $C_5H_{10}O_5$ that is not fermentable with ordinary yeasts and occurs as a constituent of xylans from which it is obtained by hydrolysis.

Wormwood: It is a European plant (*Artemisia absinthium*) having silvery, silky haired leaves and drooping yellow flower heads and yields bitter dark green oil (Plate 669).

Plate 668: Spreading of tea leaves for withering

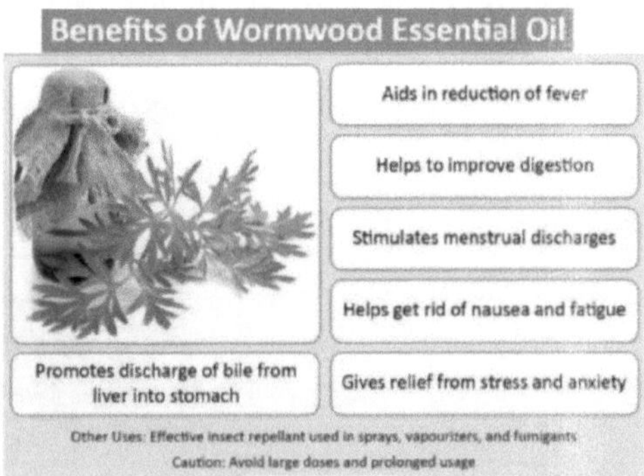

Plate 669: Wormwood benefits

Wort: It is the clear liquid obtained through the filteration of insoluble materials during mashing process of brewing. Or Wort is liquid fraction high in yeast fermentable sugars (separated form spent grain) is known as 'wort'. It is a sweet liquid drained from mash and fermented to make beer and whiskey.

Wounding: An injury operation in the plant part resulting in the accumulation of auxin, carbohydrates etc, and simultaneous increase in respiration rate at the site of wound.

Wrap: A thin flat piece of bread that is rolled around a filling of meat, fish, or vegetables is called wrap. It is similar to masala dosa or vegetable dosa.

Wraps: The wrapping machines for folding flexible material around geometrically shaped product or package is used mainly in cakes / biscuits / vermicelli packaging and its use is being extended to bread, confectionery, dairy products like cheese, butter etc. This type of packaging is called Wraps / Flow Wraps (Plate 670).

Plate 670: Fruit and vegetable Wrap

X

Xanthan gum: It is a polysaccharide that is produced by fermentation of carbohydrates by a gram-negative bacterium (*Xanthomonas campestris*) and is a thickening and suspending agent used in pharmaceuticals and prepared foods. It is also called xanthan. Xanthan is a stabilizer produced by *Xanthomonas compestris* (Plate 671). Xanthan is not hydrolysed or degraded by humans.

Plate 671: Xanthan gum

Xanthones: It is a group of yellow pigments e.g. mangiferin in mango,

Xanthophyll: Yellow hydroxyl carotene derivative which is present in all green leaves together with chlorophyll and carotene. Xanthophyll is a pigment containing O_2 derived from carotene (Plate 672).

Xerophilous: An organism which is thriving in or tolerant to a xeric environment is called xerophilous.

Xerophthalmia: A dry thickened lusterless condition of the eyeball resulting especially from a severe systemic deficiency of vitamin A is called xerophthalmia (Plate 673).

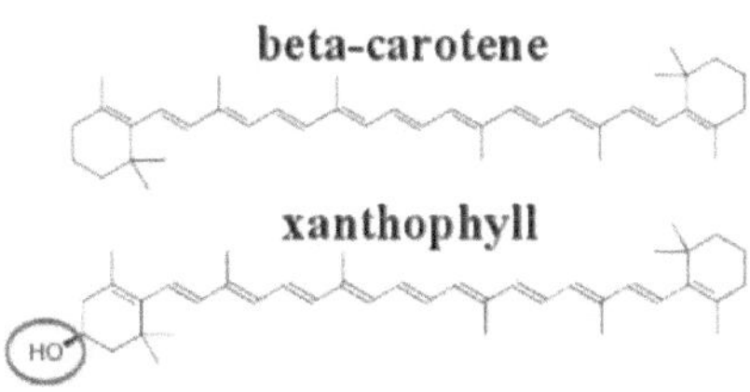

Plate 672: Xanthophyll structure

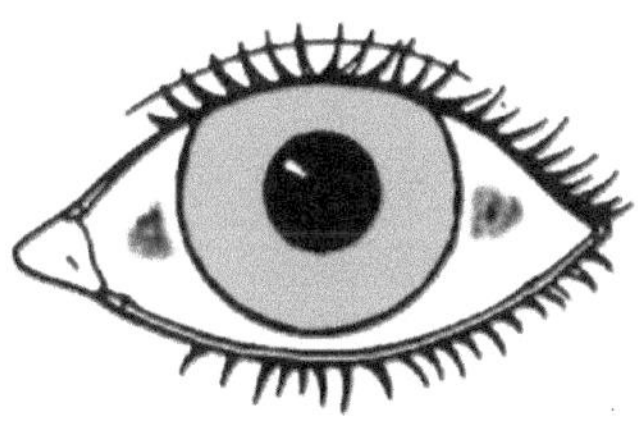

Plate 673: Xerophthalmia

X–ray: It is one of the electromagnetic radiations that have an extremely short wavelength of less than 100 angstroms and have the properties of penetrating various thicknesses of all solids, of producing secondary radiations by impinging on material bodies, and of acting on photographic films and plates as light does.

X–ray star: A luminous celestial object emitting a major portion of its radiation in the form of X-rays called X ray star. It is also called X-ray source.

XRD: It stands for X-ray Diffraction. It is a rapid analytical technique primarily used for phase identification of an unknown crystalline material e.g. minerals, inorganic compound and provide information upto one-unit cell dimension in food processing. It is a non destructive analytical technique which reveals information related to crystallographic structure, chemical composition and physical properties of the food material. XRD in done by using X-ray diffractometer which consist of three basic elements. These are X – ray tube, a sample holder and an X ray detector. It can identify the unknown material within less than 20 min.

XRF: It stands for X-ray fluorescence.

XRS: It stands for X-ray spectroscopy.

Xylan: A yellow gummy pentosan that yields xylose on hydrolysis and is abundantly present in plant cell walls and woody tissue is called xylan (Plate 674).

Xylose: A crystalline aldose sugar $C_5H_{10}O_5$ that is not fermentable with ordinary yeasts and occurs as a constituent of xylans from which it is obtained by hydrolysis. It is wood sugar and is hemicellolosic pentoses (Plate 675).

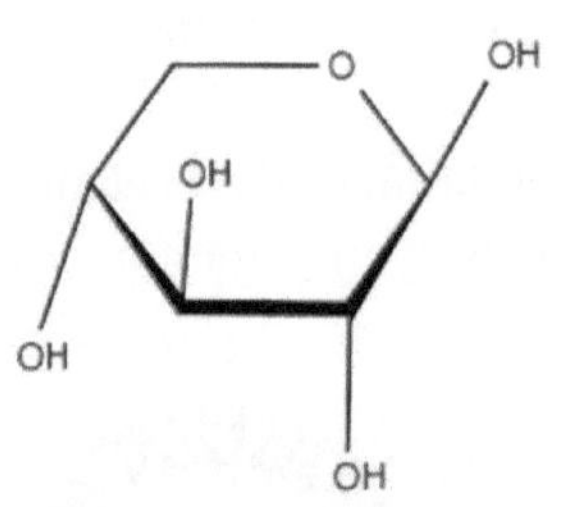

Plate 674: Xylan

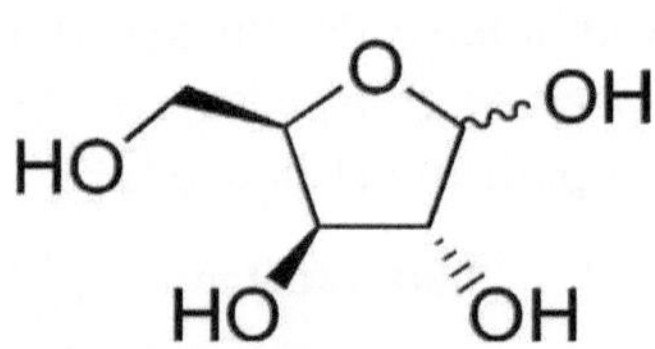

Plate 675: Xylose

Y

Yabby: It is one of the burrowing Australian crayfishes (*Cherax destructor*) that are used for food.

Yakitori: A bite-size marinated pieces of beef, seafood, or chicken on skewers is called yakitori.

Yam: It is edible starchy tuberous root of plants (genus *Dioscorea* of the family Dioscoreaceae) used as a staple food in tropical areas (Plate 676).

Yautia: It is one of the aroid plants (*Xanthosoma sagittifolium*) chiefly of tropical America with starchy edible shaggy brown tubers that are cooked and eaten like yams or potatoes (Plate 677).

Plate 676: Yam

Plate 677: Yautia

Yearling mutton: Meat obtained from 12-20 month aged sheep. It is also called Earling mutton.

Yeast: Unicellular plant body, very minute in size containing cell wall, cytoplasm with one or more vacuoles and a single nucleus, cause alcoholic fermentation (Plate 678). Yeast usually has little or no mycelium, and reproduces by budding. It results in production of yellowish surface froth or sediment that occurs in saccharine liquids (fruit juices) in which it promotes alcoholic fermentation, consists largely of cells of a fungus (as the saccharomyces, *Saccharomyces cerevisiae*), and is used in the making of alcoholic liquors and as a leaven in baking.

Yield stress: It is the energy required to flow.

Yield value: Yield value refers to the amount of the energy required to start movement of a non – Newtonian liquid.

Yogurt /yoghurt: Yogurt is a fermented, slightly acidic, often flavored semisolid food made of milk and milk solids to which cultures of two bacteria (*Lactobacillus bulgaricus* and *Streptococcus thermophilus*) have been added. Or It is fermented milk product with fine curdle gel like consistency having sour and aromatic flavour (Plate 679).

Plate 678: Yeast

Plate 679: Yogurt

Yolk: It is a yellowish internal content of an egg of the bird or reptile. It contains vitellin and other proteins and is surrounded by white substance; serve as nourishment for young one before hatching. It is rich is proteins, carotenoids, fats, lecithin and cholesterol (Plate 680).

Young's modulus: It is the ratio of the stress to strain (Plate 681).

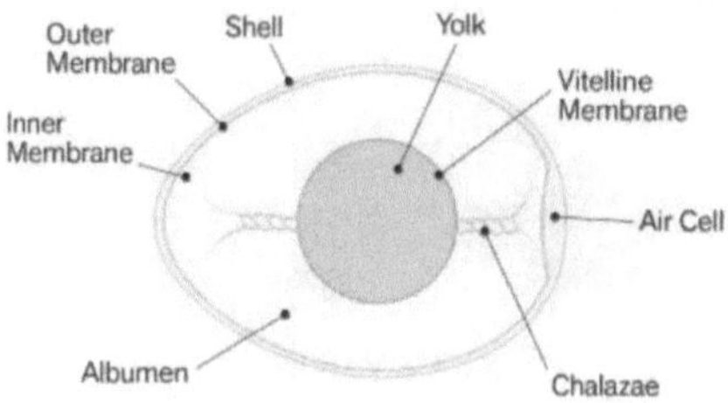

Plate 680: Different parts of egg

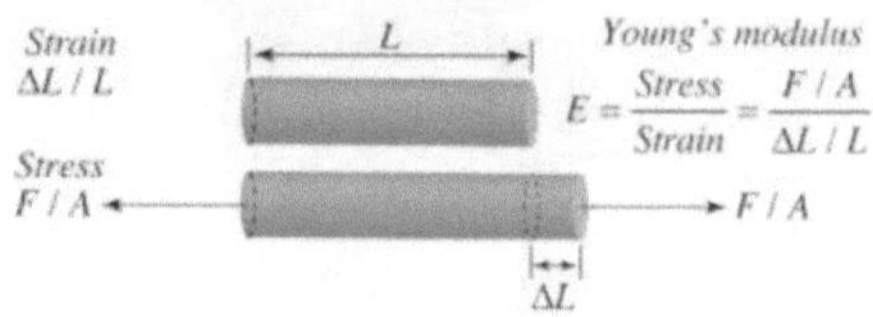

Plate 681: Young's modulus

Z

Zeatin: Zeatin is a cytokinin $C_{10}H_{13}N_5O$ first isolated from the endosperm of Indian corn.

Zeaxanthin: Zeaxanthin is an isomer of lutein occurring in fruits and vegetables (spinach and corn) (Plate 682, 683 and 684).

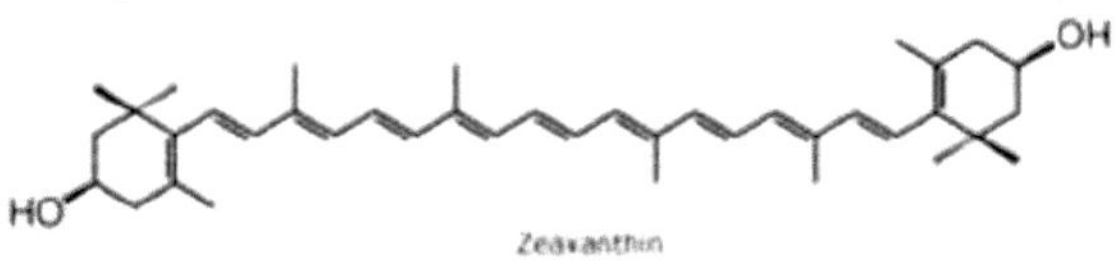

Plate 682: Chemical structure of zeaxanthin

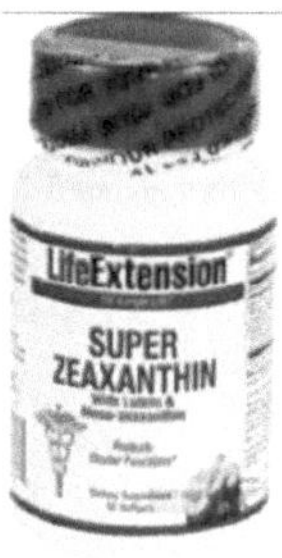

Plate 683: Commercial pack of zeaxanthin

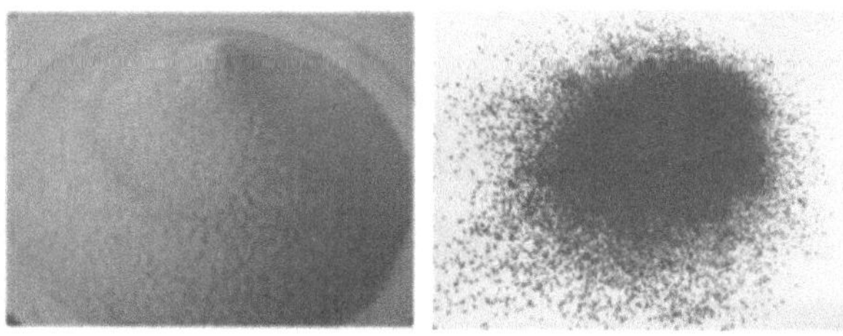

Plate 684: Zeaxanthin available in light yellow and brick red colour from different sources

Zein: Zein is a protein from Indian corn that lacks lysine and tryptophan and is used in making textile fibers, plastics, printing inks, edible coatings, adhesives and sizes.

Zeolite: It is clay embedded in the plastic film to adsorb ethylene. It is used in Japan. Zeolite is anhydrous silicates that is analogous in composition to the feldspars and used in water softening, adsorbents and catalysts. Or Zeolite is one of various hydrous silicates that are analogous in composition to the feldspars, occur as secondary minerals in cavities of lavas, and can act as ion-exchangers. These are used especially in water softening and as adsorbents and catalysts (Plate 685).

Zero 2: It is Australian oxygen scavenging plastic packaging materials.

Zero energy cool chamber (ZECC): Zero energy cool chamber (ZECC) is used to enhance shelf life of fresh fruit and vegetables by maintaining the temperature and RH during summer and winter season naturally with the application of water. It is based on the principle of direct evaporative cooling *(evaporation cause cooling effect)* and hence, does not require any electricity or power to operate and all the materials (Brick, Sand, Bamboo, Khaskha, etc.), required to make the cool chamber are available easily, cheaply and are reusable (Plate 686). Water evaporate from wall and energy required for evaporation comes from inside the room and room get cool and in this way temperature may decrease upto 10^0C i.e. 35^0C – 25^0C and in this way can store the fruit/vegetable without energy source also. This method is given by Australian and now well adopted in India. Problem is that at the bottom there is lot of water which cause PH disease and treatment with TBZ @ 1000ppm, even can solve the problem. In this way huge loss of monitory value due to power cut in storage in potato every year in India can also be minimized.

Plate 685: Zeolite powder

Plate 686: Zero energy cool chamber

Zero vacuum: It is the pressure in the headspace which is equal to atmospheric pressure whereas a vacuum of 30 inch of Hg would indicate that all gas has been removed from the container. Vacuum gauge is used for measure the pressure inside the container (Plate 687).

Zero/ Free / No Nutrition food: Foods contain no /zero or trivial (of little worth) level of nutrients like fat, saturated fat, cholesterol, sodium, sugar or calorie.

Zeroth law of thermodynamics: It states that when any two bodies are in thermal equilibrium with third body, then they are also in thermal equilibrium with each other. This law provides basis of temperature measurement (Plate 688).

Plate 687: Vacuum Pressure gauge

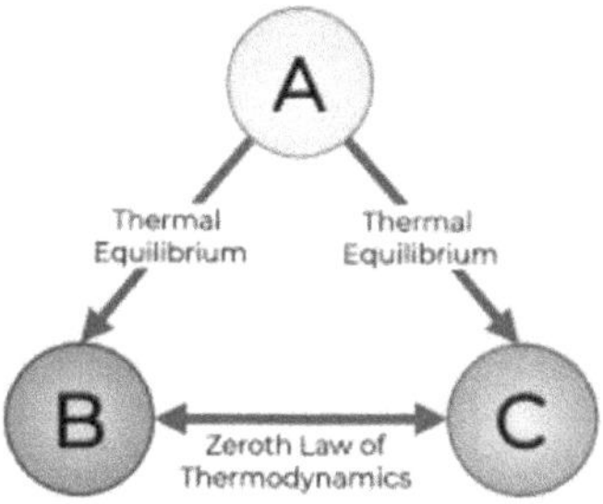

Plate 688: Zeroth law of thermodynamics

Zest: The rind of a lemon or orange grated thinly without any pith (Plate 689).

Zester: The apparatus used for the preparation of zest is called zester (Plate 690).

Zinc sulphate: It is used as a yeast food in the process of beer manufacture (Plate 691).

Zooxanthin: It is a yellow pigment found in animals.

Z-value: Z value is equal to the reciprocal of the slope (negative slope) of the thermal death time (TDT) curve; hence Z-value provides information on the relative resistance of microorganisms to different destructive temperature. Or The slope of the TDT curve is called Z- value and is defined as number of degree °C required to bring about a ten-fold change in decimal reduction time. Or It is the number of the degrees required for a specific thermal death time curve to pass through one log cycle i.e. change by a factor of 10.

Zwitterion: The dipolar ions are known as zwitterions (Plate 692).

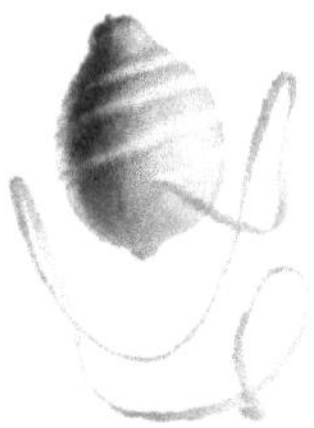

Plate 689: Grating of lemon rind

Plate 690: Different type of zester used for grating of lemon rind

Plate 691: Zinc sulphate granules and powder

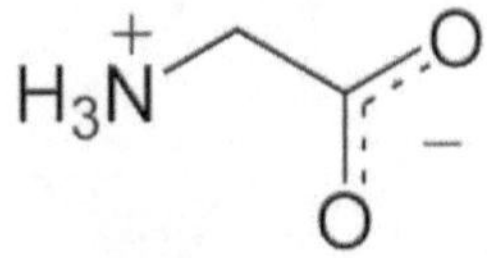

Plate 692: Zwitterion

Zymase: It is an intracellular sugar fermenting complex enzyme.

Zymogenic: It is a substance which is capable of undergoing fermentation.

Zymology: It is branch of science which deals with the study of fermentation.

Zymotechnics: It is the art of managing fermentation.

Zymotic: It is the process which causes fermentation.

Zymurgy: It is a branch of the technological chemistry that concerned with wine making, brewing, distilling and similar processes involving fermentation (Plate 693).

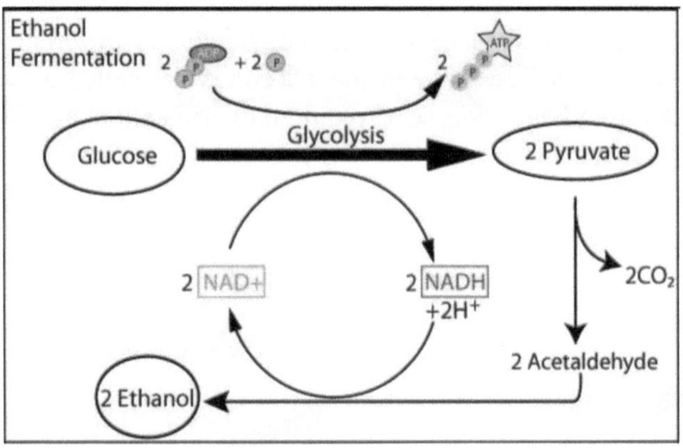

Plate 693: Chemistry of wine making i.e. zymurgy

Annexures

Annexure -1

Minimum Permissible Limits of Preservatives in Fruit and Vegetable Products (FSSAI Specifications, Earlier FPO Specifications)

Product	Minimum TSS (^{0}Brix)	Minimum fruit juice in final product	Preservatives (ppm)		Remarks
			SO_2	Benzoic acid and salt	
Fruit Syrup	65	25	350	600	
Crush	55	25	350	600	
Squash	40	25	350	600	
Cordial	30	25	350	600	
Unsweetened juice	Natural	100	350	600	
Sweetened juice	10	85	350	600	
RTS drink Aerated water	10	10	70	120	
Lime RTS	10	5	70	120	
Fruit juice concentrate	32	100	1500		
Fruit nectar	15	20	-		
Orange/ Pineapple Nectar	15	40	-		
Barley Water	30	25	-		0.25% Starch
Synthetic Sharbat/ Syrup	65	10	350	600	
Ginger Cocktail	30	-	350	600	
Jams	68	45	40	200	55% Sugar
Raspberry/ Strawberry jam	68	25	40	200	55% Sugar
Fruit Jelly	65	45	40	200	

Candy	70	Whole	150		RS(%) 25
Preserve	68	55	150		
Fruit Chutney	50	40	100	250	
Tomato Juice	5	-	-		
Tomato Soup	7	-	--		
Tomato Puree	9	-	-	250	
Tomato Paste	25	-	-	250	
Tomato Ketchup	25	-	-	750	Acidity (1%AA)
Sauce	15	-	-	750	Acidity (1.2%)
Soybean sauce	25	-	-	750	Acidity (0.6%)
Vinegar Brewed	-	-	-	-	Acidity (3.75%AA)
Synthetic Vinegar	-	-	-	-	Acidity (3.75%AA)
Pickle in Vinegar	-	-	100	250	Acidity (2%AA)
Pickle in Brine/ citrus juice	-	-	100	250	12% salt 1.2% (CA)
Sun dried Fruits	-	-	2000	-	Max. 20% moisture
Canned Fruits	-	50(drained weight)	-	-	
Canned Vegetables	-	55 (drained weight)	-	-	
Fruit Pulps			1000	-	

*AA: Acetic acid, CA: Citric acid, RS : Reducing sugars

Annexure-II

Abbreviations Related to Post Harvest Technology

APEDA	**Agricultural and Processed Food Products Export Development Authority**
2,4 –D	2,4,5-dichlorophenoxyacetic acid
2,4,5-TP	2,4,5-trichlorophenoxypropionic acid
ATP	Adenosine triphosphate
a_w	Water activity
BIS	Bureau of Indian Standard
CAC	Codex Alimentarius Commission
CACP	Commission for Agricultural Costs and Prices
CCC	Cycocyl or Chlorocholine chloride
CCFS	Central Committee for Food Standards
CCP	Critical Control Point
CFBB	Corrugated Fibre Board Box
CIPC	Isopropyl-N-Chlorophenyl carbamate
CIPHET	Central Institute of Post Harvest Technology and Engineering, Ludhiana
DON	Deoxynivalenol or Vomitoxin
EDTA	Ethylenediamine tetra acetic acid
EU	European Union
EUREPGAP	Euro-Retailer Produce Working Group on Good Agricultural Practices
FAO	Food and Agricultural Organization of the United Nations
FDA	Food and Drug Administration (USA)
FHA	Food Health Authority
FPO	Fruit Product Order (1955)
GA	Gibberllic acid
GAP	Good Agricultural Practices
GATT	General Agreement on Tariffs and Trade
GDP	Gross domestic production
GHP	Good Hygeine Practices
GM Foods	Genetically Modified foods
GMP	Good Manufacturing Practices
HACCP	Hazard Analysis Critical Control Point
IFS	International Food Standard
IPC	Isopropyl carbamate
ISO	International Organization for Standardization
KMS	Potassium Metabisulphite
LHA	Local Health Authority

MAQ	Minimum Acceptable Quality
MFPO	Meat and Food Product Order (1973)
MH	Maliec Hydrazide
MIL	Masoneilan India Limited
MMPO	Meat and Meat Product Order (1972)
MPEDA	Meat and Marine Food Products Export Development Authority
MRL	Maximum Residue Limit
MSG	Monosodium glutamate
NAA	Naphthalene Acetic Acid
NATO	North Atlantic Treaty Organization
PE	Polyethylene
PFA	Prevention of Food Adulteration Act (1954)
PHM	Postharvest Management
ppm	Parts per million
psi	Ponds per square inch
QA	Quality Assurance
QC	Quality Control
QMS	Quality Management Systems
RTE	Ready to eat
RTS	Ready to serve
SO_2	Sulphur dioxide
SOP	Standard Operating Procedures
SPS	Sanitation and Phytosanitary Measures
SQC	Statistical Quality Control
SQFI	Safe Quality Food Institute
SQFS	Safe Quality Food Standard
SRSV	Small Round Structured Viruses
SSOP	Standardized Sanitary Operation Procedures
TBT	Technical Barrier to Trade
TSS	Total soluble solids or sugars
UN	United Nations
USDA	United State Department of Agriculture
WHO	World Health Organization (UN)
WTO	World Trade Organization (UN), Geneva (1948)
WVTR	water vapour transmission rate

Annexure-III

Standard Units of Measurements

International Standards			
Quantity	**Imperical (US) unit**	**Metric unit**	**Conversion factor**
Length	Inch (in)	Centimeter (cm)	1 in = 2.54 cm
	Foot (ft)	Centimeter (cm)	1 ft = 30.48 cm
	Yard (yd)	Meter (m)	1 yd = 0.914 m
Mass	Ounce (oz)	Gram (g)	1 oz = 28.35g
	Pound (lb)	Kilogram (kg)	1 lb = 0.454 kg
	ton	Tonne (t)	1 ton = 1.02 t
Volume	Cubic inch (in^3)	Cubic centimeter (cm^3)	1 in^3 = 16.4 cm^3
	Cubic foot (ft^3)	Cubic meter (m^3)	1 ft^3 = 0.0283 m^3
	Cubic yard (yd^3)	Cubic meter (m^3)	1 yd^3 = 0.765 m^3
	Fluid ounce (fl oz)	Milliliter (ml)	1 fl oz = 28.40 ml
	Gallon (gal)	Liter (l/L)	1 gal = 3.785 L
Area	Square inch (in^2)	Square centimeter (cm^2)	1 in^2 = 6.45 cm^2
	Square yard (yd^2)	Square meter (m^2)	1 yd^2 = 0.836 m^2
Temperature	Degree Fahrenheit (° F)	Degree Celsius (° C)	° C = 5/9 (F-32)
Pressure	Atmosphere (atm)	Kilopascal (kPa)	1 atm = 101 kPa (760mm Hg)
Energy	Calorie (cal)	Joule (J)	1 cal = 4.1868 J

Conversion of Domestic Unit Standards
Liquids Measurements
1 teaspoon full = 1/8 fluid ounce
1 tablespoon full = ½ fluid ounce
8 Fluid ounce = 1 Cup
4 Cups = 1 Quart
16 Cups = 4 Quarts = 1 Gallon
Solid / Dry Measurements
3 Teaspoon = 1 Tablespoon
2 Tablespoon = 1/8 cups
1 Cup = 16 Tablespoons
Solutions Measurements
1 ppm (parts pre million) = 1 mg per litre = 0.001 g per litre = 1 mg per 1000000 μl
10 ppm = 10 mg per litre = 0.01 g per litre
100 ppm = 100 mg per litre = 0.1 g per litre
1000 ppm = 1000 mg per litre = 1 g per litre = 1 g per 1000 ml = 1 g per 1000000 μl

1 % = 1 g per 100 ml = 10 g per 1000 ml = 10 g per litre = 10000 ppm
0.1 % = 0.1 g per 100 ml = 1 g per 1000 ml = 1 g per litre = 1000 ppm
0.01 % = 0.01 g per 100 ml = 0.1 g per 1000 ml = 0.1 g per litre = 100 ppm
0.001 % = 0.001 g per 100 ml = 0.01 g per 1000 ml = 0.01 g per litre = 10 ppm
0.0001 % = 0.0001 g per 100 ml = 0.001 g per 1000 ml = 0.001 g per litre = 1 ppm
0.5 % = 0.5 g per 100 ml = 5g per 1000 ml = 5 g per litre = 5000 ppm
1 mg per ml or 100 mg per 100 ml or 1000 mg per 1000 ml = 1000 ppm = 1%
1 µg per ml or 100 µg per 100 ml or 1000 µg per 1000 ml = 1 ppm = 0.001%

Annexure IV

Sugar Requirement Calculations per 100 gm Syrup/Solution/Water

$$\text{Sugar Requirement calculations per 100 gm syrup/solution/water} = \frac{\text{TSS required in Brix}}{\text{Water present in required TSS}} \times \{(100\text{-TSS})\text{- TSS}\}$$

or

$$\text{Sugar Requirement calculations per 100 gm syrup/solution/water} = \frac{\{100}{\{(100-\text{TSS desired})} \times (\text{Initial volume})\} - \{\text{Initial wt or vol} - (\text{Initial TSS of the water / solution} \times \frac{\{100\}}{\{(100-\text{TSS desired})\}}$$

Table: Sugar required to make 70°Brix TSS

Solution TSS	Sugar required / 100 g or ml water or solution	Solution TSS	Sugar required / 100 g or ml water or solution	Solution TSS	Sugar required / 100 g or ml water or solution
0	233.3	25	150.0	51	63.3
1	230.0	26	146.7	52	60.0
2	226.7	27	143.3	53	56.7
3	223.3	28	140.0	54	53.3
4	220.0	29	136.7	55	50.0
5	216.7	30	133.3	56	46.7
6	213.3	31	130.0	57	43.3
7	210.0	32	126.7	58	40.0
8	206.7	33	123.3	59	36.7
9	203.3	34	120.0	60	33.3
10	200.0	35	116.7	61	30.0
11	196.7	36	113.3	62	26.7
12	193.3	37	110.0	63	23.3
13	190.0	38	106.7	64	20.0
14	186.7	39	103.3	65	16.7
15	183.3	40	100.0	66	13.3
16	180.0	41	96.7	67	10.0
17	176.7	42	93.3	68	6.7
18	173.3	43	90.0	69	3.3
19	170.0	44	86.7	70	0.0
20	166.7	45	83.3		
21	163.3	46	80.0		
22	160.0	47	76.7		
23	156.7	48	73.3		
24	153.3	49	70.0		
		50	66.7		

Table: Sugar required to make 65°Brix TSS

Solution TSS	Sugar required / 100 g or ml water or solution	Solution TSS	Sugar required / 100 g or ml water or solution	Solution TSS	Sugar required / 100 g or ml water or solution
0	185.7	25	120.0	51	40.0
1	182.9	26	117.1	52	37.1
2	180.0	27	114.3	53	34.3
3	177.1	28	111.4	54	31.4
4	174.3	29	108.6	55	28.6
5	171.4	30	105.7	56	25.7
6	168.6	31	102.9	57	22.9
7	165.7	32	100.0	58	20.0
8	162.9	33	97.1	59	17.1
9	160.0	34	94.3	60	14.3
10	157.1	35	91.4	61	11.4
11	154.3	36	88.6	62	8.6
12	151.4	37	85.7	63	5.7
13	148.6	38	82.9	64	2.9
14	145.7	39	80.0	65	0.0
15	142.9	40	77.1	66	
16	140.0	41	74.3	67	
17	137.1	42	71.4	68	
18	134.3	43	68.6	69	
19	131.4	44	65.7	70	
20	128.6	45	62.9		
21	125.7	46	60.0		
22	122.9	47	57.1		
23	185.7	48	54.3		
24	182.9	49	51.4		
		50	48.6		

Table: Sugar required to make 60°Brix TSS

Solution TSS	Sugar required / 100 g or ml water or solution	Solution TSS	Sugar required / 100 g or ml water or solution	Solution TSS	Sugar required / 100 g or ml water or solution
0	150.0	25	87.5	51	22.5
1	147.5	26	85.0	52	20.0
2	145.0	27	82.5	53	17.5
3	142.5	28	80.0	54	15.0
4	140.0	29	77.5	55	12.5
5	137.5	30	75.0	56	10.0
6	135.0	31	72.5	57	7.5

7	132.5	32	70.0	58	5.0
8	130.0	33	67.5	59	2.5
9	127.5	34	65.0	60	0.0
10	125.0	35	62.5	61	
11	122.5	36	60.0	62	
12	120.0	37	57.5	63	
13	117.5	38	55.0	64	
14	115.0	39	52.5	65	
15	112.5	40	50.0	66	
16	110.0	41	47.5	67	
17	107.5	42	45.0	68	
18	105.0	43	42.5	69	
19	102.5	44	40.0	70	
20	100.0	45	37.5		
21	97.5	46	35.0		
22	95.0	47	32.5		
23	92.5	48	30.0		
24	90.0	49	27.5		
		50	25.0		

Table: Sugar required to make 55°Brix TSS

Solution TSS	Sugar required / 100 g or ml water or solution	Solution TSS	Sugar required / 100 g or ml water or solution	Solution TSS	Sugar required / 100 g or ml water or solution
0	122.2	25	66.7	51	8.9
1	120.0	26	64.4	52	6.7
2	117.8	27	62.2	53	4.4
3	115.6	28	60.0	54	2.2
4	113.3	29	57.8	55	0.0
5	111.1	30	55.6	56	
6	108.9	31	53.3	57	
7	106.7	32	51.1	58	
8	104.4	33	48.9	59	
9	102.2	34	46.7	60	
10	100.0	35	44.4	61	
11	97.8	36	42.2	62	
12	95.6	37	40.0	63	
13	93.3	38	37.8	64	
14	91.1	39	35.6	65	
15	88.9	40	33.3	66	
16	86.7	41	31.1	67	

17	84.4	42	28.9	68	
18	82.2	43	26.7	69	
19	80.0	44	24.4	70	
20	77.8	45	22.2		
21	75.6	46	20.0		
22	73.3	47	17.8		
23	71.1	48	15.6		
24	68.9	49	13.3		
		50	11.1		

Table: Sugar required to make 50oBrix TSS

Solution TSS	Sugar required / 100 g or ml water or solution	Solution TSS	Sugar required / 100 g or ml water or solution	Solution TSS	Sugar required / 100 g or ml water or solution
0	100.0	25	50.0	51	
1	98.0	26	48.0	52	
2	96.0	27	46.0	53	
3	94.0	28	44.0	54	
4	92.0	29	42.0	55	
5	90.0	30	40.0	56	
6	88.0	31	38.0	57	
7	86.0	32	36.0	58	
8	84.0	33	34.0	59	
9	82.0	34	32.0	60	
10	80.0	35	30.0	61	
11	78.0	36	28.0	62	
12	76.0	37	26.0	63	
13	74.0	38	24.0	64	
14	72.0	39	22.0	65	
15	70.0	40	20.0	66	
16	68.0	41	18.0	67	
17	66.0	42	16.0	68	
18	64.0	43	14.0	69	
19	62.0	44	12.0	70	
20	60.0	45	10.0		
21	58.0	46	8.0		
22	56.0	47	6.0		
23	54.0	48	4.0		
24	52.0	49	2.0		
		50	0.0		

Table: Sugar required to make 40°Brix TSS

Solution TSS	Sugar required / 100 g or ml water or solution	Solution TSS	Sugar required / 100 g or ml water or solution	Solution TSS	Sugar required / 100 g or ml water or solution
0	66.7	25	25.0	51	
1	65.0	26	23.3	52	
2	63.3	27	21.7	53	
3	61.7	28	20.0	54	
4	60.0	29	18.3	55	
5	58.3	30	16.7	56	
6	56.7	31	15.0	57	
7	55.0	32	13.3	58	
8	53.3	33	11.7	59	
9	51.7	34	10.0	60	
10	50.0	35	8.3	61	
11	48.3	36	6.7	62	
12	46.7	37	5.0	63	
13	45.0	38	3.3	64	
14	43.3	39	1.7	65	
15	41.7	40	0.0	66	
16	40.0	41		67	
17	38.3	42		68	
18	36.7	43		69	
19	35.0	44		70	
20	33.3	45			
21	31.7	46			
22	30.0	47			
23	28.3	48			
24	26.7	49			
		50			

Annexure-V

Milk Standards

1. BIS Standards for Different Types of Milk

Milk Type	Fat (%)	SNF (%)	TS (%)	LA (%)	Sucrose (%)
Skim milk	1.5 max	8.7	10.5	0.14	-
Double toned milk	1.5	9.0	10.5	0.14	-
Toned milk	3.0	8.5	12.0	0.15	-
Standard milk	4.5	8.5	13.0	0.15	-
Full cream milk	6.0	9.0	15.0	0.16	-
Sterilized milk	3.5	8.5	14.0	0.18	6.0
Flavoured milk	3.0	8.5	12.0	0.18	-
Condensed milk (plane)	9.0	-	31.0	0.35	-
Condensed milk (Skimmed)	Not less than 9.0	-	31.0	0.35	40.0
Condensed milk (Skimmed)	Not more than 0.5	-	26.0	0.35	-
Condensed milk (Partly skimmed)	>3.0 and <9.0	-	28.0	0.35	-
UHT condensed	Not less than 8.0	-	26.0	0.35	18.0
UHT condensed and skimmed	Not more than 0.5	-	20.0	0.35	18.0

2. FSSAI Standard for Milk Products

Fat rich products

Product	Moisture (%)	Fat (%)	Curd (%)	Salt (%)
Malai	-	25	-	-
Low fat cream	-	25	-	-
Medium fat cream	-	40	-	-
High fat cream	-	60	-	-
Table butter	16	80	1.5	3.0
Cooking butter	-	86	-	-
Table spread	16-56	41-80	-	2.5

Concentrated dairy products

Product	Fat (%)	Protein (%)	Milk solids (%)	Fat (%) on db	Moisture (%)
Khoa	-	-	-	30.0	-
Channa	-	-	-	50.0	70

Paneer	20-25	-	-	50	70
Low fat paneer	15	-	-	17.0	70
Evaporated milk	8	34	26	-	-
Evaporated Partially skimmed	1-8	34	20	-	-
Evaporated Skimmmed	1	34	20	-	-
Evaporated high fat milk	15	34	27	-	-
Sweetened condensed milk	9	34	27	-	-
Sweetened condensed Partially skimmed milk	3-9	34	28	-	-
Sweetened condensed skimmed milk	1	34	26	-	-

Fermented Milk Products

Product	**Milk TS /SNF (%)**	**Fat (%)**	**Protein (%)**	**Ash (%)**	**Sugar (%)**
Chakka	TS 30.0	33	30	3.5	-
Srikand	58	8.5	9	0.9	72.5
Yoghurt	SNF 8.5	8	3.2	-	-

Frozen dairy product

Product	**Fat (%)**	**Protein (%)**	**Milk Solids (%)**	**Wt./Vol (g/l)**
Ice cream	10	3.5	36	525

Cheeses

Cheese type	**Fat (%) db**	**Moisture (%)**
Extra hard	32	36
Hard pressed	48	39
Semi hard	40	45
Soft	20	80
Mozzerella	35	60
Pizza	35	54
Processed cheese spread	40	60